Primates in Nature

Alison F. Richard
Yale University

W. H. Freeman and Company
New York

Library of Congress Cataloging in Publication Data

Richard, Alison F.
  Primates in nature.

  Bibliography: p.
  Includes index.
  1. Primates. I. Title.
QL737.P9R54   1985   599.8   84-18802
ISBN 0-7167-1487-6
ISBN 0-7167-1647-X (pbk.)

Printed in the United States of America

1 2 3 4 5 6 7 8 9 0   HA   3 2 1 0 8 9 8 7 6 5

For Bessie and Charlotte,
who screamed but rarely,
and for Bob,
who screamed not at all.

Contents

Acknowledgments

So many people have contributed to this book that it is difficult to know where to begin. I am particularly grateful to Mark Birchette, Keith Hart, David Pilbeam, Rick Potts, and John Rhoads, for discussions about the book and for their comments on the manuscript. I am constantly struck by my good fortune in having colleagues who offer support and encouragement as well as a stream of ideas and insights. Almost in the same breath I would add that past and present students of physical anthropology at Yale have also played a crucial role, and I appreciate enormously their interest, enthusiasm, and willingness to tell me when I'm talking nonsense. I thank every one of them, colleagues and students alike.

The book has benefited immeasurably from the suggestions and criticisms of colleagues and friends in the field. In particular, I thank Mike Rose, Barbara Smuts, Bob Sussman, and Richard Wrangham for valiantly ploughing through the manuscript and offering excellent advice time and again. To John Buettner-Janusch I owe a special debt. He wields a mean red pen, and his substantive and editorial comments deserve a book to themselves. To the extent that I have managed to write a book with sentences that make sense, it is in no small part due to him. My English was also improved by my mother, Joyce Richard, and I am, as ever, equally grateful for her inimitable and (she would assert) totally unbiased support. Janice Chisholm, David Chivers, Marina Cords, Tag Demment, Ken Glander, Mark Leighton, Lisa Leland, David Post, Thelma Rowell, Rich Smith, and Tom Struhsaker have all suggested improvements to the manuscript and I am most grateful for their efforts. Special thanks are due to the participants in the research on rhesus monkeys in Pakistan—Chris Burns, Jim Clapp, Serge Goldstein, Bill Keene, Don Melnick, and Mary Pearl. The inspiration for Part I came largely from their work, and they have contributed in many other ways besides. I am grateful to Mark Solomon and Annabel Keddy for their research assistance. Jeff Rogers worked long and hard on the data for Chapter 7 and many of the ideas in that chapter came out of discussions with him. Most of what I know about the evolution of primate brains and everything I know about trout feeding habits is due to Todd Preuss. My thanks to both Jeff and Todd.

Marion Schwartz came to the rescue during the final and crucial stage of preparing the manuscript for publication. I am deeply grateful for the enormous amount of work she did, helping with the basic research, laboring over the bibliography and index, reading the galleys and, not least, supplying constant encouragement. Liz Kyburg typed the manuscript with the patience and good humor of a saint and with the speed and accuracy of a true professional. My thanks to Linda Chaput, Editorial Director and President of W. H. Freeman and Company, who has nurtured this book from start to finish. Heather Wiley has been an exemplary editor and has my admiration for her transformation of the manuscript into a book and for her tolerance of an author who escaped to Madagascar just when the going got hottest. I thank Ippy Patterson for her enchanting drawings, Brenda Booth and Sally Black for the line illustrations, and John Buettner-Janusch and Ian Tattersall for overseeing the final versions of the artwork. Alison Jolly, a source of inspiration from the very outset, nobly lugged the page proof half way across the world to me, for which she has my lasting gratitude.

On a different, but to me equally important note, I would like to express my profound thanks to Marilyn Smith. It would not have been possible to write this book if she and her family had not provided such a happy home-away-from-home for our children. The term babysitter is not description for the part that she has played in our lives for the last four years. To Andrew and Amy Brownlie goes likewise a special note of thanks. They found the Vosges Mountains for me, and their friendship has done much to keep me going.

Finally, of course, there is my husband, Bob Dewar. No pithy aphorisms spring to mind to convey the central role he has played in every aspect of the whole enterprise. All I can do is to thank him from the bottom of my heart.

Alison F. Richard
Knowles Landing
Middle Haddam

PRIMATES IN NATURE

CHAPTER ONE

Studying Primates

1.1   Introduction

THE goals and methods of ecologists doing research on the primates* (Table 1.1) are often different from those of other ecologists. There are at least two reasons for this. First, most primates are large (Figure 1.1) and long-lived, whereas the field of ecology has focused primarily on small and short-lived animals. Primates, like other mammals of their size, pose different ecological questions and require different methods of study. The second reason is less straightforward. Many nonhuman primates bear an uncanny resemblance to ourselves, and throughout history they have been singled out from their fellow mammals, eliciting amusement, curiosity, and, on occasion, heated debate (Figure 1.2). The fascination of contemplating our look-alikes and, as we now accept, closest relatives in the animal world has hampered the study of primates as mammals in their own right. Until recently, behavioral research on primates was predominantly the domain of psychologists interested in applying their knowledge to the study of human nature rather than to the study of primates as members of natural communities of plants and animals.

In the last twenty-five years, however, the scope of primate field research has broadened, and this book reviews the consequences. Recent findings on the ecology of primates are brought together for comparison with material about other mammals, current theoretical problems in the field of ecology are evaluated with reference to primate data, and conversely, new directions for primatological research are suggested in light of this theory. By considering primates as mammals rather than as our closest living relatives, I hope to provide new perspectives on the ecological distinctiveness and interest of this group of animals.

---

* For convenience, we shall refer to the nonhuman primates simply as primates unless a comparison is being made between them and us.

Figure 1.1   A schematic representation showing the size of a primate, the baboon, compared with other mammals. From left to right: African elephant *(Loxodonta africana)*: height, 3.2m; weight, 5000 kg. Somali zebra *(Equus grevyi)*: head and body length, 145 cm; weight, 430 kg. Lion *(Panthera leo)*: head and body length, 200 cm; weight, 172 kg. Baboon *(Papio cynocephalus)*: head and body length, 700 mm; weight, 25 kg. Uganda grass hare *(Poelagus marjorita)*: head and body length, 470 mm; weight, 2.5 kg.

The book is divided into five parts. Part I looks at the broad sweep of primate biogeography, comparing the past and present distributions of primates with those of other mammals and exploring the possible biological and ecological factors responsible for differences in these distributions. Part II examines how a primate eats and reproduces. Part III considers various characteristics of primate populations: how do these populations fluctuate in size through time, and what causes these fluctuations? Primates do not live as isolated individuals in large, unstructured populations, for all primates maintain networks of social relationships and most spend their lives in social groups. Accordingly, Part IV examines how social life affects an individual primate's interactions with its environment and how the environment, in turn, may shape the nature of primate social organization. In Part V, we step back once more for a broader view: how do primate species interact with one another, and how do they interact with the plants and other animals that make up the natural communities to which they belong? How far can we reconstruct the evolutionary relationships of primates and other community members?

Sections 1.2 through 1.4 of this chapter consider from three perspectives the reasons why primate studies have traditionally been divorced from the rest of ecology. In Section 1.2 we briefly look at the history of ideas about and attitudes toward the primates. In Section 1.3 we review some of the historical trends that gave rise to the modern discipline of ecology. In Section 1.4 we amplify the comment that primates are "large and long-lived," examine the distinguishing characteristics of the order, and summarize ecological interpretations of these features. Finally, in Section 1.5 the conclusions reached in these earlier sections are brought together in an effort to characterize the scope of modern primate studies.

Figure 1.2   The eighteenth-century artist and social critic Richard Hogarth depicts a monkey dressed and acting like a man to satirize the politicians of the day. (Andrew Edmunds Collection.)

Table 1.1 Classification of the primates

Suborder	Superfamily	Family	Subfamily	Genus	Common name
Strepsirhini	Lemuroidea	Cheirogaleidae		*Allocebus*	Hairy-eared dwarf lemur
				Cheirogaleus	Dwarf lemur
				Microcebus	Mouse lemur
				Mirza	Coquerel's dwarf lemur
				Phaner	Fork-marked lemur
		Daubentoniidae		*Daubentonia*	Aye-aye
		Indriidae		*Avahi*	Woolly lemur
				Indri	Babakoto
				Propithecus	Sifaka
		Lemuridae		*Lemur*	Lemur
				Varecia	Ruffed lemur
		Lepilemuridae	Hapalemurinae	*Hapalemur*	Gentle lemur
			Lepilemurinae	*Lepilemur*	Sportive lemur
	Lorisoidea	Lorisidae	Galaginae	*Euoticus*	Needle-clawed galago
				Galago	Bushbaby
			Lorisinae	*Arctocebus*	Angwantibo
				Loris	Slender loris
				Nycticebus	Slow loris
				Perodicticus	Potto

Suborder	Infraorder	Superfamily	Family	Subfamily	Genus	Common name
Haplorhini	Tarsii	Tarsioidea	Tarsiidae		*Tarsius*	Tarsier
	Platyrrhini	Ceboidea	Callimiconidae		*Callimico*	Goeldi's marmoset
			Callitrichidae		*Callithrix*	Marmoset
					Cebuella	Pygmy marmoset
					Leontopithecus	Golden lion tamarin
					Saguinus	Tamarin

Suborder	Infraorder	Superfamily	Family	Subfamily	Genus	Common name
Haplorhini	Platyrrhini	Ceboidea	Cebidae	Alouattinae	*Alouatta*	Howler monkey
				Aotinae	*Aotus*	Owl or night monkey
				Atelinae	*Ateles* *Brachyteles* *Lagothrix*	Spider monkey Woolly spider monkey Woolly monkey
				Callicebinae	*Callicebus*	Titi monkey
				Cebinae	*Cebus*	Capuchin monkey
				Pitheciinae	*Cacajao* *Chiropotes* *Pithecia*	Uakari Bearded saki Saki
				Saimiriinae	*Saimiri*	Squirrel monkey
	Catarrhini	Cercopithecoidea	Cercopithecidae	Cercopithecinae	*Allenopithecus* *Cercocebus* *Cercopithecus* *Erythrocebus* *Macaca* *Miopithecus* *Papio* *Theropithecus*	Swamp monkey Mangabey Guenon Patas Macaque Talapoin Savanna baboon Gelada baboon
				Colobinae	*Colobus* *Nasalis* *Presbytis* *Pygathrix* *Rhinopithecus*	Colobus monkey Proboscis monkey Langur Douc langur Golden monkey
		Hominoidea[a]	Hominidae	Gorillinae	*Gorilla* *Pan*	Gorilla Chimpanzee
				Homininae	*Homo*	Human
			Hylobatidae		*Hylobates*	Gibbon
			Pongidae		*Pongo*	Orang-utan

Source: Adapted from Hershkovitz 1977. Classification of the Lemuroidea follows Tattersall 1982; the Lorisoidea, Charles-Dominque 1977; the Cercopithecidae, Thorington and Groves 1970; and the Hominoidea, Andrews and Cronin 1982.
[a] This classification is based on recent molecular evidence. Traditional classifications put *Gorilla* and *Pan* in a separate family from *Homo*.

1.2 The Primates:
A Historical Perspective

The complex and fascinating history of our intellectual dealings with the nonhuman primates has been considered at length by several writers, including Robert and Ada Yerkes (1929), Sir Solly Zuckerman (1932), Vernon Reynolds (1967), Donna Haraway (1978a, 1978b), Peter Reynolds (1981), and Hugh Gilmore (1981). The following account has been pieced together largely from these sources, and the reader is referred to them for a more comprehensive discussion of the changing role of primates in human science and culture.

Since ancient times, the nonhuman primates have had a special status because of their humanlike appearance. Aristotle pointed out this resemblance in 325 B.C., and a few hundred years later another Greek scholar, Galen, supported the observation with a series of dissections of animals that included the chimpanzee and a number of monkeys. Between the eclipse of the civilizations of the ancient world and the late Middle Ages in Europe, we have no record of what people thought about primates. But in the thirteenth century we find the resemblance being pondered again, this time by Albertus Magnus, a European theologian who divided up the world not into two groups, as his theological predecessors had done, but into three. To the categories of man, who has reason, and animals, who do not, he added one of creatures "similar to man," namely, the pygmies and the monkeys. Although lacking reason, these "animals," he suggested, do possess some control over their instincts through memory and imitative ability.

Albertus Magnus would probably have had additional candidates for his third category had he been familiar with the nonhuman primates most like us, the apes (Figure 1.3). But apes became known to the Western world only late in the seventeenth century, primarily through studies of the chimpanzee by the anatomists Nicolas Tulp and Edward Tyson. Tyson's work was partly inspired by the Belgian anatomist Vesalius, who challenged the prevalent view that the bodies of apes so closely resemble the human body they can be conveniently substituted for human cadavers in anatomical studies. Such a view was held by Linnaeus, author of the *Systema Naturae,* the first comprehensive classification of all plants and animals and the foundation of modern biological systematics. In the tenth (1758) edition of his work, Linnaeus grouped humans, monkeys, and apes together for the first time in the order Primates. He based his classification on anatomical criteria and, in fact, believed that humans and apes were so similar anatomically it was difficult to tell them apart. Tyson himself, we may note, concluded to the contrary that his "pygmie" (the chimpanzee) was an intermediate type, different from both humans and monkeys.

Still, scholars who described nonhuman primate morphology generally emphasized, even exaggerated, its similarity to the human form. When the subject was behavior,

Figure 1.3 Portrait of a gorilla mother *(Gorilla gorilla)*. (A. H. Harcourt/Anthro-Photo.)

however, we find more ambivalence and less data. Although most accounts discussed nonhuman primate and particularly ape behavior in much the same terms we use to describe our own, they emphasized all that was bestial, disgusting, and generally antithetical to the human condition. These accounts usually owed more to the moral attitudes and preconceptions of their authors than to knowledge of the habits of the animals. In the late Middle Ages, a number of animal species were used as religious symbols (Hutchinson 1978b), and primates in particular were frequently pictured as symbols of the devil himself or else of the poor, deluded human sinner (Figure 1.4). When their symbolic importance in Christian theology faded during the sixteenth century, they became instead figures of fun as they had been in ancient Greece, and even scientists tended to view them as grotesque and comic caricatures of people. In the mid-eighteenth century the great French natural historian Comte Georges-Louis Leclerc de Buffon published his *Histoire Naturelle*. This was the most complete and insightful account of natural history available in his day and included a lengthy description of the nonhuman primates. Yet even Buffon let a note of fascinated and unscientific disgust and fantasy creep into his description of the baboon (Reynolds 1981:39):

> He is insolently salacious, affects to show himself in this situation, and seems to gratify his desires *per manum suam,* before the whole world. This detestable action recalls the idea of vice, and renders disgustful the aspect of an animal, which nature seems to have particularly devoted to such an uncommon species of impudence; for, in all other animals, and even in man, she has covered these parts with a veil. In the baboon, on the contrary, they are perpetually naked, and the more conspicuous, because the rest of the body is covered with long hair. The buttocks are likewise naked, and of a blood red colour; the testicles are pendulous; the anus is uncovered, and the tail always elevated. He seems to be proud of all those nudities, for he presents his hind parts more frequently than his front, especially when he sees women, before whom he displays an effrontery so matchless, that it can originate from nothing but the most inordinate desire. . . .

Poor baboons—lustful, perverted, ugly, and indecent. . . .

Reynolds (1981) nonetheless dates the beginning of the modern approach to studying nonhuman primate behavior to Buffon's era. Commenting on an English traveler's account of chimpanzees, the philosopher Rousseau stressed the need for observations. It cannot be claimed that apes are animals, he argued, until it can be shown that their primitiveness is due to an absence of culture and not to inadequate training. In 1760, struck by the similarities between apes and people while writing a comparative review of their behavior, Hoppius, a student of Linnaeus, lamented the sparsity of information available about behavior.

Rousseau and Hoppius provided no new facts, but they did emphasize the importance of collecting them if the relationship of nonhuman primate behavior to human behavior was ever to be studied rationally. One hundred fifty years later the situation was in some ways little different. Even though nonhuman primates figured importantly in the debates about human evolution that raged in the second half of the nineteenth century, and even though a reasonably secure classification of the order had been achieved by the close of the century, understanding of behavior had advanced little. When the battle over evolutionary theory was won, Darwin and his allies had succeeded in extending the recognition of the morphological similarity between nonhuman primates and us to a realization that primates are our relatives and the closest living representatives of our forebears. Yet belief in the savageness and general bestiality of these

Figure 1.4 This twentieth-century political cartoon shows that old ideas die hard! (Courtesy of the Conservative Research Department, London.)

Figure 1.5   This engraving is the frontispiece of *The Malay Archipelago: The Land of the Orang-utan, and the Bird of Paradise* by Alfred Russel Wallace in 1869. One of the first proponents of the theory of natural section, Wallace dedicated his book to Charles Darwin.

relatives remained strong, particularly with regard to the great apes (Figure 1.5). For example, on a collecting mission to Africa at the turn of the century, zoologist R. G. Garner decided to spend some time observing the great apes; to this end he built a big cage in the forest—not for the apes, but for his own safety (Reynolds 1967).

The dawn of the twentieth century brought about the systematic empirical research for which enlightened thinkers of the eighteenth century had been calling. It was begun by psychologists, notably Robert Yerkes, whose interest in apes was inspired by the traditional cause, their similarity to us. For Yerkes, apes were "perfect models of human beings" (Haraway 1978a), and he believed that through his work and particularly through the study and manipulation of chimpanzee intelligence and sociosexual life, it might become possible to demonstrate the "possibility of re-creating man himself in the image of a generally acceptable ideal" (Haraway 1978a).

In a fascinating, albeit controversial, history of the relationship between primate research and theories of human nature and society, Haraway (1978a, 1978b) traced subsequent developments in the modern approach to this topic. The anatomist Sir Solly Zuckerman, the psychiatrist David Hamburg, and the anthropologists Earnest Hooton, Sherwood Washburn, and Irven DeVore must surely count as major figures in this history (e.g., Zuckerman 1932, 1933; Hooton 1954; DeVore and Washburn 1963; Hamburg and McCown 1979b). We should also take note of the role played by the social anthropologist A. R. Radcliffe-Brown, who influenced both data collection methods and the general orientation of the first physical anthropologists to do field studies on primate social behavior. These studies were predominantly descriptive and, reflecting Radcliffe-Brown's theoretical concerns, emphasized the cohesiveness and structure of the social group rather than the individual interests of its members (Gilmore 1981). But we are getting ahead of ourselves here, so let us return to Yerkes and the 1920s.

Although Yerkes's long-term concern may have been human nature or even human engineering, along the way he became increasingly interested in apes in their own right, and in 1929 he and his wife, Ada Yerkes, published an account of everything then known about apes. A recurrent theme of their book was how much remained to be found out, and they cited the lack of information about primates in the wild as one of the major gaps in current knowledge. Even though Yerkes himself never did fieldwork, he subsequently became known as the father of research on wild primates, for he trained and encouraged young researchers to go out and fill that gap.

And so the investigation of primate ecology and social behavior in the wild began in 1931 with Clarence Ray Carpenter's study of howler monkeys (Figure 1.6) on Barro Colorado Island in the Canal Zone (now Panama) (Figure 1.7). Two of Yerkes's other students, H. C. Bingham and N. W. Nissen, had left for the field a few years earlier. The subjects of their research were gorillas and chimpanzees (Figure 1.8), respectively, but in the primatological sweepstakes they lost out, as many have done since: their animals

Figure 1.6　Howler monkeys *(Alouatta palliata)* are widespread in Central and South America. (Courtesy of K. Glander.)

were obstinately shy and elusive. Although both published their findings (Nissen 1931; Bingham 1932), they were unable to paint the rich portrait of primate life that Carpenter (1934) was to do.

Carpenter (1934) set out to gather data on the howler monkey with four stated goals: (1) to shed light on problems in psychobiology, particularly those relating to behavior and social relations; (2) to aid in the interpretation of data collected under strictly controlled laboratory conditions; (3) to suggest better techniques for studying behavior and relations in the laboratory; and (4) to point to significant problems that could be investigated experimentally. Trained as a psychologist, Carpenter had done his doctoral dissertation on the sexual behavior of pigeons. His goals in doing primate fieldwork were those of a psychobiologist interested in linking behavior and physiology, and perhaps

Figure 1.7   Barro Colorado Island in Panama was the site of one of the first systematic studies of the ecology and social behavior of primates. Shaded area represents highest terrain of island. (Adapted from Carpenter 1964.)

his greatest theoretical contribution was linking the interpretations of the laboratory-based disciplines of comparative psychology and sexual physiology to the findings of evolutionary and ecological field biology (see Carpenter 1964 for a collection of his writings). Our concern here, however, is with the ecological aspect of the research tradition established by Carpenter, for it came to dominate much of the ecological field-work done on primates. This tradition was one of meticulous description but made little allusion to theoretical issues being discussed by ecologists studying other animals.

Figure 1.8   An adult female chimpanzee *(Pan troglodytes)* rests beside her two-year-old son. (Nancy Nicolson/Anthro-Photo.)

After World War II, the ranks of primate field-workers began to swell, and by 1965 "more than 50 individuals from the fields of anthropology, psychology, and zoology" (Washburn and Hamburg 1965:608) were involved. From the early 1950s, the efforts of researchers in the West were complemented by a strong tradition of primatological research initiated in Japan by Imanishi and his colleagues (see S. A. Altmann 1965 for translations of this work). The number of species about which information was available increased accordingly, and much of it was summarized in a volume edited by Irven DeVore (1965) entitled *Primate Behavior: Field Studies of Monkeys and Apes.* Two features of this volume are particularly noteworthy here. First, Washburn and Hamburg (1965:621) restated the importance of the link forged by Carpenter: "The most fundamental reason for field studies is that monkeys and apes have evolved as a result of selection pressures under natural conditions. The field studies are essential to an understanding of the way structure and behavior are adapted to various ways of life." The goal of fieldwork, in other words, is to establish a theory of the evolution and adaptiveness of nonhuman primate behavior. The possible contribution of nonhuman primate studies to the understanding of ourselves goes unmentioned here and, with few exceptions, in the volume as a whole. A second notable feature of the book is that the accounts of ecology by zoologists and anthropologists alike are mainly descriptive, covering the size and composition of social groups, foods eaten, ranging and activity pat-

Figure 1.9   An infant chacma baboon *(Papio cynocephalus ursinus)* suckles while its mother is groomed by an adult male, beside whom an older sibling is huddled. (Sherwood L. Washburn/ Anthro-Photo.)

tern and so forth. The only ecological problem discussed at length is the way in which the environment shapes these aspects of primate life. *Primate Behavior* is, in my estimation, an accurate reflection of "state-of-the-art" primate field research in the mid-1960s.

One of the earliest explicit hypotheses about the relationship between the environment and social behavior was based on the observations of an English psychologist, K. R. L. Hall. In the 1950s Hall embarked on a series of studies of African monkeys and baboons, and in a study of the chacma baboon *(Papio cynocephalus ursinus)* (Figure 1.9), he found a correlation between group size and habitat type: groups were generally smaller in impoverished environments. This, he suggested, illustrates a more general relationship whereby variations in group size represent adaptations to differing environmental conditions (Hall 1962, 1963). Taking advantage of the tremendous increase in information about primates over the next ten years, Hall's student John Crook and his student Steve Gartlan made the first attempt to develop this idea formally. They postulated five adaptive grades among the primates, each an interrelationship of behavioral and environmental features (Crook and Gartlan 1966). This classification stimulated a wave of research and syntheses that continues yet and will be discussed in detail in Chapter 9.

Over the past 15 years, the theoretical content of primate ecology and the diversity of problems considered have grown enormously. The issues that initially preoccupied researchers, namely the adaptiveness of social behavior on the one hand and its importance in understanding our own species on the other, are now complemented by many others. For example, we shall discuss the findings of researchers interested in the links between a species' biology and its diet, primate population dynamics, and the ecological relationships between primates and other animals and between primates and the plants on which they feed. We shall also look at the work of those trying to reorient the study of environmental influences upon social organization. In almost all instances, the exchange of ideas between these researchers and those studying other animals, particularly other mammals, is clear. In short, ecological research on the primates has entered a new and particularly productive era.

We began by stating that primates have traditionally received special treatment, and in this section I have tried to show how and why. To substantiate the claim further, however, we now turn to the history of ecological research on other animals, which long followed a rather different pathway.

1.3 Notes on the History and Antecedents of Ecology as a Discipline

Like the history of primate studies, the multifaceted history of the discipline of ecology has been discussed in depth by several writers, and once again the reader is referred to these sources for a more comprehensive account than that presented here (Allee et al. 1949; Cole 1957; Doutt 1964; Odum 1964, 1971; Egerton 1968a, 1968b, 1969; Krebs 1972).

The word *ecology* was coined in 1868 by Ernst Haeckel, the great German biologist. Charles Elton's *Animal Ecology,* published in 1927, was the first generally accepted synthesis of knowledge in this area. With it, ecological research acquired the status of a science concerned, in Haeckel's words, with the "total relations of the animal to both its organic and its inorganic environment" (Egerton 1968b). A hundred years later, this definition is still in use (Odum 1971), although many others have been offered, from Elton's "scientific natural history" to Krebs's (1972) recent rendering, "the scientific study of the interactions that determine the distribution and abundance of organisms." Despite its fledgling status as a science, the roots of animal ecology are deeply buried in diverse traditions, and modern definitions of the field reflect this fact. Elton, for example, emphasizes the importance of the age-old pursuit of natural history while Krebs stresses *demography*, the statistical study of populations.

Ancient accounts of natural history commonly dwelt upon the sudden and dramatic increases in population that characterize many animal species, and which were clearly a source of alarm to people in those days. Egyptians, Babylonians, and Romans alike accorded special significance to plagues of animals, often interpreting them as signs of divine wrath. Yet a few thinkers did propose natural causes for these events. For example, Aristotle suggested that explosions of the vole population were related to high reproductive rates and that this population was ultimately curbed not by predators ("they make no way against the prolific qualities of the animal") but by climate ("nothing succeeds in thinning them down except the rain") (Egerton 1968a).

At the heart of Aristotle's writing (indeed, at the heart of Greek science) was the belief that all life is maintained in a constant and harmonious state designed for the benefit of the human species. We find this belief revived when, following almost a thousand-year gap, interest in natural history was rekindled in the Western world during the twelfth century. For hundreds of years thereafter, the writings of the ancient Greeks were enthusiastically pored over as infallible sources. By the seventeenth century, however, biologists had once more begun to study nature for themselves. The resulting increase in knowledge allowed Linnaeus to embark on his celebrated classification of plants and animals. Like the Greeks, Linnaeus believed that the natural world was designed to function harmoniously and that all species were created independently of one another, yet he was one of the earliest biologists to stress the importance of external conditions in regulating the lives of animals.

Less interested in structure and systematics than Linnaeus, his contemporary Buffon focused more on the behavior of animals and on the factors regulating their numbers under natural conditions. In the *Histoire Naturelle* he considered a wide range of animals in addition to primates and pondered many ecological phenomena; he suggested, for example, that Aristotle's voles were killed not by rain but by "contagion, a necessary consequence of too great a mass of living matter assembled in one place" (Cole 1957).

Earlier ideas developed by students of human demography began to influence natural historians toward the end of the eighteenth century. In 1798, Malthus published a work in which he argued that although populations can in principle increase at a geometric rate, their food supplies increase only arithmetically. He concluded that because of the vast discrepancy between these two rates, the availability of food must ultimately restrict increases in a given population.

Malthus said little that was not already known, but his work helped trigger a major debate in biology that continued for much of the nineteenth century. At the core of the debate lay a 2000-year-old belief that the natural world was designed to be stable and to function harmoniously for the benefit of the human species. In questioning this belief, scientists began to argue about the possibilities of some species becoming extinct and others evolving. The climax of the debate came in 1859 with the publication of Charles Darwin's treatise *On the Origin of Species,* but much of Darwin's reading material during his voyage on the *Beagle* was provided by turn-of-the-century scholars such as Humboldt, Candolle, and Lyell, who were caught up by the winds of change that had even then begun to blow.

In a discussion of plant geography in 1807, Alexander von Humboldt stressed the importance of *physical factors,* especially climate, topography, and soil, for plant distribution. In 1820, Augustin Pyrame de Candolle took up the theme but went beyond Humboldt by introducing *competition* as an important determinant of plant distributions. The most significant attempt at synthesis during the early years of the nineteenth century was Charles Lyell's *Principles of Geology,* published between 1830 and 1833. Lyell broached many questions concerning the nature of species, most of which led to some discussion of population dynamics. While he inclined toward the idea of the supernatural creation of species and was clearly influenced by the concept of a providential ecology, Lyell was nevertheless the first to emphasize the changing nature of animal populations; he described population size as fluctuating in response to many factors and saw changes in population distribution as an active response to the pressure of numbers. He understood the importance of competition in nature, and he accepted the proposition that species can become extinct.

When Darwin expounded his theory of evolution in 1859, the curtain was finally brought down on the 2000-year-old belief in the harmony of nature; in Darwin's vision, individuals compete for survival in a constantly changing world. Although the notion that we share a common ancestor with other primates gave rise to much colorful and often emotionally charged debate, Darwin's theory depended heavily on integrating his observations of many different animals with concepts of demography developed for human populations and reapplied to animals. This integration of natural history and population studies was both a cause and an effect of Darwin's ideas. It was in that process that ecology was born.

If natural history and demography are the parents of ecology, the role of midwife belongs to the applied research of the past two hundred years in fisheries, medicine, and agriculture. Many tenets of modern ecology derive from work done in these areas. Let us briefly consider two illustrations of this point, both from agriculture.

Crop depredation by a wide assortment of pests is a perennial problem for farmers, and the fact that certain insects prey upon these pests has probably been well known since ancient times (Doutt 1964). By the eighteenth century, a growing literature documented the effects of insect-eating insects, and in the last hundred years biological and agricultural knowledge has been applied to produce biological control, the introduction of predators to control the populations of potentially devastating pests.

Successful application of biological control involves considerable understanding of population dynamics and of the factors regulating the abundance of organisms in nature. For example, it requires knowing not only that members of species A eat members of species B but also that they limit the population size of B in the process. These relationships were known by the early nineteenth century through the practical observations of farmers and an increasing interest in population dynamics triggered by Malthus's work, but the efficacy of biological control became widely established only in 1889. In 1887 the young citrus industry of California was threatened with bankruptcy by a massive infestation of cottony-cushion scale. Late in 1888, as a last resort, the vedalia beetle was introduced to California from Australia, and it proceeded to devour cottony-cushion scale with such relish that by 1889 the epidemic was over.

The practical concern of farmers with pest control provided the impetus for research that later spawned within the discipline of ecology a host of theoretical models of predator–prey relationships. The preoccupation of farmers with crop production was of major importance in the development of a central concern of community ecologists, the measurement of biological growth in different habitats (Section 11.2). In fact, the modern framework of this area of ecological research can be traced back to an eighteenth-century botanist and horticulturalist named Richard Bradley (Egerton 1969). Bradley evaluated agricultural productivity as a ratio of financial costs to benefits, and he used financial profit as a comparative measure of the relative productivity of organisms ranging from grapevines to crayfish. Today biological productivity is evaluated in terms of the rate and efficiency of energy flow through the environment, about which little was known in Bradley's time. Yet his recognition that production rate is important and his attempt to find a common denominator, a currency, to permit comparisons between diverse forms provided an approach to the problem that was to be adopted and developed by theoretical ecologists two centuries later.

The definition and position of modern ecology in relation to the other biological sciences have been a subject of much discussion, but today the prevailing view is that ecology is concerned primarily with the principles governing particular levels of biological organization, namely individual organisms, populations, communities, and ecosystems (e.g., Miller 1957, Odum 1971, Krebs 1972) (Figure 1.10). At one end of this spectrum, ecology merges into the study of individual physiology; at the other, into the study of the earth as a single unit, the biosphere. Characteristic properties arise at each

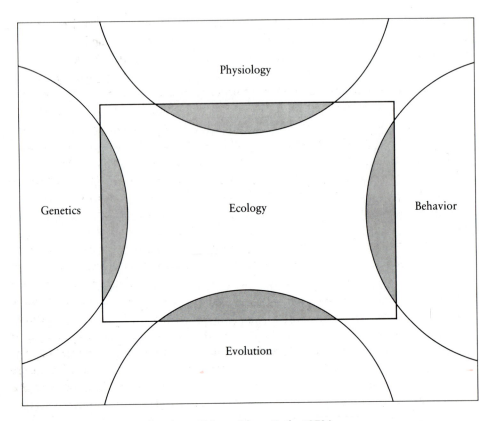

Figure 1.10   The scope of ecology. (Adapted from Krebs 1972.)

level of organization and are usually referred to as *emergent features*. Although research at one level can help in explaining a feature of another level, it cannot provide a complete explanation. For example, if we want to characterize the dynamics of a population, we must study all its members over time: knowledge of the life histories of a few individuals will not be sufficient. Emergent properties and, more generally, approaches to the study of hierarchical systems are currently subjects of heated debate (Wimsatt 1980). Ecologists study higher levels of biological organization and oppose the *reductionist* view that research at any level higher than the cell is a waste of time until the organization of the cell is fully understood.

Today our knowledge of animal ecology is concentrated at the levels of the individual and the population. Plant ecologists, in contrast, have traditionally directed their attention to the community level. There are good reasons for this difference of emphasis. The complexity of organization and the difficulty of studying animal communities under natural conditions are daunting, and such communities are hard to replicate within the confines of a laboratory. Plant ecologists, on the other hand, deal with stationary organisms with less diversified roles than animals; individual plants are often hard to differentiate, they may live for hundreds of years, and many produce dormant seeds, so that recognizing populations is a problem in itself. One result of these differences is that many of the terms used by animal ecologists to describe processes at levels higher than the population were developed by plant ecologists. For example, the term *ecosystem,* referring to a community of plants and animals and their physical environment, was originally proposed by a noted plant ecologist, A. G. Tansley (1935).

Ecologists differ not only in what they study but also in how they study it. First historically are classical ecologists who uphold the long tradition of meticulous description and classification characterizing natural history. These ecologists inventory the flora and fauna of a particular area and describe their interrelationships, basing conclusions upon extensive observations under field conditions. Next are theoretical ecologists, who seek to identify rules governing the operation of ecosystems. Theoretical ecology came into its own in the twentieth century. Emlen (1973) distinguishes between the goals and methods of theoreticians interested in explaining short-term change and those interested in evolutionary change. Ecologists studying short-term change develop models that describe the feedbacks, or "inter-locking cause–effect pathways," present in complex systems (Watt 1966). In contrast, models of evolutionary change "sacrifice precision in an effort to grasp at general principles" (May 1973). Far fewer factors are taken into account, less complex cause–effect relationships are invoked, and the model's predictive accuracy is more limited. On the other hand, such models attempt to describe changes occurring on a much larger scale over a much longer time period.

All these approaches are fruitful, and all have their drawbacks. Ideally they should be complementary, but in practice they tend to remain separate (Fretwell 1972) and modern ecology lacks as yet a unifying theme or theoretical framework to accommodate and integrate all the work being carried out under its flag. Krebs (1972:11) has summarized the current situation well, with a blend of realism and optimism: "The theoretical framework of ecology may be weak at the present time, but this must not be interpreted as a terminal condition. Eighteenth-century chemistry was perhaps in a comparable state of theoretical development as ecology at the present time. Sciences are not static and ecology is in a strong growth phase."

1.4   The Primate Pattern

While nonscientific considerations have undoubtedly played their part in separating the study of the primates from studies of other animals, there are also practical, if not exactly scientific, reasons why primates have been excluded as subjects from the mainstream of ecological research. These reasons have to do with certain primate characteristics. In particular, a long life span makes primates impractical subjects for research requiring knowledge of the life histories of large numbers of individuals, and their large size makes it difficult to simulate a natural environment for them in the laboratory, where controlled experiments can most readily be performed. (Note, however, that other mammals of comparable body size pose similar difficulties as research subjects.)

In this section we begin by showing the distinctiveness of primate life history patterns compared to other mammals. Then we shall consider another important distinguishing feature of primates, namely a large brain relative to body size. Finally, we shall discuss other characteristics of the order and how these characteristics have been interpreted ecologically.

To say that primates are long-lived compared with many other animals is to describe one aspect of their *life history patterns*. A complete description of a life history pattern must answer a number of questions including the following: at what age does the individual first reproduce? Does it reproduce once or many times in its life? If it reproduces repeatedly, what is the interval between births? How many offspring does it give birth to at a time? How long does pregnancy, or *gestation,* last? How long are offspring dependent on either or both parents? For primates, the general answer to these questions goes something like this: they are long-lived, have a long period of immaturity, and begin to reproduce relatively late. Usually, only one large offspring is born at a time. Extended periods of gestation and nursing of the newborn by females insure a long interval between births in most species, and seasonal changes in the environment may limit breeding to certain times of year. Females are involved in reproductive activities during much of their adult life, while direct male involvement in reproduction is much more intermittent. In only a few species do males assist in infant care, and social factors may limit a male's access to fertile females for much of his life.

Body size, a recurring theme in this book, is important in understanding variations in life history patterns. Its relationship to these patterns among a variety of mammals has been examined by Western (1979), and in order to place the primate pattern in broader perspective we turn now to his work.

Table 1.2 lists life history data for selected primates, carnivores, and artiodactyls (even-toed ungulates). *Gestation time* is the number of days between conception and birth. *Age at first reproduction* is defined as the age at sexual maturity plus gestation time, since there are very few data on actual age at first breeding. *Life span* is the maximum recorded longevity. Such records are usually for captive animals, partly because the age of adult animals in the wild is rarely known and partly because animals often live longer in captivity than in the wild.

This observation about life span leads to some general notes of warning. First, Table 1.2 combines information from many sources. In several life history characteristics, animals living in the wild may differ significantly from those maintained in zoos or labora-tories, and the data presented are mostly based on so few individuals that they cannot be considered representative of the species. Second, the list of species is far from complete. Primates are long-lived, the time units for many reproductive variables are relatively great, and individuals are rarely studied systematically in either the lab or the field for years at a stretch. Thus there are few species for which any, let alone reliable, informa-tion is available. Third, the body weights listed are averages for both sexes combined, and this can be misleading.* Many mammalian species are *sexually dimorphic*: males and females differ in morphology, particularly in body size (Figure 1.11). In such species, there may be no adults that weigh the average computed for the two sexes together.

Figure 1.12 plots gestation time, age at first reproduction, and life span against body weight for the species listed in Table 1.2. Each life history variable shows an absolute increase from small species to large ones, but each differs in the *rate* of increase. Specifi-cally, in all three groups of mammals, the larger an animal is, the shorter its gestation time and the more delayed its age at first reproduction are relative to its size. In other words, big animals have longer gestation periods and take longer to mature than small animals, but their gestation period is not as long as would be expected if there were a one-to-one relationship between it and body size, and maturation takes longer than would be expected if there were a one-to-one relationship between it and size.

The three groups of mammals differ markedly in two respects. First, gestation time for a carnivore of a given weight is significantly shorter than for a primate or an artio-dactyl of equivalent weight. Second, both life span and age at first reproduction are greater in primates than in artiodactyls and carnivores: a primate of a particular weight takes longer to mature and lives longer, on average, than a carnivore or artiodactyl of similar size.

In sum, there is a relationship between the characteristic life history of a mammal and its body weight. Primates are no exception. The biological mechanisms involved are not understood. It has been proposed that metabolic rate, itself closely predicted by body weight, regulates reproductive variables, but no one knows just how. Besides, the relationship between body size and life history is not completely deterministic. Figure 1.12 shows that animals of similar size may differ significantly in some respects. How can this be accounted for?

The partial answer to this question introduces another feature of the primate pat-tern, namely a large brain relative to body size (Jerison 1973). The slower growth rates of primates both before and after birth and their long lives are no longer anomalous com-pared with other mammals if viewed from the perspective of their large brains. Sacher

---

* However, when slopes for primates were recomputed using female weights, they were found not to differ significantly from those calculated by Western.

Table 1.2   Life history data for various mammals

Species	Body weight[a] (kilograms)	Gestation time (days)	Age at first reproduction (days)	Life span (days)
Arriodactyls				
Hippopotamus (Hippopotamus amphibius)	1,000	437		10,230
Giraffe (Giraffa cameleopardalis)	750	339	1,450	7,665
African buffalo (Syncerus caffer)	450	263	1,150	7,665
Eland (Taurotragus oryx)	340	234	910	6,205
Gnu (Connochaetes taurinus)	165	257	980	5,000
Waterbuck (Kobus defassa)	160	219	690	7,413
Hartebeest (Alcelaphus buselaphus)	125	219	775	6,205
Waterbuck (Kobus leche)	72	224	630	6,205
Kob (Kobus kob)	58			
Wart hog (Phacochoerus aethiopicus)	45			
Grant's gazelle (Gazella grantii)	40		724	4,570
Impala (Aepyceros malampus)	40	191	660	3,800
Bushbuck (Tragelaphus scriptus)	30	178		3,650
Gray duiker (Sylvicapra grimmia)	17	210		3,980
Thomson's gazelle (Gazella thomsonii)	15	191	550	4,015
Oribi (Ourebia ourebia)	15			
Dik-dik (Madoqua kirkii)	6	174	355	3,470
Primates[b]				
Orang-utan (Pongo pygmaeus)	40.0	275	4,290	11,680
Chimpanzee (Pan troglodytes)	30.0	224	5,110	14,965
Baboon (Papio cynocephalus)	13.0	175	2,365	10,460
Langur (Presbytis entellus)	11.6	168	1,445	7,300
* Siamang (Hylobates syndactylus)	10.4	233		5,900
Rhesus macaque (Macaca mulatta)	6.9	168	1,345	9,125
Spider monkey (Ateles geoffroyi)	5.8	139		7,300
Woolly monkey (Lagothrix logotricha)	5.8	180	1,640	4,380
Howler monkey (Alouatta palliata)	5.7	140	1,508	
Patas monkey (Erythrocebus patas)	5.6	165	1,447	7,360
Gibbon (Hylobates lar)	5.5	210	2,947	12,045
Long-tailed macaque (M. fascicularis)	4.1	165	1,734	9,855
Ring-tailed lemur (Lemur catta)	2.7	132	1,095	9,885
Slow loris (Nycticebus coucang)	1.1	193		4,620
Saki (Pithecia pithecia)	1.0	225		4,985

Species	Body weight[a] (kilograms)	Gestation time (days)	Age at first reproduction (days)	Life span (days)
Primates[b] (cont.)				
Bush baby (*Galago crassicaudatus*)	1.0	133		5,110
Squirrel monkey (*Saimiri sciureus*)	0.6	175	1,270	7,660
Tamarin (*Saguinus oedipus*)	0.5	140	685	2,795
Golden marmoset (*Leontopithecus rosalia*)	0.4	133	498	5,475
Titi (*Callicebus jacchus*)	0.3	145	715	5,840
Bush baby (*G. senegalensis*)	0.2	120	488	3,285
Slender loris (*Loris tardigradus*)	0.2	167		2,555
Tarsier (*Tarsius bancanus*)	0.1	180		4,380
Bush baby (*G. demidovii*)	0.1	112	477	
Carnivores				
Tiger (*Panthera tigris*)	209	99	8,868	6,935
Lion (*Panthera leo*)	120	108	1,205	8,760
Spotted hyena (*Crocuta crocuta*)	70.5	110	1,185	9,125
Puma (*Felis concolor*)	70	90	750	7,300
Cheetah (*Acinonyx jubatus*)	57.5	93	1,085	5,657
Coyote (*Canis latrans*)	53	63	806	5,840
Hunting dog (*Lycaeon pictus*)	28.4	76	704	3,650
Caracal (*Felis caracal*)	17	74	418	
Old World badger (*Meles meles*)	13	53	353	5,475
Red fox (*Vulpes fulva*)	9	53		4,380
American badger (*Taxidea taxus*)	6.7	42	343	4,745
Arctic fox (*Alopex lagopus*)	6	51	351	5,110
Kit fox (*Vulpes velox*)	5.5	51		4,380
White-tailed mongoose (*Ichneumia albicaudata*)	4.5			3,650
Tayra (*Tayra barbara*)	4.4			4,380
European wildcat (*Felis silvestris*)	4	63	363	
Striped skunk (*Mephitis mephitis*)	1.6	64	429	3,650
Fennec fox (*Fennecus zerda*)	1.5	51		4,015
Ring-tailed cat (*Bassariscus astutus*)	0.9			2,920
Eastern mink (*Mustela vison*)	0.7	49	149	
Black-footed ferret (*Mustela nigripes*)	0.6	42	162	

Source: Data for artiodactyls and carnivores are from Western 1979. Data for primates differ from Western 1979 when more recent and reliable information is available.

[a] Artiodactyl and carnivore weights are averages for males and females combined. Primate weights are averages for females alone.

[b] Findings based on the revised primate data set presented here do not differ significantly from those based on Western's (1979) data set.

Figure 1.11   In many primates, such as the baboons pictured here, males are much larger than females. In only a few are females the same size as or larger than males. (Courtesy of R. Johnson.)

and Staffeldt (1974; Sacher 1970, 1974) began a series of studies by suggesting that differences in fetal growth rates might be attributable to a common limiting factor, the maximum growth rate of neural tissue, which is known to be slower than that of other tissues. If this suggestion is correct, they reasoned, the maximum rate of fetal growth would be constrained by the maximum growth rate of the brain. In other words, brain size would limit the rate of development: the larger the brain, the slower would be the rate of growth. Sacher and Staffeldt used data from 91 species of mammals to test this idea and found that neonatal brain weight does closely predict gestation time: embryonic development takes longer in species in which the adults have a large brain. Further, variations among different groups of mammals in the life spans of similarly sized species such as those already noted are much reduced when life span is related to *brain* weight instead of body weight: animals with big brains live longer. Variations are best accounted for when both brain weight and body weight are taken into account.

These findings indicate tantalizing links between some of the features unique to primates and also show the biological continuity between primates and other mammals.

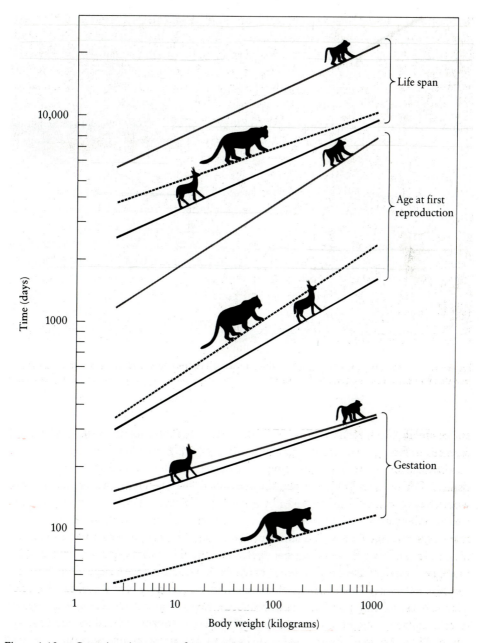

Figure 1.12   Gestation time, age at first reproduction, and life span in relation to body weight in selected primates (grey lines), carnivores (dashed lines), and artiodactyls (solid lines). (Adapted from Western 1979.)

However, the issues *not* tackled are as interesting as those that are. For one thing, a correlation between brain weight on the one hand and gestation length and life span on the other is a long way from being a complete explanation of the relationship. How, for example, does the growth rate of neural tissue constrain that of other tissues, and why are neural tissue growth rates slower than those of other tissues? Second, is the acquisition of a large brain a way of achieving a longer life or, contrarily, is long life an incidental consequence of having a large brain? In other words, has natural selection favored a change in primate life history patterns, with increased brain size the means by which this was achieved, or has it favored larger brains themselves because of the advantages they confer, with the change in life history pattern occurring as a secondary effect? Is it reasonable to try to distinguish cause from effect in the first place?

The selective pressures favoring an increase in brain size interested Darwin a hundred years ago and continue to attract attention today (e.g., Jolly 1966a; Jerison 1973; Charles-Dominique 1975; Humphrey 1976; Eisenberg and Wilson 1978, 1981; Clutton-Brock and Harvey 1980; Martin 1981; Passingham 1982). A review of ideas about the evolution of primate brains would demand a volume to itself, however, and here we consider only explanations of the ecological and evolutionary significance of primate characteristics taken together.

No single distinguishing feature or set of features characterizes all living primates. However, primates can be viewed as products of a number of distinctive evolutionary trends (Le Gros Clark 1962). The presumed ancestral mammalian condition provides the starting point for defining these trends, and animals are said to be *specialized* in features that have undergone major evolutionary change away from the corresponding feature of this condition and *primitive* in those that have changed little. At best, it would be possible to trace each evolutionary trend through the fossil record. In practice, trends are more often inferred from comparative studies of living primates. In these studies, primate species are arranged in a graded series that suggests an evolutionary sequence. There is, of course, a danger in this since every species is itself the product of evolutionary processes and a set of living species cannot, therefore, be taken to represent an evolutionary sequence. However, the fossil record is incomplete and some of the features with which we are concerned do not fossilize in any case; thus, a comparative approach is often the only possible one.

Take the primate brain, for example. A large brain relative to body size characterizes all living primates to some extent. Primate brains also show varying degrees of increased elaboration compared with other mammals, particularly in the neocortex (Figure 1.13), the area concerned with "higher functions" such as the integration of sensory messages and voluntary control of movement. The expansion of primate brains can be traced with some precision in the fossil record. In contrast, stages in the elaboration of the

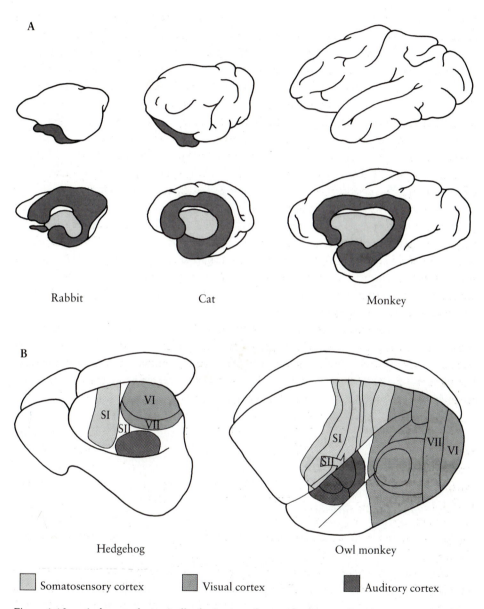

Figure 1.13   **A** shows schematically the increased area of neocortex in primates compared with other mammals. (Adapted from MacLean 1954.) **B** shows the addition of cortical areas that accompanied this expansion of neocortex. The hedgehog has only a few cortical areas, including visual areas VI and VII and somatosensory areas SI and SII. In contrast, the owl monkey has several "new" areas, particularly in the visual cortex. (Adapted from Merzenich and Kaas 1980.)

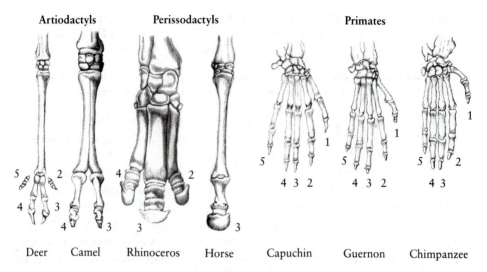

Figure 1.14   The first mammals had five digits on each of their hands and feet. Unlike many other mammals, primates have retained this feature. (Not to scale.) (Adapted from Kent 1954 and Hershkovitz 1977.)

brain are much more difficult to pinpoint and are usually inferred from living forms. The other six primate evolutionary trends listed below are delineated using a similar combination of methods (Le Gros Clark 1962):

1.   Retention of an unspecialized postcranial skeleton. For example, primates generally have five digits on each of their hands and feet (Figure 1.14); they also have a collarbone, or clavicle, a bone that is reduced or absent in some groups of mammals.

2.   Increased mobility of the digits, especially the thumb and big toe.

3.   Replacement of pointed claws by flattened nails. This probably occurred along with the evolution of a more discriminating sense of touch through the development of sensitive tactile pads on the digits.

4.   Shortening of the snout and, more generally, reduction of the apparatus of smell.

5.   Elaboration of the visual apparatus with the development of binocular vision.

6.   Reduction in the number of teeth, with preservation of a primitive cusp pattern on the molar teeth.

Together, these features add up to a pattern that Le Gros Clark (1962:41–42) characterized as follows:

While many other mammalian orders can be defined by conspicuous specializations of a positive kind which readily mark them off from one another, the Primates as a whole have preserved rather a generalized anatomy and, if anything, are to be mainly distinguished from other orders by a negative feature—their lack of specialization. . . . The evolutionary trends associated with the relative lack of structural and functional specialization are a natural consequence of an arboreal habitat, a mode of life which among other things demands or encourages prehensile functions of the limbs, a high degree of visual acuity, and the accurate control and co-ordination of muscular activity by a well-developed brain.

In this statement, Le Gros Clark was echoing a theme that dated back almost to the turn of the century (Cartmill 1972). In his presidential address to the British Association for the Advancement of Science in 1912, G. Elliot Smith argued that many of the characteristic traits of primates had evolved as adaptations to an arboreal existence. For instance, up in the trees vision and touch are more important than a sense of smell, hence the elaboration of primates' visual apparatus and the atrophy of the olfactory apparatus; exceptional agility is needed to negotiate unstable, discontinuous, and three-dimensional networks of branches, hence the elaboration of areas of the primate brain concerned with locomotor control. A few years later, Frederic Wood Jones developed these ideas further; for example, he suggested that selection for efficient tree climbing had favored the evolution of grasping hands and feet (Figure 1.15). Elliot Smith and Wood Jones both considered the primate trend toward frontal orientation of the eye sockets to be a by-product of the reduction in snout size. In 1921, Treacher Collins proposed that *orbital convergence* had been selected because it permitted overlap and stereoscopic integration of the two visual fields; this would be a major asset for animals that must judge distance accurately when leaping from one branch to another.

Over the next 50 years, certain weaknesses were pointed out in these ideas, which were collectively labeled the arboreal theory of primate evolution, but they remained generally accepted as the most plausible ecological and evolutionary interpretation of the traits distinguishing members of the order. In 1972, however, Cartmill expounded the theory's fatal flaw at length and proposed a rather different interpretation of the basic primate pattern. The main problem with the theory is that several kinds of mammals have adapted to life in the trees without evolving a primatelike set of characteristics. Squirrels, for example, are highly arboreal and yet they have eyes that face sideways, a well-developed sense of smell, and hands and feet with little manipulative ability (Figure 1.16). Yet if abundance is a measure of success, squirrels have done very well for themselves.

Cartmill's (1972) revised interpretation of primate traits was based on an examination of the distribution of these traits among other animals, with particular attention given to the life-styles of these animals. He found prehensile, or grasping, hands and feet to be universal among mammals that forage for insects or vegetable foods in the terminal branches of shrubs; such animals suspend themselves by their hindlimbs and use their forelimbs to reach and manipulate food items. Outside the primates, orbital con-

Figure 1.15   The grasping hands and feet of primates help them hang from small branches and twigs while feeding in the periphery of trees. (Photography by A. Richard.)   Figure 1.16   The grey squirrel *(Sciureus carolinensis)*. Claws, rather than grasping hands and feet, help this animal keep its balance and move around in the trees. (Animals Animals/©Leonard Lee Rue III.)

vergence occurs in predators that rely on vision to detect their prey (Figure 1.17), and contrary to the expectations aroused by the arboreal theory, there is no correlation between this trait and the degree of acrobatic skill demonstrated by arboreal mammals. In fact, orbital convergence is not merely unnecessary for gauging distance in arboreal leaps but probably as much a handicap as an advantage. Among the living primates, orbital convergence may be most advantageous for those that feed on easily alerted and fast-moving insect prey. Stereoscopic vision improves the accuracy of a primate's final strike, and increased overlap of the visual field makes it easier to compensate for evasive action taken by prey. Finally, there is no evidence that an arboreal life-style encourages

Figure 1.17   Forward-facing eyes are found not only in primates but also in predatory animals such as cats, owls, and hawks.

the loss of olfactory acuity. Reduction of the olfactory apparatus in primates, contended Cartmill, had more to do with a rearrangement of the facial skeleton resulting from the increase in orbital convergence than with some trend toward "perfected adaptation to arboreal life" (Cartmill 1972:115).

In summary, Cartmill argued that many of the features characterizing living primates were originally adaptations to a life-style involving nocturnal, visually directed predation on insects in the terminal branches of the lower strata of tropical forests. The earliest mammals to show this hypothesized basal primate adaptation lived about 50 M.Y.B.P. (million years before the present) (Kay and Cartmill 1977) (Figure 1.18). Szalay and Delson (Szalay 1968, 1972; Szalay and Delson 1979), in contrast, believe that the fossil evidence indicates an earlier evolutionary origin for the primates, marked by a shift from insect eating to a more vegetarian diet. This debate concerns us here only insofar as it provides a background for Sussman's (in press) suggestion that the basal adaptation of the primates involved feeding on both insects and plant material in the terminal branches of trees. Sussman's argument is as follows. The location of prey by sight does not adequately explain the visual adaptations of the early primates, since there is increasing evidence that the living *strepsirhines* (i.e., lemurs and lorises) rely heavily on hearing and smell when they hunt. Besides, all the species in this suborder of the primates have substantial vegetable components in their diets. The radiation of the early primates coincided with the radiation of many modern groups of flowering plants, and increased visual acuity and other characteristic primate features were more likely a means of capitalizing upon a newly available array of resources including flowers, fruits, gums, nectars, and insects rather than upon insects alone. In Sussman's scenario, the early primates fed on and manipulated small food items in the terminal branches of trees under the low light conditions of the nocturnal world. This necessitated acute powers

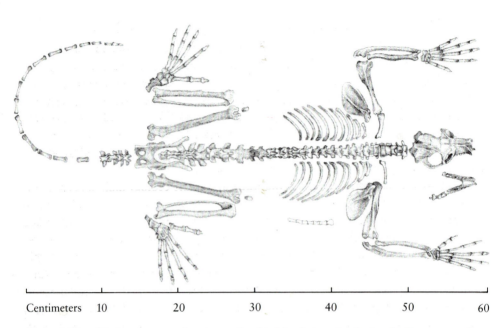

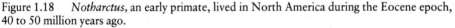

Centimeters   10   20   30   40   50   60

Figure 1.18   *Notharctus,* an early primate, lived in North America during the Eocene epoch, 40 to 50 million years ago.

of visual discrimination and precise hand–eye coordination, as well as the grasping hands and feet characteristic of terminal-branch feeders generally.

While this discussion of the likely life-style of the earliest primates may seem tangential in a book concerned primarily with the ecology of their living descendents, I believe it is in fact highly relevant. We shall return to Sussman's notion of early primates as discriminating eaters of insects and high-quality plant foods in Part I.

1.5   Overview

Our knowledge and understanding of primate ecology come almost entirely from the labors of psychologists, anthropologists, and zoologists over the last half century. These researchers initially approached the study of wild primates with somewhat different orientations, reflecting the interests of their own disciplines. For the psychologist, nonhuman primates provided access to the human mind and served as good subjects for studying links between physiology and behavior in complex social animals. For the physical anthropologist, nonhuman primates presented reflections of ourselves or our past selves; in particular, nonhuman primate societies provided an opportunity to study possible antecedents of our own social structures. For the zoologist, primates were a group of mammals about whose natural history little was known, and their relatively large size and social complexity, though disadvantages in the eyes of many, made them particularly interesting and challenging subjects for those who did elect to study them.

Yet despite the differences, the fieldwork of psychologists, anthropologists, and zoologists alike had several features in common. First, it was highly descriptive, at its best a meticulous natural history. Second, it considered ecology primarily with regard to how the environment might shape the nature of primate social groups. Third, most researchers stressed the adaptiveness of social organization and tended to overlook the individual's interests and, indeed, the individual as a unit of selection.

Why, we may ask once more, was the study of primate ecology limited in scope for so long, even though from the 1950s the ranks of primate field-workers included zoologists presumably well trained in the discipline of ecology? In Section 1.3 we identified three historical trends affecting this discipline, namely the development of applied research into medical and agricultural problems, the development of demographic methods and theory, and the emergence of a rigorous tradition of descriptive natural history. The primates figured little in the first two of these trends and this helps account for their anomalous position. To the applied scientists of Europe and the United States, primates were of small interest. They never threatened the California orange crop; indeed, for the most part they lived in areas so remote to pre-twentieth-century Europeans as to be practically irrelevant. Besides, primates do not occur in life-threatening plagues, nor are they promising agents of biological control. As a result, they generated little of the research interest that was brought to bear so successfully on marine life, insects, and a variety of small mammals.

Regarding demography, it is ironic that its basic tenets were derived from the study of a long-lived species, *Homo sapiens,* for in modern times the demography of other long-lived animals, including nonhuman primates, has been little studied. The reason, of course, is that demographers require data on the life history patterns of large numbers of individuals. We are the only long-lived species obliging enough to provide verbal or written information about our lives. In other species, it is necessary to identify individuals and then study them for years on end just to chart the history of a single generation; long-lived animals are also large-bodied and live at low densities compared with small animals, so that locating an adequate number of subjects is itself a problem. The population density of mice is high by comparison, and their generation span pleasingly short; those of fruit flies even more so. The demographer's choice of subjects follows accordingly.

For practical reasons, then, primates played little part as subjects in two fields that contributed to the modern discipline of ecology. For these and additional practical reasons, they continued to play little part even after the birth of the discipline. Yerkes's lament of many years ago gives the psychologist's version of the problem (1916:231):

Figure 1.19   A rare photograph of the pygmy chimpanzee *(Pan paniscus)*. This shy ape lives deep in the West African rain forest. It is extremely difficult to observe and only recently has it become the subject of a major study. (Courtesy of R. Malenky, Lomako Forest Pygmy Chimpanzee Project.)

Most investigators are either impelled or compelled by circumstances to work on easily available and readily manageable organisms. Many of the primates fail to meet these requirements, for they are relatively difficult and expensive to obtain by importation or breeding, and to keep in normal condition. It is clear from an examination of the literature on these organisms and a survey of the present biological situation that the neglect by scientists of all the primates excepting man is due, not to lack of appreciation of their scientific value, but instead, to technical difficulties and the costliness of research.

For the ecologist bent on experimentation in the laboratory, prohibitive costs are compounded by the difficulty of re-creating within the confines of the laboratory an environment resembling the wild. For the field ecologist, the problems arising from the expense of reaching a study site in the tropics when one could be studying squirrels on one's own temperate-zone doorstep are compounded by conditions at the end of the trip. Primates, tree-living forms in particular, are often elusive, and the tropical and subtropical forests where most are found tend to be inhospitable as well as inaccessible (Figure 1.19). The difficulties of locating and following arboreal animals living at low densities, let alone of doing field experiments in a capture–release program, mean that built-in methodological constraints generally limit the kinds of questions one can hope to answer.

Finally, the ecology of primates is not easily modeled. In this they are little different from other large, social mammals that eat a range of plant foods. Since we shall repeatedly return to this point in later chapters, one brief example will suffice here for illustration. A large chunk of the ecological literature is concerned with models predicting the diet and foraging behavior of animals under a variety of circumstances. Most of these models assume that the nutritional value of food items in the diet is constant, and only the distribution and abundance of these items change. This assumption is reasonable for an animal feeding on, say, a range of insects. It is not reasonable for an animal that eats a range of plant parts, for plant parts vary in nutritional quality not only from species to species, but also from individual to individual within a species and even from one hour to the next within a single individual. Faced with complexity of this kind, it is small wonder that until recently most ecologists chose to concentrate on species that preyed on insects.

The third historical influence on ecology is natural history. In Section 1.2 we saw that primates have not been ignored by natural historians but rather singled out for special treatment because of their similarity to ourselves. This, I believe, is the key to another less practical but nonetheless powerful reason for their exclusion until recently as subjects from much of ecological research. Nonscientific fascination with the resemblance between us and our closest relatives translated in the twentieth century into an emphasis on the scientific study of nonhuman primates as models for *Homo sapiens,* the goals being otherwise unobtainable insights into human psychology, behavior, and evolution. Today the widespread use of nonhuman primates in biomedical research bears witness to their continuing value to scientists interested in the biological bases of human behavior.

For scientists interested in the functioning of natural communities or parts of those communities, I believe that the resemblance that renders nonhuman primates such an attractive model to others has represented instead a threat to objectivity, and the findings of primate field-workers have been treated with corresponding skepticism (Figure 1.20). Most wildlife ecologists are little interested in the similarities or lack thereof between animals and humans; rather, they are interested in the relationships among the animals themselves and their environment. The striking and sometimes poignant parallels of human behavior seen among the nonhuman primates are thus distracting and present a potentially dangerous source of distorted and anthropomorphic interpretation, pitfalls that have not, in fact, been universally avoided (Pilbeam 1972).

This book is primarily about the findings of the growing number of researchers more interested in the nonhuman primates as mammals than as relatives. If a range of factors discouraged this approach in the past, then what precipitated the change in the last 15 years or so? I would suggest two contributing factors, one affecting primarily those trained as physical anthropologists, whose numbers have increased dramatically in the last 20 years (Gilmore 1981), and the other affecting those trained as ecologists in departments of biology, who make up most of the balance of primate field-workers. The first factor results from the change in how we think about our ancestors. Where once

Figure 1.20   Because we resemble other primates, particularly apes such as this young orang-utan *(Pongo pygmaeus)*, there is a tendency to ascribe to them the feelings we experience ourselves. (Courtesy of A. Mitchell.)

even our remotest ancestors were envisioned as humanlike in many features, today the fossil evidence suggests that during much of their evolution, our ancestors were more like other mammals than like us (Pilbeam 1980) (Figure 1.21). This changed perception makes the study of nonhuman primates as mammals in their own right much more compelling for physical anthropologists, who have turned to the discipline of ecology and increasingly contributed to research on the ecology of medium- and large-bodied mammals.

The second factor has been a change in the discipline of ecology itself. It was initially assumed that models developed for animals exhibiting a relatively simple pattern of behavior could subsequently be applied to animals with more complex behavior, but rarely, in practice, has this progression proved possible. This has brought increasing recognition of the need for data and models developed explicitly for these latter animals, particularly large, social, plant-eating mammals. Primates represent a substantial element in this general category. They also make up a major proportion of the arboreal herbivores present in most tropical forests in the world (Chapters 2 and 11); as such, they warrant systematic ecological study. More thorough and patient surveying has located forested environments where primates can be consistently if not continuously observed, the ease with which members of many species can be individually identified has been recognized, and lately successful capture–release programs have been shown to be possible.

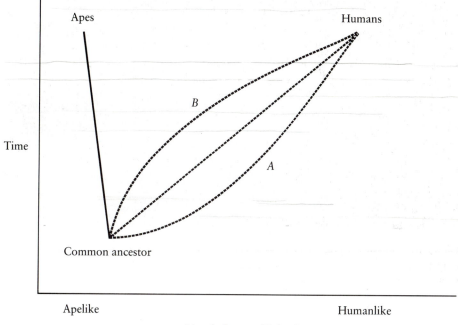

Figure 1.21   Two ways of viewing the trajectory of human evolution. According to the traditional view, *B*, most humanlike features emerged early in our evolution. Recent evidence suggests that *A* may be a more accurate representation, with humanlike features emerging late in our evolutionary history. (Adapted from Pilbeam 1980.)

Today, the primates probably remain the best models we have when trying to flesh out the fossil record or, more generally, to reconstruct the stages in human evolution. However, inferences from living primates to long-departed ancestors are made more cautiously than once they were (Landau, Pilbeam, and Richard 1982; Pilbeam 1984). At the same time, primates have acquired a new significance as important subjects in ecological and sociobiological research, a status achieved independently of their relationship to us. It is with their significance in ecological research that this book is concerned.

PART I

PRIMATES AT LARGE

Introduction

IT has been said that the primates are an unusual order because most primates are limited to areas with a warm climate (Simons 1972). It has also been said that they are one of the most successful groups of mammals that has ever evolved (Buettner-Janusch 1966). Are both these statements accurate? How can a group of animals have an unusually limited distribution and still be called successful?

It depends on what you mean by *success*. Reflecting upon his own earlier comments about the success of the hominoids, Pilbeam (1980:266) noted that "low diversity with the few remaining species mostly on the verge of extinction could surely only connote 'success' in a group that included humans." There is undoubtedly an element of truth in the suggestion that hominoids in particular and primates in general are considered successful simply because we ourselves are hominoids and primates. However, I believe there are other reasons for claiming that the primates are at least moderately successful, their limited distribution notwithstanding. In Part I we explore these reasons, looking at success from an ecological and biogeographic perspective. More generally, the goal of Part I is to provide a broad ecological characterization of the order Primates that distinguishes it from other mammalian orders.

In Chapter 2 we survey the kinds of environments that primates occupy and do not occupy. Then, with much less confidence, we reconstruct their distribution over the last 50 million years in relation to prevailing environmental conditions. This allows us to answer the question, have primates always had a primarily tropical distribution? These data form the basis for a preliminary discussion of both the successfulness of primates and the factors that may limit their distribution.

We move on in Chapter 3 to compare the distribution of primates with the distribution of other orders of land mammals, thus seeing whether the primates are indeed unusual as Simons suggests. Returning to a consideration of limiting factors, we look at ways in which other mammals solve the problems presented by life at high latitudes and the characteristics that distinguish these mammals from primates. We then examine more closely the few primates who have managed to colonize temperate zones. Do these species have features that set them apart from the rest of the order, and do they recall features of other temperate-zone mammals? Finally, based on this discussion and on that of the early primates presented in Section 1.4, Part I concludes with a broad ecological characterization of the whole order of living primates.

CHAPTER TWO

Primate Distribution: Past and Present

2.1 Biomes of the World

IN this chapter we consider the distribution of primates among *biomes,* "the largest land community unit which it is convenient to recognize" (Odum 1971:378). A biome is a grouping of natural communities that are broadly similar in vegetation structure. Whittaker (1975:1) has characterized a *natural community* as "an assemblage of populations of plants, animals, bacteria, and fungi that live in an environment and interact with one another, forming together a distinctive living system with its own composition, structure, environmental relations, development, and function."*

Although it is convenient to divide the natural world into various categories, it can also be misleading, for the world does not really consist of discrete geographic units, each with characteristic climate, vegetation, soil conditons, and so forth. The categories we create intergrade in nature. An *ecocline* is a gradual change through space, or a gradient, in the characteristics of a community that results from a gradual change in the abiotic environment. In many cases an ecocline passes through adjacent communities, making it impossible to define community boundaries clearly (Whittaker 1967). For example, in tropical Africa an ecocline is formed by the climatic moisture gradient (Figure 2.1). Tropical rain forests in the Zaire River basin give way to mixed evergreen and deciduous forests in the north, where the climate is drier and rain falls seasonally. In these forests, the proportion of deciduous trees increases as the average annual rainfall decreases. Further along the same gradient, seasonal forests disappear altogether and are replaced by woodlands, which in turn become thornwoods. Then the size of trees dwindles until the thornwoods give way to low thorn scrub or grassland. Finally, at the end of the line lies the Sahara Desert.

---

* Whittaker uses the term *environment* to refer to *abiotic,* or nonliving, features, such as climate, soil conditions, and topography. In this book the term will refer to *biotic* (living) as well as abiotic features of an animal's surroundings except when a more limited meaning is specified.

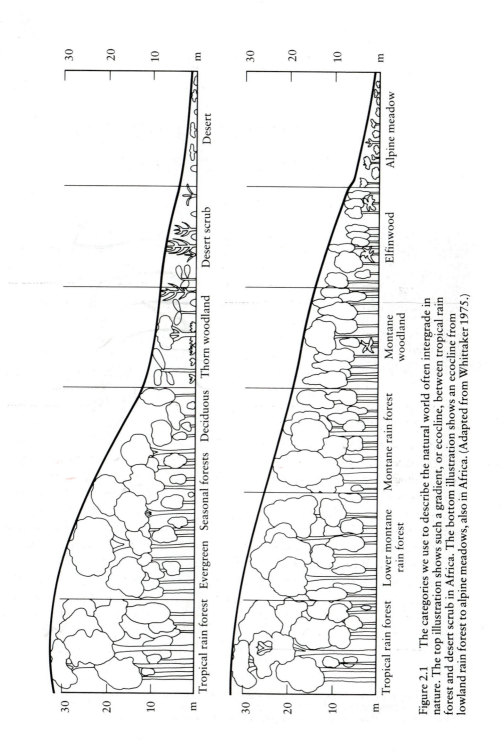

Figure 2.1   The categories we use to describe the natural world often intergrade in nature. The top illustration shows such a gradient, or ecocline, between tropical rain forest and desert scrub in Africa. The bottom illustration shows an ecocline from lowland rain forest to alpine meadows, also in Africa. (Adapted from Whittaker 1975.)

In tropical Africa, we also find ecoclines associated with increasing altitude (Figure 2.1). For example, the tropical rain forest that predominates at low altitudes in Zaire is replaced in the east by the smaller trees of montane forest on the lower slopes of the Virunga volcanoes. As altitude increases, trees become smaller still, and the vegetation becomes more of a dense thicket than a forest; thickets then give way to the small, gnarled trees and shrubs of elfinwoods, and finally elfinwoods are replaced by alpine shrubland and grasslands seasonally covered by snow (Whittaker 1975; Schaller 1963).

A community is not simply a cross section through a single ecocline, however, for living things respond simultaneously to many features of the abiotic environment. Take the climatic moisture gradient in West Africa, for example. At many points on this gradient the annual rainfall may be able to support more than one type of vegetation, but local topography, soil conditions, or the frequency of fires is likely to determine which type actually occurs.

In short, it is not easy to impose spatial boundaries and discrete categories on a world that varies continuously and in complex ways. The difficulty is compounded because the environment varies through time as well as through space. For example, when a farmer in the northern United States abandons a field, it is first colonized by annual weeds, then by perennial weeds and grasses, shrubs, and trees. Over time, the initial species of trees are replaced by others until finally the farmer's field is covered with full fledged forest that will not change further if left undisturbed. This gradual transformation from one vegetation type to another is called *succession,* and the vegetation of the final, steady state is called the *climax.* How often in reality is the succession completed and a climax community achieved? We do not know, and the answer probably differs from one environment to another, depending on the frequency of such catastrophic events as fires, hurricanes, and floods (Bormann and Likens 1979). It is important to recognize, however, that what we label as woodland today may be forest in a decade or two.

Because of these problems, classifying the natural world into communities is unavoidably arbitrary. Even though biome is a more inclusive concept than community, similar problems arise when classifying the world by biomes. Still, despite discrepancies in various biome classifications, there is enough agreement for the biome to be a useful tool in describing the general range of conditions under which primates live (and do not live) both at present and in the past.

Four major structural classes are represented in the thirteen biomes listed in Table 2.1: (1) forest; (2) woodland grading into shrubland; (3) savanna, grassland, and savanna-mosaic; and (4) desert and semidesert scrub. Each of these classes includes a wide range of vegetation, and the boundaries between them are not clear-cut. Tall trees are the dominant plant form in *forests,* their crowns providing a continuous ceiling of

Table 2.1   World biomes

Tropical	Temperate	Subalpine and alpine	Subarctic and arctic
Rain forest	Rain forest	Elfinwood	Taiga
Seasonal or monsoon forest	Woodland, deciduous, and evergreen forest	Shrubland and meadow	Tundra
Woodland, thornwood, and shrubland	Grassland		
Savanna and savanna-mosaic	Desert and semi-desert scrub		
Desert and semi-desert scrub			

branches and leaves. Small, more widely spaced trees and an abundance of undergrowth characterizes *woodland*. In drier areas, many of the trees are spiny and have small, drought-resistant, deciduous leaves; others are succulent, conserving water in their trunks and branches. Such vegetation is often called thornwood. *Shrubland,* in which at least half the vegetative cover is made up of shrubs, is also included in the woodland category. *Grassland, savanna,* and *savanna-mosaic* are dominated by grasses or grasslike plants. (The term *savanna* is conventionally used in the tropics, and the term *grassland* in temperate regions.) Savanna-mosaic is more heterogeneous than pure savanna, being composed of a patchwork of woodland, wooded grassland, grassland, and shrubland (Van Couvering 1980). *Semidesert scrub* comprises sparse, low-growing shrubs and other plants interspersed with large patches of exposed soil.

The biomes in Table 2.1 are broadly grouped according to latitude and altitude, and their approximate distribution over the world's landmasses is shown in Figure 2.2. Again, the boundaries drawn here between climatic zones should not be taken too literally; for example, an area with a temperate climate may lie within the tropics, defined according to latitude. In general, though, tropical climates and *tropical biomes* exist at low altitudes in the climatic zone straddling the equator between the Tropic of Cancer (23°28′ north latitude) and the Tropic of Capricorn (23°28′ south latitude). *Warm and cool temperate biomes* exist at low altitudes in two climatic zones circling the earth from the Tropic of Cancer to about 50° north latitude and from the Tropic of Capricorn to about 50° south. Most temperate biomes are characterized by cold winters, with

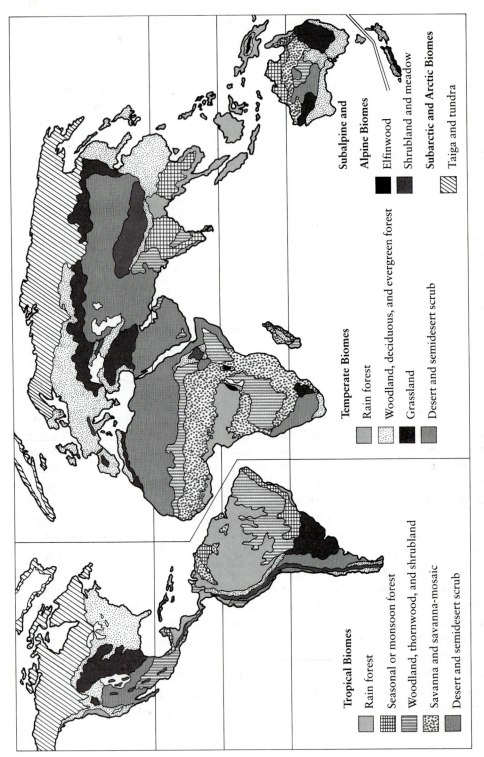

Tropical Biomes

Rain forest

Seasonal or monsoon forest

Woodland, thornwood, and shrubland

Savanna and savanna-mosaic

Desert and semidesert scrub

Temperate Biomes

Rain forest

Woodland, deciduous, and evergreen forest

Grassland

Desert and semidesert scrub

Subalpine and Alpine Biomes

Elfinwood

Shrubland and meadow

Subarctic and Arctic Biomes

Taiga and tundra

Figure 2.2   The approximate distribution of biomes over the world's landmasses.

marked seasonal variations in temperature and day length. *Subalpine and alpine biomes* are found from about 3000 m above sea level in the tropics and at lower altitudes in temperate regions; these biomes literally stand out from those closer to sea level and are cold and damp year-round. *Subarctic and arctic* biomes surround the two poles to about 50° north and south latitudes; temperatures are extremely low in these regions, and light conditions vary from months when the sun hardly rises above the horizon to months when it never sets.

Biomes differ in many attributes other than general climate and vegetation. Two features commonly used to distinguish biomes are the biomass and productivity of the vegetation. *Biomass,* or standing crop, is the amount of organic matter present at a given time per unit of the earth's surface; the term can be applied equally well to all the living matter in an area or to a subset of that matter, such as the primates, birds, or leaves on the trees. The biomass of vegetation tends to decrease with declining moisture and temperature. The biomass of climax forests, whether in the tropics or temperate regions, generally ranges from about 10 to 60 kg/m². In contrast, the biomass of woodlands is usually only in the range of 4 to 20 kg/m²; of shrublands, 2 to 10 kg/m²; of savannas, 0.5 to 3 kg/m²; and of deserts and tundra, 0 to 2 kg/m². The biomass of savanna is so low because aboveground plant parts are frequently eliminated by fire and grazing animals (Whittaker 1975).

Plants harness only a small fraction of incoming solar energy (Figure 2.3). The rate at which energy is bound or organic material is created by photosynthesis is called *primary productivity.* The total energy bound by green plants is their *gross primary production.* The energy that is left over after respiration by these plants is their *net primary production.* Productivity, like biomass, generally decreases along moisture and temperature gradients. However, the proportion of gross production used up in respiration by plants in tropical forests is usually higher than in temperate forests (70–80% compared with 50–60%), so that in practice the net productivities of tropical and temperate forests are often similar, though maximum values are higher for the tropics. These values, between 2 and 3 kg/(m²)(yr), are characteristic of some rain forests and successional communities with favorable environments. Values of 1 to 2 kg/(m²)(yr) characterize a variety of tropical and temperate forests and savannas. Many nonforest communities (woodlands, shrublands, and some savannas) are limited by drought, low temperatures, or nutrient availability and have net productivities ranging from 0.25 to 1 kg/(m²)(yr). Finally, the least productive biomes are deserts, semideserts, and parts of the arctic tundra, which have net productivities ranging from 0 to 0.25 kg/(m²)(yr) (Whittaker 1975). It should be emphasized, however, that while the net primary production of plants is potentially available for harvest by animals, in practice it is often unsuitable as food.

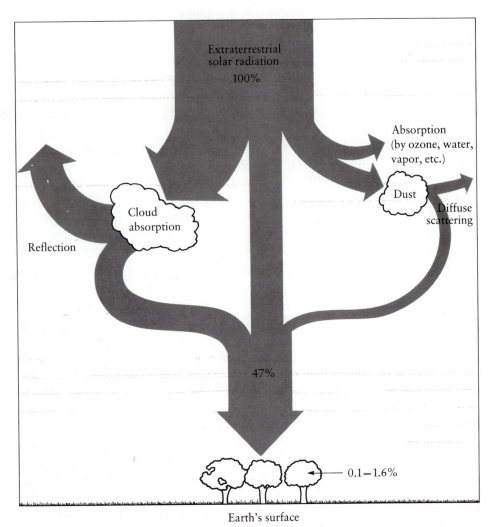

Figure 2.3   Less than half the incoming solar energy actually reaches the earth's surface, and of this only a small fraction is harnessed by plants. (Adapted from Kormondy 1976.)

2.2   Distribution of Primates among Biomes Today

The following account focuses on tropical and temperate biomes, since primates are found almost exclusively in or near these biomes.

2.2A   Tropical Biomes

Rain Forest

Rain forest covers three major regions of the world, (1) the Amazon and Orinoco basins in South and Central America; (2) the Zaire, Niger, and Zambezi basins of Central and West Africa; and (3) Southeast Asia, through Indonesia to northeastern Australia (Figure 2.4). Temperatures in these areas are high and vary more between night and day than between summer and winter. Heavy rain (over 200 cm/yr) falls fairly evenly throughout the year, and there is little variation in day length. These conditions permit trees to keep their leaves year-round, and most rain forest trees are broad-leaved evergreens. Flower and fruit production is also possible during much of the year, and there is often little synchrony in these cycles among different species and sometimes even among individuals of a single species (but see Raemaekers, Aldrich-Blake, and Paynes 1980). Despite the overall stability of climate, the immediate microclimate to which a plant is exposed varies according to its position in the forest (Richards 1957). On or near the forest floor it is constantly warm, dark, and humid, whereas higher up it is sunnier, drier, and hotter by day and cooler by night.

Of all biomes, tropical rain forest contains the greatest array of plant species. The forests of the Malay Peninsula, for example, contain on the order of 2500 species of trees and another 5400 species of seed-producing plants such as shrubs, herbs, vines, and epiphytes. This makes them the richest forests in the world. The forests of Great Britain, in contrast, contain only 35 native tree species, which constitute 2.4% of 1430 seed plant species in over twice the area of peninsular Malaysia (Raemaekers, Aldrich-Blake, and Paynes 1980). The species richness of the vegetation in rain forest is matched by its immense volume, or biomass (see Section 2.1).

Rich in species and luxuriant in growth, rain forest is also architecturally complex. From a bird's-eye view, the most striking feature is the horizontally continuous layer of the forest, the *main canopy,* made up of a leaf-covered network of branches of large trees. These trees support a host of *epiphytes,* plants that grow on their branches and catch nutrients from the rain and the air. Scattered at intervals above this sea of green are the crowns of soaring trees called *emergents* (Figure 2.5). Up to 90 m high, these trees are so tall that a vertical gap often separates their crowns from the vegetation of the canopy below.

In the shady world under the canopy grows an assortment of young saplings and woody climbers, woody shrubs, and shade-adapted herbs, which together make up the *understory* and *ground cover.* In most rain forests, the understory is dense and junglelike only where tree falls have made a hole in the canopy through which the sun can stream, producing a dense patch of luxuriant young growth. Such gaps provide many trees and climbers with the only chance they ever get to grow beyond the seedling stage, for less than 1% of the sunlight that strikes the canopy filters down to the forest floor (Rae-

Figure 2.4 (left)   Rain forest is difficult to photograph well, but this picture of the Kibale forest, Uganda, gives a good sense of its great height, luxuriance, and complex vertical structure. (Courtesy of T. T. Struhsaker.)
Figure 2.5 (right)   Emergents stand out above the canopy as the mist rises at dawn in the Kibale forest, Uganda. (Courtesy of R. A. Mittermeier.)

maekers, Aldrich-Blake, and Paynes 1980). Because of this, the species richness of rain forests is most readily apparent up in the canopy, where epiphytes, climbers, and the trees themselves vie for a place in the sun (Figure 2.6).

So far we have described the rain forest as though it were a homogeneous vegetation type, but in reality there are many variants (e.g., Gentry 1982), and some primates strongly prefer particular components within a given block of vegetation broadly characterized as rain forest (e.g., Gautier-Hion 1978; Struhsaker and Oates 1975). For example, gallery, or riverine, forest grows along the banks and flood plains of rivers. It is particularly dense and lush, and it often contains early successional species that grow only where gaps have been created by tree falls in forest further from water. Swampy

Figure 2.6   This photograph of rain forest in Itatiaia National Park, Brazil, shows the diversity of plant species living high above the ground on the branches of huge trees. (Courtesy of R. A. Mittermeier.)

areas not associated with rivers also tend to have characteristic species, and in some instances these species provide primates with crucial nutrients not readily available in other parts of the forest (Oates 1977a).

All primate families have tropical rain forest representatives (Table 2.2). The only major rain forest blocks from which primates are absent are in New Guinea and northeastern Australia. These landmasses were cut off from Southeast Asia by deep, wide sea channels many millions of years before primates evolved, forming effective barriers to the spread of many life forms, including primates.

The number of primate species in a few hundred square kilometers of rain forest can be high, although not all of them are necessarily *sympatric* in the sense of belonging to the same community. Different preferences for particular components of the forest's

Table 2.2   Distribution of primates over biomes

Primate family	Tropical biomes					Temperate biomes				Subalpine and alpine biomes		References
	Semidesert scrub	Savanna	Woodland	Seasonal forest	Rain forest	Semidesert scrub	Grassland	Deciduous and evergreen forest	Rain forest	Meadow	Elfinwood	
Strepsirhines												
Cheirogaleidae			x	x	x							Charles-Dominique et al. 1980
Daubentoniidae				x	x							Petter 1977
Indriidae			x	x	x							Richard and Sussman 1975
Lemuridae			x	x	x							Richard and Sussman 1975
Lepilemuridae			x	x	x							Richard and Sussman 1975
Lorisidae			x	x	x							Roonwal and Mohnot 1977; Doyle and Bearder 1977
Haplorhines												
Callimiconidae				x	x							Hershkovitz 1977
Callitrichidae				x	x							Hershkovitz 1977
Cebidae			x	x	x							Freese 1976
Cercopithecidae	x	x	x	x	x	x	x	x		x	x	Napier and Napier 1967; Hall 1965; Roonwal and Mohnot 1977; Dunbar 1977b
Hominidae[a]		x	x	x	x	x	x			x	x	Napier and Napier 1967; Fossey 1974; Schaller 1963; Fossey and Harcourt 1977; Suzuki 1969
Hylobatidae				x	x							Chivers 1977
Pongidae					x							Wolfheim 1983
Tarsiidae					x							Fogden 1974

vegetation often ensure that the home ranges of animals of different species overlap little, even though the overall distribution of the various species may be similar. New
World rain forests contain more primate species then any other region in the world,
with as many as 22 present in just 500 km² in Upper Amazonia (Hershkovitz 1977; Mittermeier and Coimbra-Filho 1977). In southern Colombia alone there are 14 species
(Hernandez-Camacho and Cooper 1976). In West Africa, 17 primate species, including
monkeys, apes, and nocturnal strepsirhines, live in the forests of northeastern Gabon
(Charles-Dominique 1977). Looking east, we find 10 species in the forests of the Malay
Peninsula (Chivers 1980). Despite the extinction of the larger lemurs over the past 2000
years (Dewar 1984), the primates of Madagascar still present an impressive array, with at
least 9 species to be found in a single patch of rain forest (Richard and Sussman 1975).

The estimated biomass of primates varies considerably from one rain forest to another (Table 2.3). To estimate biomass accurately, the density of animals and the
weights and relative numbers of individuals of different age and sex classes must be
known. These features are rarely well documented for even one species, let alone a
whole primate community, and for this reason the figures in Table 2.3 should be viewed
with caution.

Table 2.3   Primate genera, species, and biomass in different habitats

	Rain forest					Seasonal forest	Woodland
Characteristic	Makokou, Gabon	Kibale, Uganda	Kuala Lompat, Malaysia	Barro Colorado Island, Panama	Manu National Park, Peru	Polonnaruwa, Sri Lanka	Ankazobe, Madagascar
Number of genera	11	4	3	5	9	3	7
Number of species	15	7	6	5	11	4	7
Total estimated biomass (kg/km²)	652[a]	2345[b]	834[c]	550[d]	749	2754	600–1000

Sources: Data for Gabon from Gautier-Hion 1978; Uganda, Struhsaker 1978; Malaysia, Raemaekers and
Chivers 1980; Panama, Hladik and Hladik 1969; Peru, Terborgh 1983; Sri Lanka, Hladik and Hladik 1972;
Madagascar, Charles-Dominique et al. 1980.

[a] *Colobus, Pan, Gorilla,* and *Madrillus* not included.
[b] *Pan* and *C. lhoesti* not included.
[c] *M. nemestrina* not included.
[d] *Aotus* and *Saguinus* not included.

Figure 2.7   Monsoon forest in the Kalakkudu Sanctuary in the Western Ghat Mountains, South India. (Courtesy of J. Oates.)

Seasonal Forest

Seasonal forest includes monsoon forest and forest vegetation ranging from partly evergreen to largely or even completely deciduous (Figure 2.7). It occurs in humid tropical climates where the amount of rainfall varies during the year. Some or many trees lose their leaves in the dry season, and canopy height is lower than in the rain forest and does not provide such a thick, carpet effect. There is extensive seasonal forest in India and Southeast Asia, as well as in Central and South America, the West Indies, and northern Australia.

Most primate families have representatives in seasonal forests (Table 2.2), and the total number of species present in a particular forest block can be quite high. For example, the monsoon forests of western Thailand harbor eight species, including an ape, six monkeys, and one strepsirhine (Eudey 1980). The species are not as diverse in seasonal forest in India, where three or four species are typical (Roonwal and Mohnot 1977). While the number of species is in general lower in seasonal forests than in rain forests, the highest estimate of primate biomass comes from a seasonal forest in Sri Lanka (Table 2.3).

Figure 2.8   Woodland is found in tropical regions that receive too little rainfall to support forest vegetation. Note the Masai encampment in the center of photograph. (Photography by A. Richard.)

Woodland, Thornwood, and Shrubland

Woodland, thornwood, and *shrubland* typically exist in areas between the moist conditions of rain forest and the dry expanses of savanna or semidesert scrub (Figure 2.8). While total annual rainfall may be high, it is distributed unevenly through the year, and there is a long, hot, dry season. Soil conditions in this biome are less favorable than in areas supporting lusher forests. Woodland vegetation consists of small (4–7 m) trees that remain in full or partial leaf throughout the year and can withstand long periods of drought. Along a gradient of decreasing rainfall, these trees are replaced by more *xerophytic,* or drought-preferring, plants that form thornwoods. Thornwoods in turn give way to shrublands. This range of vegetation characterizes regions of South America north and south of the Amazonian forest block, interior southern Africa, extensive areas in southern and southwestern Madagascar, and drier parts of South and Southeast Asia.

Woodland is occupied by one New World species and several Old World families, although the distribution of most woodland-dwelling species encompasses forest or savanna and savanna-mosaic as well. In other words, no primates are adapted uniquely to this biome. Woodland rarely supports more than two or three species in sympatry except in Madagascar. There two *diurnal* (active by day) species and up to three *nocturnal*

Figure 2.9   Ninety percent of the plant species are endemic, or unique, in the spiny forests of southern Madagascar; spines help prevent water loss in these desertlike areas, and they also deter animals in search of vegetable food. (Photography by A. Richard.)

(active by night) species live in the spiny forest (Figure 2.9) unique to the south and southwest of the island. No biomass estimates are available, but woodland primates are thin on the ground, and their total biomass is almost certainly a small fraction of that in either rain forest or seasonal forest. Certainly, in the woodlands of western Madagascar, the estimated biomass of the seven primate species present is very low (Table 2.3).

Savanna and Savanna-Mosaic
Savanna and *savanna-mosaic* exist in areas with 100–150 cm of rain that falls in one or two clearly delineated wet seasons separated by dry seasons (Figure 2.10). In this respect, they differ little from woodland; indeed, it is not always clear why one or the

Figure 2.10   Savanna and savanna-mosaic are found in dry tropical areas. It is often unclear why a particular region is covered by savanna rather than by woodland; natural fires and the grazing and trampling of ungulates may prevent the spread of woodland into savanna areas. This photograph was taken in the Masai Mara National Park, Kenya. (Courtesy of C. First.)

other vegetation type predominates in a particular area. Fires, set naturally as well as by people, can eliminate woodland, thereby allowing grasses to establish themselves. Large herbivorous mammals may also destroy woodland by trampling on or eating the leaves, shoots, and young woody parts of trees and shrubs; these areas then become available for colonization by grasses able to survive being burned, trampled, and cropped. The number of plant species in a tropical savanna is much lower than in a tropical rain forest, perhaps partly because savanna plants are subject to such stringent selection pressures. In the tropics, savanna-mosaic is common, and extended tracts of treeless grassland are rare (Van Couvering 1980).

Figure 2.11   Members of these primate species can be seen out on the savanna in some parts of the species range, but only baboons and patas monkeys forage there extensively. Members of all four species return to sleeping in trees at nightfall.

No primates occupy this biome today in the New World, Asia, or Madagascar, but in Africa baboons (*Papio* spp.), patas monkeys *(Erythrocebus patas),* vervet monkeys *(Cercopithecus aethiops),* and chimpanzees *(Pan troglodytes)* live in savanna-mosaic (Figure 2.11). Baboons and vervets have particularly extensive distributions within this biome. Yet the home ranges of these open-country primates always contain trees, which often provide an important source of food as well as a refuge from predators.

Semidesert Scrub
Semidesert scrub is found under very arid conditions. Average annual rainfall is low, rain falls only during a short wet season, and it may fail altogether in some years. Vegetation, rarely more than waist high, is thinly spread over the ground. Macaques and baboons are the only primates that exploit this biome, where they manage to survive at very low population densities.

2.2B   Temperate Biomes

Rain Forest
Rain forests are found on the Pacific coast of North America from Washington to central California. Despite its temperate-zone location, this region has a climate in many ways more characteristic of the tropics, with relatively high temperatures that vary little seasonally, heavy winter rain (76–130 cm), and thick summer fogs and clouds. In summertime, tall trees intercept as much as 130 cm of rain in the form of condensed fog dripped down their branches (Oberlander 1956). In contrast to tropical rain forest, these temperate forests are dominated by a few gigantic coniferous tree species, such as the magnificent Californian redwood *(Sequoia)*. There are no primates here or anywhere in North America today except possibly the Bigfoot (Napier 1973)!

Woodland and Deciduous and Needle-Leaved Evergreen Forest
Woodland and *deciduous* and *needle-leaved evergreen forest* once covered eastern North America and formed a belt through Eurasia from western and central Europe, across the southern foothills of the Himalayas (Figure 2.12), and through to northern Japan. Today many of these tracts have been destroyed by human activities. In the Southern Hemisphere, temperate forests are found in Australia south of the great deserts of the center and west, in Tasmania, and also in southern South America. They occur in areas with plenty of rain (75–130 cm) and no long dry season but where warm summer weather gives way to cold winters.

Broad-leaved trees such as the oak *(Quercus)*, beech *(Fagus)*, and maple *(Acer)* dominate deciduous forests, which sometimes reach heights comparable to those of the tropical rain forest. Absence of a continuous canopy allows more light to reach the forest floor than in a rain forest, so that the understory is denser and more diverse. Growth and reproduction take place in warmer months when days are longer, and cycles of leaf, flower, and fruit production tend to be much more closely synchronized between and within species than they are in the tropics. The reproductive organs of many plants are close to or under the ground to protect them from winter cold.

According to local climatic, topographic, soil, and fire conditions, needle-leaved evergreens mix with deciduous species; in areas where the range between winter and

Figure 2.12   Many tracts of temperate forest have been destroyed by human activities, but primates are still found in some of those that remain in Eurasia, such as this forest in the Himalayan foothills of northern Pakistan. (Courtesy of D. Melnick.)

summer temperatures is extreme, these evergreens replace deciduous species entirely. Woodland generally exists in areas too dry for real forest and, with increasing aridity, it gives way to temperate grassland or semidesert scrub. Temperate woodlands are communities of small trees spanning a range from transitional forest vegetation to transitional grassland, and from needle-leaved evergreens to broad-leaved deciduous trees.

Only in the Old World do primates occupy temperate forests. Macaques *(Macaca spp.)* (Figure 2.13) range from eastern Afghanistan through northern Pakistan, India, and Tibet to China, Japan, and Taiwan, as well as south into tropical India, the Malay Peninsula, and the islands of Southeast Asia (Roonwal and Mohnot 1977). There is still a small, relict population of rhesus monkeys *(Macaca mulatta)* west of Beijing, and the distribution of rhesuslike fossils in China suggests that this modern isolate may represent the last vestiges of a distribution pattern that once extended much further north

Figure 2.13 (left)   Macaques live in temperate forests from Afghanistan to China and Japan.
This photograph of rhesus macaques *(Macaca mulatta)* was taken in northern Pakistan just after
a snowstorm. (Courtesy of W. Keene.)
Figure 2.14 (right)   Langurs *(Presbytis* spp.) are sympatric with macaques in many northern
forests. (Courtesy of N. Bishop.)

than it does today (Section 2.3). Even now, at the northern limits of their distribution,
macaques live in forests covered by several meters of snow for three or four months in
winter, and at higher altitudes in the Atlas Mountains of Morocco and Algeria, Barbary
macaques *(Macaca sylvanus)* weather the winter in snow-carpeted cedar and oak forests
(Deag 1977; Taub 1977).

Langurs *(Presbytis* spp.) (Figure 2.14) and golden or snub-nosed monkeys *(Rhino-*
pithecus roxellanae) are sympatric with macaques in some northern forests. Like macaques, langurs live in forests seasonally blanketed by snow. In fact, they range up to
3500 m, even higher than macaques, in the Nepalese Himalayas (Bishop 1979). Golden
monkeys are found only in China and northern Vietnam. In China, some parts of their
range are covered by snow for up to six months of the year (Roonwal and Mohnot
1977; Happel and Cheek forthcoming).

Grassland
Grassland covers vast expanses of interior North America (the prairies) and Eurasia (the steppes). There are also temperate grasslands in southern Africa (veldt) and in South America (pampas). In these regions "true" grassland predominates, and tree cover is rare or absent over large tracts. The major characteristics of the climate in this biome are hot summers and long, bitterly cold winters with relatively low annual rainfall. There are no primates in temperate grassland.

Cool Semidesert Scrub
Cool semidesert scrub covers much of the Great Basin, between the Rocky Mountains and the Cascades in the western United States. Similar communities are found in Central Asia, South America (in Patagonia), and Australia. No primates occupy this biome.

2.2C   Subalpine and Alpine Biomes

Elfinwood
Elfinwood, "rain forest in miniature" (Whittaker 1975), is found at high altitudes in tropical areas (Figure 2.15). For example, it covers the upper slopes of the Virunga volcanoes and the Ruwenzori Mountains in Africa and parts of the highlands on the island of Sri Lanka, off the southern coast of India. New Guinea's central highlands also support elfinwood in some areas. The climate in all these regions is continuously cold, rainy, foggy, and windy.

Only two primates exploit this biome. Gorillas forage through elfinwoods on the Virunga volcanoes (Schaller 1963), but they always include other kinds of vegetation in their home range. In Sri Lanka, there are populations of purple-faced leaf monkeys *(Presbytis senex)* on the high central plains covered by grassland and oases of elfinwood. In contrast to gorillas, these monkeys live wholly in the elfinwoods and apparently do not forage in the surrounding grassland (Hladik and Hladik 1972; G. H. Manley, pers. com., 1970).

Alpine Shrubland and Meadows
Alpine shrubland and *meadows* are found above the timberline in mountainous areas at both tropical and temperate latitudes. In Africa, gelada baboons *(Theropithecus gelada)* are confined to a highland plateau in Ethiopia, where they depend heavily on grass for food (Figure 2.16). Hamadryas baboons *(Papio hamadryas)* are also found in alpine meadows, but their density is much lower, perhaps because they digest grass leaves less efficiently and are able to use fewer parts of the grass plant than geladas (Dunbar 1977a, b). Gorillas enter this biome but spend little time in it. In Eurasia, langurs have been reported to make seasonal forays into alpine meadows in Nepal, but they also need other types of vegetation in their home ranges.

Figure 2.15   Elfinwood such as this, photographed in the highlands of Sri Lanka, is found at high elevations in just a few places in the world. (Photography by A. Richard.)

Figure 2.16   Gelada baboons *(Theropithecus gelada)* feeding on grassland and drinking at a pool in the Ethiopian highlands. (Courtesy of R. I. M. Dunbar.)

2.2D   Subarctic and Arctic Biomes

Taiga
Taiga, or needle-leaved forest, is the biome of the cold edge of the climatic range of
forests. It stretches in broad belts across North America and Eurasia where humidity is
relatively low and winters long and cold. The most common trees in these forests are
spruce *(Tsuga),* fir *(Abies),* and pine *(Pinus).* There are no primates in this biome.

2.3   Past Distribution of Primates
among Biomes

Today most primates live in tropical rain forest, woodland, and savanna-mosaic. Has
this always been so? We look now at changes in climate and vegetation and in primate
distribution since the start of the Eocene epoch 55 M.Y.B.P., when primates of "modern
aspect" (in other words, mammals with a full set of primate characteristics) first appeared (Simons 1972). This period makes up most of the Cenozoic era, often called the
Age of Mammals because much of mammalian evolution has taken place during it. The
Cenozoic era is conventionally divided into two periods, the Tertiary and Quaternary,
which are themselves divided into five and two time blocks, or *epochs,* respectively (Figure 2.17). Today, most researchers use major changes in marine invertebrate life forms
to define the start and end of epochs, but in the past they used a variety of criteria and
the dates assigned to particular epochs in early papers cited here may vary as a result.

World climates have shown a general cooling trend during the Cenozoic. How do we
know this? The reconstruction of ancient climates relies heavily upon indirect evidence
provided by plants and animals associated with differing climatic conditions. Plants
making up a particular vegetation type millions of years ago can sometimes be identified
from pollen preserved in mud at the bottom of an ancient lake. Small marine organisms
are extremely sensitive to changes in the ocean's temperature, and different species are
associated with water at different temperatures. Changes in the kinds of marine microorganisms deposited on the ocean floor during millions of years thus provide an indirect
record of changes in water temperature and hence the general climate of a region. (For
reviews of the techniques used in climatic reconstruction, see Kennett 1977; Wolfe
1978.)

Why do climates change? Many factors contribute, none of them completely understood. One is the continually changing position of the world's seas and continents,

Figure 2.17   The last 65 million years, or Cenozoic era, are usually divided up into seven time
blocks, or epochs. This figure shows associated events or phases in primate evolution discussed
in the text.

Era	Period	Epoch	Africa	Europe	Asia	North America	South America
Cenozoic	Quaternary (12,000yr)	Holocene	Essentially modern fauna			No primates	Primates present
	Quaternary (1.6)	Pleistocene	*Papio* largely replaces *Theropithecus* Homo evolves	Cercopithecines present until last interglacial Colobines disappear	Cercopithecines and colobines present	No primates	Primates present
	Tertiary (5)	Pliocene	*Theropithecus* baboons widespread Homininae present in east and south	Colobines and cercopithecines present	Colobines and cercopithecines present	No primates	Primates present
	Tertiary (22)	Miocene	Fossils resembling modern strepsirhines rare but present. Colobines from at least 9 M.Y.B.P. Sivapithecids present by 17 M.Y.B.P. Dryopithecids appear early in epoch	Colobines spread from Africa. Sivapithecids present 7–13 M.Y.B.P. Dryopithecids appear about 16 M.Y.B.P.	Colobines spread from Europe and Africa. Sivapithecids present 7–13 M.Y.B.P. Dryopithecids appear about 16 M.Y.B.P.	No primates	Fragmentary remains tentatively attributed to lineages of living New World primates
	Tertiary (38)	Oligocene	Primates with monkeylike skeletons in the Fayum: probably direct ancestors of haplorhines	Primates disappear	?	Primates disappear	First appearance of primates
	Tertiary (55)	Eocene	A few fragments of (?) primates	Strepsirhine-grade primates present	Strepsirhine-grade primates by mid-epoch. Haplorhinelike fragments from Burma at 40 M.Y.B.P.	Strepsirhine-grade primates present	Primates not present
	Tertiary (65)	Paleocene	?	Small mammals with primatelike features present	Probably no primatelike mammals	Small mammals with primatelike features present	No primatelike mammals

M.Y.B.P.

which has important consequences not only for climates but also for the ability of plants and animals to spread over the world's land surfaces (Simpson 1953). The earth's crust is made up of huge plates that are constantly in motion. New crust emerges slowly but steadily, spreading away from oceanic *ridges* on the ocean floor, so that plates on either side of a ridge are gradually pushed away from one another. Old crust disappears under the surface of the earth through *trenches* that exist on land as well as under the sea (Figure 2.18) (Wilson 1976; Tarling 1978). These processes are called *plate tectonics,* and the movement of the land masses that form part of the plates is referred to as *continental drift.* As a result of these geophysical processes, the size of continents and their positions with respect to the poles change. Fluctuations in sea level further vary the world area of land.

Changes in the world's continents influence climate through two important effects that the continents have on the heat balance of the earth's surface (Barron, Sloan, and Harrison 1980). First, they alter patterns of heat transport in the oceans and atmosphere. This is important because, other things being equal, the net loss of solar energy from the atmosphere is almost the same the world over, but the net gain of energy by the earth's surface is greatest in tropical regions. For this reason, temperatures are usually highest in the tropics, and if heat were not transported poleward by large, warm bodies of air and water moving away from the equator, the tropics would become progressively hotter and the poles progressively cooler. When continents are arranged so that sea and

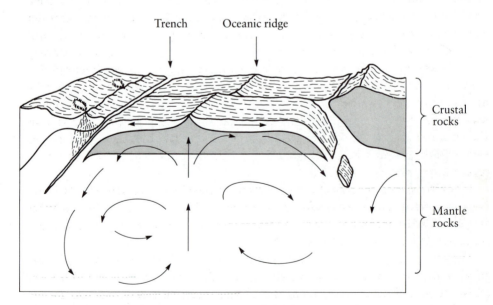

Figure 2.18   A schematic illustration of the processes that give rise to continental drift. (Adapted from Tarling 1978.)

air currents tend to circulate around the world rather than north and south away from the equator, the transport of heat from the equator toward the poles is reduced and the temperature gradient increases (Crowell and Frakes 1970).

The second effect produced by changes in the size and disposition of continents is modification of the surface *albedo*, or reflectivity, of the earth with respect to incoming solar radiation (Barron, Sloan, and Harrison 1980). The albedo of oceans is lower than that of land and considerably lower than that of land covered by desert or snow. An increase in the ratio of land to ocean area thus increases the amount of radiation reflected back into the atmosphere. Because high latitudes receive less radiation from the sun than the tropics, an increase in the area of land at high latitudes increases the amount of incoming radiation there that is reflected, and the temperature gradient between the equator and the poles increases. Large landmasses close to the equator are often covered by desert, and an increase in the amount of land in this region tends to increase the area of desert. Because deserts are highly reflective, their expansion in regions of the globe where solar input is high tends to produce worldwide cooling. Indeed, Barron and associates suggest that the transformation of the Mediterranean Sea into a vast desert about 5 M.Y.B.P. has helped trigger the ice ages of the last 2 or 3 million years.

Despite great strides in the last 20 years or so, our knowledge of world climates, vegetation, and continental positions over the past 60 million years remains far from complete, and the primate fossil record itself is patchy in time and space. In Africa, for example, about 85 fossil sites containing mammals are now known, but over half of them date to the last 5 million years, and most sites are located around the Mediterranean rim and in South Africa, Kenya, and Tanzania. Almost nothing is known about the fossil record of the west and southwest of the continent. The earliest mammal finds date only to the Eocene epoch (55–38 M.Y.B.P.) and come from just seven localities. Knowledge of the Oligocene (38–22 M.Y.B.P.) rests almost exclusively on a single site in Egypt and a handful of others scattered throughout northern Africa. There are over 20 Miocene sites, but they are concentrated in the Mediterranean area and East Africa. The 55 or so sites in Africa from the last 5 million years (Pliocene and Pleistocene epochs) are more evenly distributed over the north, east, and south of the continent, but the west is still a big question mark (Cooke 1978).

In South America, the record is temporally more complete for most mammals. However, few primates have been found there (Simons 1972; Patterson and Pascual 1972; Simpson 1980), probably because most sites are located in temperate regions of the continent, where we would least expect to find primates.

Turning to Madagascar, the lemurs apparently evolved there in isolation, for Madagascar is now believed to have separated from mainland Africa as early as 120 M.Y.B.P. (Rabinowitz, Coffin, and Falvey 1983). But the only traces we have of their past are not more than 3000 years old (Dewar 1984). All these problems must be kept in mind while reading the following account.

2.3A   Eocene Epoch (55–38 M.Y.B.P.)

Although a cooling trend began in the early Eocene (Shackleton and Kennett 1975), the world climate was still warm and wet, with much less difference than exists today between the climate of the middle latitudes and polar regions and that of the tropics (Frakes and Kemp 1972; Barron, Sloan, and Harrison 1980). The locations of the continents were also quite different. South America was attached to Antartica throughout the Eocene, and Australia was attached to Antarctica until about 53 M.Y.B.P., when it started moving north. However, by the epoch's close the two landmasses were separated only by a narrow channel. Early in the Eocene, Europe was joined to North America and separated from Asia by the Turgai Strait. The Tethys Sea, larger than the Mediterranean today but in about the same location, isolated Africa from Europe and Asia (Figure 2.19, top left). By the middle of the epoch, North America and Europe had divided; it is likely that a land bridge briefly connected Europe and Asia about this time. Toward the epoch's end, Europe and Asia became joined as they are today, and a land bridge at the western end of the Tethys Sea formed a connection between Africa and Europe (Figure 2.19, top right). The late Eocene also saw the initial uplift of the mountains of southern Europe—the Alps and the Pyrenees.

Reconstructions of vegetation covering the Eocene world reflect the milder climates prevailing then. For example, temperatures in parts of Antarctica were warm enough to support cool, temperate vegetation; if there were ice caps at all, they were restricted to the western part of the continent (Kennett 1977). In the Northern Hemisphere, vegetation right up to the Gulf of Alaska was of a type associated with subtropical conditions today (Wolfe and Hopkins 1967).

In the early Eocene, most primates lived in the warm, hospitable lands of Europe and North America, and the predominant biome throughout their range was rain forest or seasonal forest. Fragments from northern Africa suggest that a few may have crossed from Europe over a land connection that formed briefly during the Paleocene (65–55 M.Y.B.P.). The temporary land bridge from Europe to Asia in the middle Eocene allowed primates to spread east, and a few fragments are known from the late Eocene of East Asia (Simons 1972).

All but one of the Eocene primates resemble living members of the suborder Strepsirhini more than they do tarsiers, monkeys, or apes (members of the suborder Haplorhini), and so they are said to be of a strepsirhine-grade of evolution. The exception is *Pondaungia*, represented by a small collection of fragmentary remains, mainly teeth, found in Burmese deposits about 40 million years old. This animal had some features of modern haplorhines, although the primates it resembles most are those that lived in the Fayum region of Egypt about 35 M.Y.B.P. More fossils are needed before it can be described in detail and its relationships to later forms firmly specified.

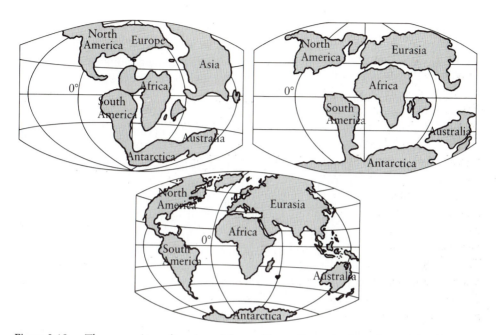

Figure 2.19   The approximate location of the earth's continents in the early Eocene (top left), in the late Eocene (top right), and today (bottom).

2.3B   Oligocene Epoch (38–22 M.Y.B.P.)

The Cenozoic cooling trend accelerated at the beginning of the Oligocene, increasing the temperature gradient between the equator and the poles and producing widespread glacial conditions in Antarctica. There were still no ice caps on the polar landmass, however, except perhaps at the end of the epoch (Kennett 1977). As Australia moved north during the Oligocene, the ocean separating it from Antarctica widened (Figure 2.19, top right). South America probably moved north away from Antarctica in the middle to late Oligocene, triggering the establishment of a strong oceanic current around the polar landmass. This, in turn, would have reduced the transport of heat toward the poles by the world's oceans. All things are relative, however, and cool, temperate vegetation persisted on the margins of Antarctica until the end of the epoch, when it disappeared (Kennett 1977).

Primates vanished from Eurasia and North America during the Oligocene, even though lush forests still covered the southerly parts of these continents. The most interesting fossil material comes from northern Africa. Where today the Fayum region of Egypt is a desert badland, 35 M.Y.B.P. there was a mosaic of tropical forest and woodland in areas seasonally flooded by the Nile, with tropical savanna covering higher ground beyond the floodplain (Butzer 1977). In this environment lived an assortment of mammals, including primates that may be the ancestors of Old World monkeys, apes, and *Homo* (Kay, Fleagle, and Simons 1981). These primates had monkeylike skeletons and were probably arboreal, restricted to the forest belt straddling the great, slow-moving river (Fleagle 1980b).

Where did these primates come from? We do not know. Perhaps they were the descendents of strepsirhines that entered Africa from Europe over a land bridge in the Paleocene. Alternatively, they may have evolved in Europe and, like two or three other mammalian groups, crossed to Africa over a temporary land bridge in the late Eocene or early Oligocene (Kennedy 1980; Coryndon and Savage 1973).

In South America, primates first appear in the fossil record about 35 M.Y.B.P. Their affinities with living New World monkeys are a matter of debate (Simpson 1980; Ciochon and Chiarelli 1980). Although they show some resemblances to the Fayum primates, their origins and the route they took to South America remain obscure. Both a northern route over North America and an eastern route across the Atlantic from Africa have been proposed (Delson and Rosenberger 1980).

2.3C   Miocene Epoch (22–5 M.Y.B.P.)

Temperatures rose again in the early Miocene (22–14 M.Y.B.P.) for reasons that are still unclear. Continents were approaching the positions they occupy today (Figure 2.19, bottom) and as the African–Arabian plate jostled against its Eurasian neighbor, land bridges formed between Africa and Europe, opening up new opportunities for the interchange of plants and animals. In the New World, ocean still separated North and South America.

The middle and late Miocene (from about 14 M.Y.B.P.) marked a renewal of the Cenozoic cooling trend and conditions simultaneously became drier. Major ice caps developed on eastern Antarctica for the first time (Shackleton and Kennett 1975), and the steep temperature gradient existing today between the equator and the poles was established. The factors precipitating this renewed cooling are unclear (Kennett 1977; Crowell and Frakes 1970). By 10 to 12 M.Y.B.P., the buildup of the Antarctic ice sheet was essentially complete, although toward the end of the epoch it expanded again, becoming thicker than it is today. This expansion was associated with further global cooling and a drop in sea level, probably because more and more water was being locked up in the form of ice. The close of the Miocene was also marked by the drying up of the Tethys Sea, and what are today the Mediterranean and Red seas turned into a vast inland desert (Hsu, Ryan, and Cita 1973; Ross et al. 1973). This transformation was brought about both by the global fall in sea level and by the formation of a land link between Morocco and Spain, which plugged the entrance to the sea.

World vegetation reflected these changes. The temperate and tropical forests that had blanketed much of the earth's landmass were transformed into more heterogeneous vegetation, often into a mosaic of woodland and grassland. In the region of the Tethys Sea, for example, woodland seems to have become the predominant biome in the middle Miocene; by the end of the epoch, savanna had taken over and trees were left only along riverbanks (Delson 1975b). In Pakistan just south of the Himalayan foothills, reconstructions of a large, late Miocene drainage basin conjure up a landscape laced by meandering river channels and covered by a mosaic of vegetation types that included forest, woodland, and grassland (Badgley and Behrensmeyer 1980).

In East Africa, as in the Mediterranean region, global effects were compounded by dramatic local events. Starting about 16 M.Y.B.P., north–south rifting of the African plate began, accompanied by great volcanic and mountain-building activity. Increased amounts of high land caused *rain shadows* to form, areas with little or no rainfall. Prior to this time, rain forest had flourished in much of eastern Africa and perhaps all the way to the Indian Ocean. But from 16 M.Y.B.P. on, there was a slow shift toward a much more seasonal climate, and by 3 M.Y.B.P. savanna-mosaic had largely replaced rain forest in East Africa. The middle and late Miocene was a period of slow transition between the two biomes, with woodland gradually taking over from forest before giving way to grassland (Van Couvering 1980; Andrews and Van Couvering 1975; Van Couvering and Van Couvering 1976). In fact, savanna-mosaic reached its greatest world distribution late in the Miocene, when it covered much of North America, Asia, South America, southeastern Europe, the Middle East, sub-Himalayan Asia, and several parts of Africa (Van Couvering 1980). In areas that continued to support forest, such as northwestern Europe, deciduous and broad-leaved evergreen forests were increasingly mixed with coniferous species. By the close of the epoch, these forests had become real taiga, dominated by pine, hemlock, and spruce (Hammen, Wijmstra, and Zagwijn 1971).

During the Miocene, the higher primates of the Old World diversified greatly. In contrast, although fossils resembling modern strepsirhines in many morphological features have been found in African deposits, they are not nearly as abundant or as diverse as the Eocene strepsirhine-grade primates. In the New World, the record is sparse. Fossils dating from the late Oligocene and early and middle Miocene have been attributed to the lineages of living New World monkeys by some (Rosenberger 1979; Delson and Rosenberger 1980), but these links are still a matter of debate (Hershkovitz 1970, 1982).

The first primates with apelike characteristics come from early Miocene deposits in Uganda and Kenya, where at least six species have been identified. These forms belong to the family Dryopithecidae and are placed in the Hominoidea, the same superfamily to which the living apes are assigned (Simons and Pilbeam 1965). The dryopithecids are often referred to as *dental apes,* for their tooth morphology is distinctively apelike. Postcranially (from the neck down), however, they exhibit monkeylike features as well

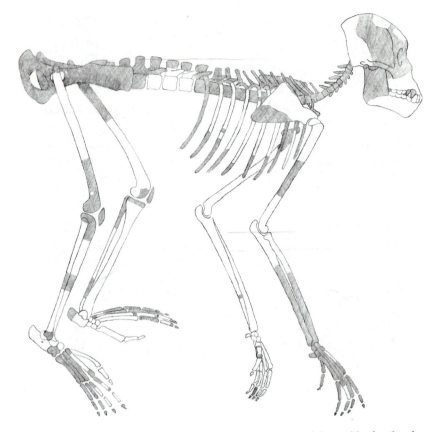

Figure 2.20   Reconstruction of *Proconsul africanus*, a member of the apelike family of Dryopithecidae that was widespread in Africa, Europe, and Asia in the early Miocene. (Adapted from D. Pilbeam and *Scientific American*.)

as apelike features, in addition to features unlike those of any living primate (Figure 2.20) (Rose 1983). The various species ranged widely in size, the smallest weighing about 2 kg and the largest as much as 65 kg. Environmental reconstructions strongly indicate that all of them were forest dwellers, and the morphology of their teeth suggests that all of them ate a substantial amount of fruit. Beginning 16 or 17 M.Y.B.P., they spread out of Africa into Eurasia, from Spain to China. Two species are known from western Europe, where they ranged as far north as the Rhine Valley of Germany. In Eurasia as in Africa, they lived in tropical forests and woodlands, and they are never found in association with plants or animals typical of open-country conditions. About 13 M.Y.B.P., they may also have occupied temperate deciduous forests at the northern limit of their distribu-

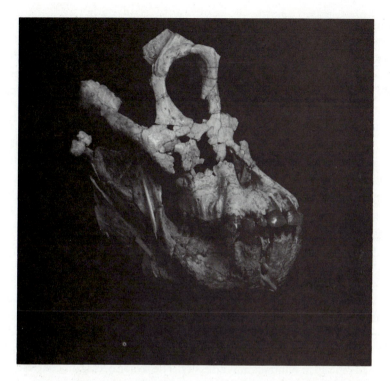

Figure 2.21   This 7-million-year-old partially complete skull of *Sivapithecus indicus* from Pakistan shows striking resemblances to the modern orang-utan. (Photography by W. Sacco/specimen courtesy of D. R. Pilbeam.)

tion (Von Koenigswald 1962), but the evidence for this is tenuous (D. R. Pilbeam, pers. com., 1984).

A second distinctive group of primates, the Sivapithecidae, appears in Miocene deposits in Europe and Asia between about 13 and 7 M.Y.B.P. Remains of this group, which includes *Ramapithecus, Sivapithecus,* and *Ouranopithecus,* have been recovered from Greece, Turkey, India, Pakistan, and China. More fragmentary evidence suggests that similar forms were present in East Africa at the same time or even as early as 17 M.Y.B.P. (Pickford 1982; A. C. Walker, pers. com., 1984). Several sivapithecid features distinguish them from the dryopithecids, particularly their large, thick-enameled molar teeth and heavily buttressed jaws. These features formerly led researchers to believe that a sivapithecid was the earliest direct ancestor of the hominids, for similar characteristics are found in hominids living in East Africa about 3.5 M.Y.B.P. With the addition of new finds, particularly the partially complete skull of a *Sivapithecus indicus* from Pakistan (Figure 2.21), it now looks as though the closest affinities of the Eurasian forms, at least, are with the living orang-utan *(Pongo)* (Pilbeam 1982; Ward and Pilbeam 1983; Preuss

1982; Andrews 1982). *Sivapithecus indicus* appears to have been quite dimorphic sexually, with males weighing about 45 kg and females less. It lived in woodland and forest, and probably moved and fed on fruit and leaves both in the trees and on the ground. Its life-style is difficult to reconstruct, though, for postcranially it does not closely resemble any of the living hominoids, and none of the living primates have such thick-enameled teeth (Pilbeam 1984; Kay 1981).

The third group to blossom in the Old World during the Miocene were colobine monkeys, which make up one of the two subfamilies within the family Cercopithecidae, to which all Old World monkeys belong (Table 1.1). The earliest unquestionably colobine fragments, from Kenya, are dated at about 9 M.Y.B.P. In the late Miocene, colobines spread into southern and central Europe, where they diversified further, and by 6 to 7 M.Y.B.P. they were in Asia (Simons 1970; Delson 1975a; Delson and Andrews 1975). The morphological and paleoecological evidence suggests that the colobines of the late Miocene occupied not only woodland and forest biomes, like modern colobines, but more open habitats as well, and were precursors of the highly terrestrial open-country forms that evolved in the Pliocene (Birchette 1982).

What was the fate of these three Old World groups? The question cannot be answered fully until deposits from 4 to 8 M.Y.B.P. are found, for at present the fossil record of this period is almost a complete blank. We do know that about 10 M.Y.B.P. dryopithecid fortunes turned, for from then until the hiatus in the record 2 million years later, they declined rapidly in numbers and diversity. The sivapithecids, in contrast, were still widespread when we lose sight of them 8 M.Y.B.P. When the fossil record resumes in East Africa at about 4 M.Y.B.P., it contains at least one species of hominid. As we noted, the material assigned to this species recalls in some respects sivapithecid morphology, but the exact relationship between the two groups remains a mystery. The fossil record of the African apes is virtually nonexistent, although it seems likely that they too evolved in their modern form from dryopithecids or sivapithecids in late Miocene or Pliocene times (Pilbeam 1984). The colobine group is the least problematic, since there are close resemblances between the Miocene and Pliocene fossils from Europe and Asia and between some of these fossils and the living colobines.

In sum, three distinctive groups of primates became widespread in the Old World during the Miocene. They were limited, probably without exception, to regions broadly describable as tropical in climate and vegetation. The relationships of two of the three groups to living primates are unclear.

2.3D   Pliocene (5–1.8 M.Y.B.P.) and
Pleistocene (1.8 M.Y.B.P.–12,000 B.P.) Epochs

Despite a retreat of the Antarctic ice sheets from their late-Miocene–earliest-Pliocene maximum, other indicators show that the Cenozoic cooling trend surged on. Major ice sheets in southernmost South America formed about 3.5 M.Y.B.P., and 2.5–3 M.Y.B.P. marks the formation of an ice sheet at the North Pole for the first time. It is not yet known why ice sheets formed about 10 million years later in the Northern Hemisphere than in Antarctica, although it probably had to do with hemispheric differences in plate tectonics and heat transport to the poles (Kennett 1977). Whatever the cause, formation of the northern ice sheet seems to have triggered a series of alternating warm and cold phases in world climate, with a steadily sinking mean temperature. Over the past 800,000 years and probably longer, cold phases have been marked by large ice sheets in the middle latitudes of the northern continents as well as by a massive expansion of the polar ice caps. These *glaciations* have been punctuated by *interglacials,* or warm phases, when ice sheets recede and sea level rises. We are in an interglacial today. However, temperatures fluctuate slightly even during glaciations. The warm peaks during these fluctuations are called *interstadials.* We know that there have been 10 or more glacial cycles in the last 800,000 years, but their chronology is not completely known; the further back in time we go, the harder a chronology is to reconstruct. We can say with some confidence that the most recent glaciation began at least 65,000 B.P. and ended about 10,000 B.P., and the one before that began about 200,000 B.P. and lasted until about 130,000 B.P. (Butzer 1977).

Glaciations transform temperate biomes into arctic desert and tundra, but their effect on the tropics is not so clear. Over the past million years, the rain forests of Africa and South America have repeatedly shrunk, leaving small, scattered pockets of forest surrounded by savanna (Figure 2.22). These retreats were followed by expansions back to the forest's former size. It is likely, though not certain, that retreats coincided with glaciations at higher latitudes and expansions with interglacials (Hammen 1972).

At the end of the Miocene, as we noted, the Mediterranean Sea was sealed off and turned into a vast desert. Early in the Pliocene, the Atlantic Ocean broke through at the Strait of Gibraltar and refilled the landlocked basin, and the boundaries of the new sea became approximately those of the Mediterranean today. The moderating effect of this sea transformed woodland and semidesert scrub into monsoon forest in the surrounding region. The other major geographic change during the Pliocene was the formation about 3 M.Y.B.P. of the Panamanian land bridge between North and South America, finally bringing all the oceans and continents of the globe approximately into their present relationships.

Formation of the land bridge between North and South America marked the beginning of a major exchange of land mammals between the two continents (Marshall et al. 1982), but so far as we know, South American primates never spread further than southern Mexico, the northernmost limit of their modern distribution.

In the Old World, the Pliocene saw the spread of the second subfamily within the Cercopithecidae. Eight genera make up the subfamily Cercopithecinae (Table 1.1). Members of four, macaques *(Macaca), Papio* baboons, *Theropithecus* baboons, and the

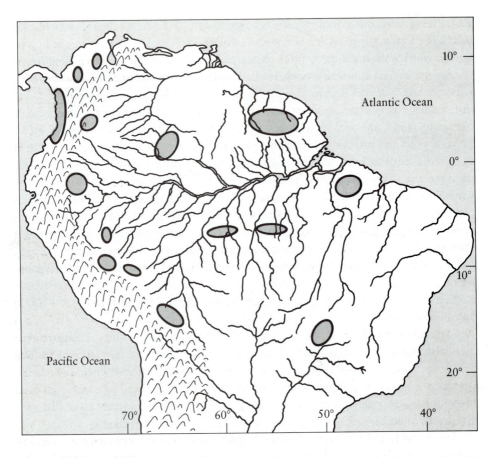

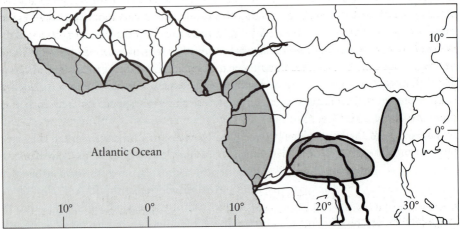

patas monkey *(Erythrocebus patas)*, spend much of their time on the ground, while the others are primarily arboreal, forest-dwelling monkeys.

The only cercopithecines outside Africa today are the macaques, which live almost exclusively in Asia. Ancestors of the modern Barbary macaque *(Macaca sylvanus)* were the first cercopithecines to appear in Europe, and although colobines continued to dominate Pliocene primate assemblages in that region for some time thereafter, macaques became more and more abundant and by the Pleistocene they were the only cercopithecid present. From Europe they spread east into Asia during the late Pliocene and Pleistocene (Melnick and Kidd forthcoming). In interglacials and interstadials, macaques dispersed widely through Europe; however, their range apparently contracted during glaciations, because they are never found in association with animals considered to be really cold adapted. Macaques lived in European temperate forests and woodlands until the last interglacial, perhaps persisting into the early stages of the glaciation that followed (Delson 1975a, 1975b). Thereafter, except for the possibly relict population still on Gibraltar, the westernmost limit of their Eurasian distribution became Afghanistan.

The first baboons to appear in Africa belonged to the *Theropithecus* group. Three and possibly four species ranged from the Mediterranean coast to the tip of South Africa from the middle Pliocene (C. J. Jolly 1972; Dunbar 1977a). Only one species, the gelada baboon *(T. gelada)*, survives today, and it is found only on the mountainous plateau of Ethiopia. The other species became extinct in the late Pleistocene, their disappearance attributed by some (e.g., Tappen 1960) to the spread of *Papio* baboons through Africa at about that time. Others (e.g., C. J. Jolly 1972; Dunbar 1977a) have suggested that early hominids may have had a hand in these extinctions.

The mammalian fossil record for West Africa is a virtual blank throughout the Cenozoic, and the Pliocene and Pleistocene are no exceptions. However, it is probably that the arboreal, forest-dwelling cercopithecines diversified at this time. We noted earlier that "islands" of forest were repeatedly created during the periodic contractions of the Old and New World rain forests. This process may have contributed to the high number of species not only of primates but also of other animals found in these regions today. The following scenario is suggested (following Moreau 1969). When the forests receded, the arboreal cercopithecines were trapped in remaining pockets of forest, sometimes separated by hundreds of kilometers of open country. In these refuge areas they remained isolated from one another for tens of thousands of years. During that time, random genetic changes and perhaps differing local selection pressures caused the pop-

Figure 2.22   Location of postulated forest refuge areas in South America (top) and West Africa (bottom) during periods of contraction of tropical forests in the last 2 million years (Adapted from Vanzulini 1973 and Laurent 1973.)

ulations to diverge. When the forests expanded once more and contact between populations was reestablished, animals had diverged sufficiently not to interbreed. In other words, these populations had evolved into separate species. (For a discussion of the impact of forest refuges on the New World primates, see Kinzey 1982.)

Our account ends with the appearance of hominids in East Africa between 3 and 4 M.Y.B.P. The number of species present in East Africa at that time is still not settled (Leakey and Lewin 1977; White, Johnson, and Kimbel 1981), although it is likely that at most there were two. Interpretation of the hominid fossils dating to about 2 M.Y.B.P. is equally problematic; some scientists propose one and others as many as four species (Walker and Leakey 1978). Irrespective of species, these East African hominids of the Pliocene and Pleistocene all walked upright and had small, bladelike canines and large, thick-enameled cheek teeth, and the paleoecological evidence suggests that they lived in savanna-mosaic. About 1 M.Y.B.P., one species of hominid *(Homo erectus)*, our own immediate ancestor, moved out of Africa and spread rapidly through the Old World. However, the reconstruction of hominid evolution claims whole books to itself and does not directly concern us here.

2.4   Discussion

Throughout their evolution, primates have lived primarily in the tropics. Today, approximately 170 species occupy all the major tropical biome types, from rain forest to woodland, shrubland, savanna, and semidesert scrub. If primates are successful, it is a tropical success. Can we say, then, that within the tropics primates are more successful in some biomes than in others? At this point, we may ask once more what we mean by success. What is it we want to measure? At least three possibilities spring to mind. One is to compare the sheer bulk, or biomass, of primates in different biomes. A second is to compare the numbers of species. A third and I believe most useful approach is to compare the numbers of genera. Let us consider each of these possibilities.

The highest estimate of primate biomass is for dry seasonal forest at Polonnaruwa, Sri Lanka. The conclusion that primates are most successful in seasonal forest must be viewed with great caution, however. First, quite apart from the doubtful accuracy of most biomass estimates, there are simply not enough of them, accurate or no, upon which to base broad generalizations. Second, many of the estimates come from areas in which there has been substantial human interference. This can influence biomass in many ways. Degradation of the forest, for example by felling trees or grazing cattle, decreases the abundance of many species, though it may actually have the opposite effect on some (Goldstein 1984). If primates are hunted, their biomass will be below its natural level. If other species are hunted, this too may indirectly affect primate biomass, because primates may normally compete for food or other resources with these species or be preyed upon by them. Thus, if such species are eliminated or reduced in abundance, the biomass of primates may become unusually high. It should be noted that at Polonnaruwa few mammals are left except primates, dogs, and people.

A third factor must also be considered when using biomass as a criterion to measure primate success in different biomes. The biomass of animals in a particular environment is strongly influenced by the primary productivity of that environment and by the form in which plant matter is produced. For example, the biomass of mammals is higher in tropical savanna than rain forest because much more of the primary production is available as food. In this respect, the apparent success of primates as measured by biomass may simply reflect differences in primary productivity that affect all mammals equally.

In Section 2.2 we noted that the greatest number of primate species are found in the rain forests of Africa and South America. If we use species richness as an indicator of success, then primates are most successful in rain forest. Most of the problems with using biomass as a measure apply here, too. In addition, and most important, the high number of species in New and Old World rain forests today may be a temporary phenomenon reflecting the repeated retreats of these forests over the last several million years rather than real adaptive radiations. An *adaptive radiation* is a set of species that exploit an environment in many different ways, having evolved from a single ancestral species.

The third comparison involves the more inclusive taxonomic unit of the genus. A genus is traditionally defined as a species or group of related species with a presumed common phylogenetic origin (Buettner-Janusch 1966). (In some recent classifications, no inferences are made about common origin, and species are assigned to a single genus purely on the basis of similarity.) Members of a single genus are usually similar to one another and clearly distinct from other genera in morphology and ecology. The number of genera in a biome is thus likely to be a better measure of an order's ecological diversity in that biome than is the number of species. Table 2.3 shows that the number of primate genera is highest in rain forest, specifically in the rain forests of South America and West Africa. The low diversity of primates in Malaysia is probably due to several factors, including competition with other animals and the relatively recent arrival of at least some primate groups in this outpost of Eurasia.

In summary, if primates can be said to be successful at all, the theater of this success is the tropics. When we try to assess the success of primates in different tropical biomes, however, the difficulty of actually measuring success becomes apparent. Of the three measures discussed, the least ambiguous and the one most indicative of my conception of evolutionary and ecological success is the number of genera that a biome supports, for this provides at least a general indication of ecological diversity. Using this criterion, we can say that primates are most successful in rain forest but not, it should be emphasized, in all rain forests: life is not that simple.

Although primates are most widespread in the tropics, their distribution today does not coincide perfectly with the limits of tropical climates and vegetation, nor, so far as we can tell, did it coincide completely in the past. For example, the distribution of a few species extends into temperate regions. These we shall consider further in Chapter 3. In other cases, primates have disappeared from or failed to colonize seemingly accessible tropical biomes. We end this chapter by reviewing these cases briefly. This and especially the next chapter are concerned with the ways in which biology of the primates constrains their ability to exploit highly seasonal environments. But additional factors, in particular competition with other animals, likely play crucial roles as well. The importance of competition is considered at length in Part V. The purpose of the quick review that follows here is to emphasize that it may not be possible to explain the distribution of the primates simply as a consequence of their biology and the distribution and accessibility of certain climates and vegetation.

Our first case is that of the Eocene strepsirhines. Widespread in the northern continents throughout that epoch, they virtually disappeared from Eurasia early in the Oligocene. Although world climates cooled considerably during this epoch, at least the southern parts of Eurasia remained covered by the tropical vegetation in which strepsirhines had thrived. The second case is that of the dryopithecids, which disappear from the fossil record during the Miocene after being both abundant and diverse for 10 million years. Although large tracts of tropical forest were giving way to savanna-mosaic during this epoch, the tropical forest biome by no means disappeared.

A third puzzle is the failure of primates to invade the tropical savannas of the New World. Although cebus monkeys (Cebus capuchinus) occasionally cross open areas from one patch of woodland to another in Central America (Freese 1976), there are no New World monkeys adapted primarily to life in open country, nor have there ever been, so far as we can tell from the sparse fossil record. In contrast, several groups moved out into this biome in Africa and today baboons (Papio spp.), patas monkeys (Erythrocebus patas), and vervets (Cercopithecus aethiops) range through open country south of the Sahara. Africa and South America were widely separated throughout the Cenozoic and have quite distinctive faunas, but as in Africa, savanna-mosaic gradually replaced rain forest in South America during the Miocene, and there were several periods in the Pleistocene when Amazonian rain forest occurred only in small, isolated patches. Today, the fringes of this forest give way to woodland and savanna as in Africa. Thus, while rain forest covers 20% of South America and only 6–7% of Africa in modern times (Keast 1972), it cannot be argued that open country is or was unavailable for colonization by South American primates.

Competition with other mammals has been cited as the reason for the disappearance or nonappearance of primates in each of these cases. The decline of the Eocene strepsirhines has been attributed to competition resulting from the diversification of rodents (Romer 1966) and bats (Sussman and Raven 1978). The decline of the dryopithecids, starting about 10 M.Y.B.P., coincided with the first major radiation of monkeys, and it has been argued that the two events are related, with monkeys outcompeting their ape-like relatives in the forests of Africa and Eurasia (Temerin and Cant 1983). Finally, in the case of the New World monkeys, it has been postulated that the early colonization of the South American savannas by ungulates, sloths, and rodents led to the competitive exclusion of monkeys (Keast 1972; Patterson and Pascual 1972).

Because competition is difficult to document in the present and even harder to sub-stantiate in the past (Part V), propositions about competition are largely speculative. Moreover, competitive exclusion by other animals and the constraints of species biol-ogy are not mutually exclusive explanations. The outcome of competition between two species is likely to be different depending on the environment in which the competition occurs and, conversely, the ability of a species to survive in a particular environment may often be affected not only by its biological makeup but also by the intensity of competition. As we return in Chapter 3 to a more straightforward account of the biol-ogy of primates compared with that of other mammals, we should keep this complexity in mind.

CHAPTER THREE

Primate Distribution: A Broader Perspective

3.1   The Distribution of Mammals

JUST how unusual among mammals is the tropical distribution of the primates? This question leads to a broader inquiry than in Chapter 2, where we considered the primates alone. Here we compare the distribution by family of the primate order with the distributions of other mammalian orders, and then we investigate the implications of these comparisons. Our survey includes only the 14 land-living eutherian orders. The class Mammalia (Table 3.1) is made up of two subclasses, the *Prototheria*, whose members produce eggs from which their young hatch, and the *Theria*, whose members give birth to live young. The subclass Theria is in turn composed of two infraclasses, the *Eutheria*, or placental mammals, and the *Metatheria*, mammals in which most of the development of the young takes place in a pouch. Only a single order, the Marsupialia, is classified in the Metatheria, while the order Primates is one of 17 in the infraclass Eutheria. Our discussion is limited to other members of this infraclass excluding the three marine orders (Cetacea, Pinnipedia, and Sirenia), so that the comparison is among land forms with broadly similar reproductive biology.

There are eutherians in every biome type listed in Table 2.1. Table 3.2 provides an overview of their distribution by latitude and altitude. Darlington (1957:341) has summarized the global picture thus:

> Zonation of land mammals is well marked but is mostly a matter of subtraction. North-temperate mammal faunas are essentially tropical faunas from which much has been subtracted. And the arctic mammal fauna is essentially a temperate fauna from which much more has been subtracted. Southward, too, mammal faunas are reduced by subtraction.

As we leave the treeline of the North American boreal forest, for example, bats (order Chiroptera) disappear, leaving five orders still represented: insectivores (Insectivora), rodents (Rodentia), carnivores (Carnivora), hare and rabbits (Lagomorpha), and even-toed ungulates, such as the caribou (Artiodactyla).

Table 3.1   The class Mammalia

Taxon	Number			Common name
	Families	Genera	Species	
Subclass Prototheria				
Infraclass Ornithodelphia				
Order Monotremata	2	3	6	Monotreme
Subclass Theria				
Infraclass Eutheria				
Order Artiodactyla	9	75	171	Even-toed ungulate
Order Carnivora	7	96	253	Carnivore
Order Cetacea	8	32	69	Whale
Order Chiroptera	17	176	883	Bat
Order Dermoptera	1	1	2	Flying lemur
Order Edentata	3	14	31	Anteater, sloth
Order Hyracoidea	1	3	11	Hyrax
Order Insectivora	8	77	406	Insectivore
Order Lagomorpha	2	9	63	Rabbit, hare
Order Perissodactyla	3	6	16	Odd-toed ungulate
Order Pholidota	1	1	8	Pangolin
Order Pinnipedia	3	20	31	Seal, walrus
Order Primates	14	55	166	Primate
Order Proboscidea	1	2	2	Elephant
Order Rodentia	33	352	1685	Rodent
Order Sirenia	2	3	5	Sea cow, dugong
Order Tubulidentata	1	1	1	Aardvark
Infraclass Metatheria				
Order Marsupialia	8	81	242	Marsupial

Source: Adapted from Anderson and Jones 1967.

Moving yet further poleward, from the icy cold tundra of the continental arctic to the Arctic archipelago, the northernmost landmass on earth, we leave behind the shrew family (Soricidae), last representatives of the Insectivora. Our list of mammals is now limited to a single rodent, the collared lemming (Dicrostonyx); a single lagomorph, the arctic hare (Lepus); three carnivores, the wolf (Canis), arctic fox (Alopex), and polar bear (Ursus); and two artiodactyls, the musk ox (Ovibos) and caribou (Rangifer). All these animals except the musk ox are also represented in northern Eurasia by the same or closely related species. In the Southern Hemisphere, there are no land mammals on Antarctica, but the tip of South America (55° south latitude), the next most southerly

Table 3.2   Mammalian distribution over biomes

Taxon	Common name	Tropical	Temperate	Subarctic, subalpine, arctic, and alpine
Artiodactyla				
Antilocapridae	Pronghorn		x	
Bovidae	Buffalo, bison, antelope, gazelle, cow, goat, sheep	x	x	x
Camelidae	Camel, llama	x[a]	x[a]	x[b]
Cervidae	Deer	x	x	x
Giraffidae	Giraffe, okapi	x		
Hippopotamidae	Hippopotamus	x		
Suidae	Pig, hog	x	x	
Tayassuidae	Peccary	x		
Tragulidae	Chevrotain, mouse deer	x		
Carnivora				
Canidae	Dog, wolf, fox, jackal	x	x	x
Felidae	Cat	x	x	x
Hyaenidae	Hyena	x		
Mustelidae	Badger, otter, skunk, weasel	x	x	x
Procyonidae	Raccoon	x	x	
Ursidae	Bear	x	x	x
Viverridae	Civet, genet, mongoose	x	x	
Chiroptera				
Desmodontidae	Vampire bat	x		
Emballonuridae	Ghost bat	x	x	
Furipteridae	Smoky bat	x		
Megadermatidae	False vampire bat	x		
Molossidae	Mastiff bat	x	x	
Mormoopidae	Leaf-chinned bat	x		
Mystacinidae	Short-tailed bat	x		
Myzopodidae	Golden bat	x		
Noctilionidae	Bulldog bat	x		
Nycteridae	Hollow-faced bat	x		
Phyllostomatidae	American leaf-nosed bat	x		
Pteropididae	Flying fox	x		
Rhinolophidae	Horseshoe bat	x	x	
Thyropteridae	Disk-wing bat	x		
Vespertilionidae	Common bat	x	x	
Dermoptera				
Cynocephalidae	Flying lemur	x		

Taxon	Common name	Tropical	Temperate	Subarctic, subalpine, arctic, and alpine
Edentata				
Bradypodidae	Tree sloth	x		
Dasypodidae	Armadillo	x	x	
Myrmecophagidae	Anteater	x		
Hyracoidea				
Procaviidae	Hyrax, dassie	x		
Insectivora				
Chrysochloridae	Golden mole	x	x	
Erinaceidae	Hedgehog	x	x	
Macroscelididae	Elephant shrew	x		
Solenodontidae	Solenodon	x		
Soricidae	Shrew	x	x	x
Talpidae	Mole		x	
Tenrecidae	Tenrec	x		
Tupaiidae	Tree shrew	x		
Lagomorpha				
Leporidae	Rabbit, hare	x	x	x
Ochotonidae	Pika		x	x
Perissodactyla				
Equidae	Horse, zebra, ass	x	x	
Rhinocerotidae	Rhinoceros	x		
Tapiridae	Tapir	x		
Pholidota				
Manidae	Pangolin	x		
Primates				
Callimiconidae	Goeldi's marmoset	x		
Callitrichidae	Marmoset	x		
Cebidae	New World monkey	x		
Cercopithecidae	Old World monkey	x	x	
Cheirogaleidae	Dwarf lemurs	x		
Daubentoniidae	Aye-aye	x		
Hominidae[c]	Chimpanzee, gorilla	x		x[b]
Hylobatidae	Gibbon	x		
Indriidae	Indri	x		
Lemuridae	Lemur	x		
Lepilemuridae	Sportive lemurs	x		
Lorisidae	Loris	x		
Pongidae	Orang-utan	x		
Tarsiidae	Tarsier	x		

[a] Desert biome.
[b] Subalpine biome.
[c] Excluding *Homo*.

(Continued on page 88)

Taxon	Common name	Tropical	Temperate	Subarctic, subalpine, arctic, and alpine
Proboscidea				
Elephantidae	Elephant	x		
Rodentia				
Abrocomidae	Chinchilla rat		x	
Anomaluridae	Scaly-tailed squirrel	x		
Aplodontidae	Mountain beaver		x	
Bathyergidae	Mole rat	x		
Capromyidae	Hutia	x		
Castoridae	Beaver		x	x
Caviidae	Guinea pig	x	x	
Chinchillidae	Chinchilla			x[b]
Cricetidae	Hamster, gerbil, vole, lemming	x	x	x
Ctenodactylidae	Gundi	x		
Ctenomyidae	Tuco-tuco		x	
Dasyproctidae	Agouti, paca	x		
Dinomyidae	Paca-rana	x		
Dipodidae	Jerboa	x[a]	x[a]	
Echimyidae	Spiny rat	x		
Erethizontidae	New World porcupine	x	x	
Geomyidae	Gopher	x	x	x
Gliridae	Dormouse	x	x	
Heteromyidae	Kangaroo rat		x[a]	
Hydrochoeridae	Capybara	x		
Hystricidae	Old World porcupine	x	x	
Muridae	Old World rat, mouse	x	x	
Myocastoridae	Nutria, coypu		x	
Octodontidae	Hedge rat		x	
Pedetidae	Spring hare	x		
Petromyidae	Dassie rat	x		
Platacanthomyidae	Spiny dormouse	x		
Rhizomyidae	Bamboo rat	x		x[b]
Sciuridae	Squirrel, marmot	x	x	x
Seleviniidae	Dzhalman		x[a]	
Spalacidae	Mole rat		x	
Thryonomyidae	Cane rat	x		
Zapodidae	Jumping mouse		x	
Tubulidentata				
Orycteropodidae	Aardvark	x		

Source: Adapted from Walker et al. 1964; Anderson and Jones 1967; Burt and Grossenheider 1976; Gunderson 1976; and Hagmeir and Stults 1964.
[a] Desert biome.
[b] Subalpine biome.

Table 3.3   Distribution over biomes of extant mammalian orders

Primarily tropical	Tropical and temperate	Tropical, temperate, arctic, and alpine
Dermoptera	Chiroptera	Artiodactyla
Edentata	Perissodactyla	Carnivora
Hyracoidea		Insectivora
Pholidota		Lagomorpha
Primates		Rodentia
Proboscidea		
Tubulidentata		

landmass, is occupied by various rodents, four carnivores, the mountain lion *(Felis),* skunk *(Conepatus),* fox *(Dusticyon),* otter *(Lutra),* and two artiodactyls, the llama *(Lama)* and deer *(Pudu)* (Osgood 1943).

Five orders of mammals are represented in tropical, temperate, and arctic zones (Table 3.3). A sixth, the Chiroptera, is found from the equator to the treeline in both hemispheres. The Perissodactyla are today limited to the New and Old World tropics and to the temperate grasslands of Mongolia and Chinese Turkistan, although they were more diverse and more widely distributed in the past (Simpson 1951; Romer 1966). Few families within these orders are distributed worldwide. Three bat families probably come closest to a worldwide distribution, and rabbits (Leporidae), squirrels (Sciuridae), dogs (Canidae), weasels (Mustelidae), and cats (Felidae) are found on all the major land-masses except Australia. At the level of genus, the most widespread forms are bats *(Myotis),* otters *(Lutra),* cats *(Felis),* hares *(Lepus),* weasels *(Mustela),* and wolves *(Lupus).*

Seven orders, including the primates, are limited primarily to the tropics. On the face of it, this seems sufficient to dismiss the claim that primates have an unusual distribution pattern, but let us look at these other orders more closely.

3.1A   Flying Lemurs

Flying lemurs (Dermoptera) are today represented by only one genus and two species in southern China and the islands of Java, Borneo, and the Philippines (Figure 3.1). Simpson (1945) tentatively ascribes to this order one extinct group containing two genera, dating from the North American Paleocene and Eocene. Flying lemurs are in no way related to the lemurs of Madagascar—aside from both being eutherian mammals—and they do not fly, although they are skillful gliders. Primarily nocturnal, they live on a diet of leaves and fruit (Gunderson 1976).

Figure 3.1   This drawing of the Philippine flying lemur *(Cynocephalus volans)* shows the membrane that extends between its limbs and tail, which allows it to glide up to 70 m in a single flight.

3.1B   Edentates

Edentates (Edentata) are a more diverse order, comprising 14 genera and 31 species grouped into three families: anteaters, armadillos, and tree sloths (Figure 3.2). Edentates are found almost exclusively in tropical South and Central America. Anteaters have long noses, long, thin, sticky tongues, and strong front claws for digging into ant and termite nests. They are climbers as well as diggers and live in forests and more open country. They range in size from tiny ratlike to giant boar-sized animals. Armadillos have a similar size range, but they are terrestrial, digging forms that feed on fallen fruit, invertebrates and, when they find it, carrion (Hershkovitz 1972). Sloths are highly arboreal, leaf-eating specialists that move so little and so slowly it is not uncommon for them to be covered with moss! At least seven other families belonging to this order evolved and became extinct over the past 65 million years (Patterson and Pascual 1972), so that today the edentates are but a shadow of their former selves.

3.1C   Pangolins

Pangolins (Pholidota), represented by two genera and seven species (Patterson 1978), are the anteaters of sub-Saharan Africa and South and Southeast Asia (Figure 3.3). They have been studied in Africa, where there are two arboreal and two terrestrial species.

Figure 3.2   Today there are only three families in the order Edentates, anteaters (middle), armadillos (bottom), and sloths (top), but in the past the order was more highly diversified.

Figure 3.3   Pangolins (*Manis* spp.) have a coat of armor and curl up in a well-protected ball when disturbed.

Figure 3.4   The aardvark *(Orycteropus afer),* sole living representative of the order Tubuliden-tata, lives only in dry open country in southern Africa. In the past, members of this order were distributed throughout Africa, Madagascar, and southern Eurasia, but the fossil record indicates that the order's diversity has always been low. (Animals Animals/©Patti Murray.)

The arboreal forms are sympatric in lowland forest, but one is nocturnal while the other is diurnal; of the two terrestrial forms, one is primarily savanna dwelling and the other forest dwelling (Rahm 1960; Bigalke 1972). Early members of the order are represented by a few bones from the Miocene and Oligocene of Europe. Pangolins probably reached Africa from Eurasia late in the Oligocene, although the only African fossils are a handful of fragments from the Pliocene and Pleistocene.

3.1D   Aardvarks

Aardvarks are the sole living representatives of the order Tubulidentata (Figure 3.4). The range of this species, which looks like a "pig on a diet" (Gunderson 1976), is today limited to southern Africa. Aardvarks are found primarily on the savanna, where they use their considerable digging skills to tear open termite mounds, whose occupants are their main food. Aardvarks also construct burrows that provide them with shelter from heat and predators. The order is represented in the fossil record by three genera in addition to the genus surviving today, with a geographic range that included Madagascar, Africa, and southern Eurasia (Patterson 1978).

Figure 3.5   African and Indian elephants are the last representatives of the once diverse and widely distributed order Proboscidea. Extinct species in this order, such as the mastodon and the mammoth, are among the largest mammals that ever lived.

3.1E   Elephants

Elephants (Proboscidea), even more than edentates, are remnants of a once much more diverse and widely distributed order (Coppens et al. 1978) (Figure 3.5). Today the order is represented by one family comprised of two genera, each containing one species, the African and Indian elephant, respectively. Both species are vegetarians, eating grass, leaves, shoots, fruit, and woody material. The first elephant ancestors appear in late Eocene and early Oligocene assemblages of the Fayum in northern Egypt. Their later history is one of successive movements out of Africa to the north and east. By the Pliocene there were six families, and twelve different species had crossed into North America over the land bridge intermittently linking it to Asia. In the late Pliocene, at least three reached South America, so that 2 million years ago there were Proboscidea on all the major landmasses of the world except Australia, occupying biomes from the tropics to the Arctic Circle.

3.1F   Hyraxes

Hyraxes (Hyracoidea) are today represented by three genera and 11 species, none of them bigger than rabbits (Figure 3.6). Most hyraxes live in tropical Africa, but members of one species live in the Middle East. The order includes nocturnal, diurnal, arboreal, and terrestrial forms, but all are primarily herbivores. The earliest hyraxes are found in

Figure 3.6   Female rock hyrax (left) *(Procavia capensis)* with three infants. These animals take refuge in cracks in rocks to avoid predators and also to avoid the heat of the midday sun. (Animals Animals/©B. G. Murray, Jr.)

early Oligocene deposits from the Fayum. At that time they constituted a more diverse group than today and were represented by at least six genera ranging from the size of a big rabbit to that of a tapir (Meyer 1978). Hyraxes appear in China and southern Europe in Miocene fossil deposits, but the rise of ungulates during the Miocene signaled the beginning of a decline in the abundance and diversity of the order, which nonetheless has continued to the present.

3.1G   Commentary

With the addition of information about fossil forms, the picture presented by Table 3.3 is changed, and Simons' claim for the primates appears more accurate. Three groups can be distinguished in the category *primarily tropical* in Table 3.3. The first includes hyraxes, flying lemurs, pangolins, and aardvarks. These orders were limited to tropical zones in the past, as they are today, and although they contained more genera and species in the past, none of them ever approached the diversity of the primates. The second group contains just the Proboscidea. Today this order is small and limited to the tropics, but in the past it was highly diversified and as recently as 12,000 B.P. its members were distributed over all the world's landmasses except Australia, from polar regions to the tropics. The third group contains the primates and, tentatively, the edentates. The primate order is unique today in that it is diverse and rich in species compared with other orders limited to the tropics, and yet has never had members able to penetrate deep into temperate, let alone arctic, biomes. Macaques and langurs alone have made inroads into temperate areas. The edentates are tentatively included with the primates because the fossil record indicates that in the Miocene, Pliocene, and Pleistocene epochs, this order too was diverse and yet limited mainly to tropical areas. Today, an armadillo (*Dasypus novemcinctus*) is the only edentate whose range, from Nebraska to Patagonia, extends beyond the tropics.

In summary, the order Primates is unusual among living mammals in that it contains a substantial number of genera but its members are almost all confined to tropical areas. With the possible exception of the edentates, other orders that have always had a uniquely tropical distribution are of much lower diversity. What is the biological and ecological significance of the primate distribution pattern? The rest of Chapter 3 considers this question.

3.2   Stresses at High Latitudes

As latitude increases, seasonal variation increases and mean annual temperature declines. These factors act in complex ways to limit the spread of all life forms. For example, mean annual temperature, maximum summer temperature, or minimum winter temperature may critically affect the ability of animals to reproduce, raise young successfully, maintain body temperature as adults, or find sufficient food to avoid starvation. In the following review, we shall consider the morphological and behavioral responses of mammals to two kinds of stress, cold and food scarcity. While it is useful to treat these stresses separately since they call for different responses, it is important to recognize that they have strongly interacting effects. Thus, an animal may grow a thick, insulative coat or adopt behaviors such as huddling and sunning to keep warm in winter, but an adequate supply of food is also essential: the animal will freeze to death if it no longer has the fuel to produce sufficient body heat. Conversely, an animal weakened by using up most of its energy reserves in an effort to keep warm may lack the strength to forage.

Food scarcities can be caused in various ways. First, snow cover may act as an impenetrable barrier to resources on or close to the ground that would otherwise be available. Second, snow may prevent animals moving from one food source to another. Formozov (1946) has considered these and other physical properties of snow in detail, showing how the ability of animals to dig out food, move, and protect themselves from predators and cold is affected by whether the snow is formed into compacted crusts and layers or into windblown banks of fine, dry flakes. A third cause of food scarcity can be too few hours of daylight for a diurnal animal to eat enough, even if in principle resources are in sufficient supply.

Finally, food may be scarce in the absolute sense. Those plant parts upon which an animal habitually feeds may be absent in winter, and the plant parts that are present may provide too little nutrition to be even a temporary substitute for animals not specialized to exploit these parts. Water is also likely to be in short supply when temperatures are low, with much of it locked up in the form of snow or ice and the remainder inaccessible under sheets of ice. Eating snow is a poor alternative for most animals because so much energy is needed to melt it. In most cases, several of these factors will act together to create food scarcity, and their separate effects may be difficult or impossible to distinguish. The distinctions are important, however. Where possible, they will be described in 3.3, because each factor represents a separate problem requiring a separate solution if animals are to survive temperate or arctic winters.

3.3   Mammals at High Latitudes

In this section we look at ways in which mammals of different size solve some of the problems presented by life at high latitudes, particularly the harsh winters. Our survey focuses on New World mammals in the Northern Hemisphere. Many New World genera have representatives in the Old World, however, and conclusions based on the former generally hold good for the latter. We shall not consider dietary carnivores, animals that feed largely or exclusively on meat, since we are the only primates with a major meat component in our diet. Meat eaters face different problems at high latitudes than animals with a mixed or entirely vegetable diet, because they depend for survival on capturing prey that may be not only scarce, like plants, but also alert, mobile, cryptically colored, and sometimes fierce. Note that the term *carnivore* refers to an order as well as to a set of dietary habits. All dietary carnivores (except a few human populations) are members of the order Carnivora, but not all members of the order Carnivora are dietary carnivores. Certain species, particularly smaller ones, feed heavily on fruit, and others eat both meat and fruit. Some of these forms are included in our survey.

We consider animals in the following three weight classes:

1.   *Small mammals* weighing less than 1 kg. This weight class includes *herbivores,* species with a primarily vegetarian diet, and *insectivores,* species that feed mainly on insects. Like *carnivore, insectivore* is also used to describe a member of an order, the Insectivora. Unlike the order Carnivora, which includes species with vegetable as well as animal diets, all members of the Insectivora are insectivorous. In further contrast, whereas dietary carnivores are not found outside the Carnivora, dietary insectivores are found in other orders.

2.   *Medium-sized mammals* weighing 1 to 20 kg. This class includes herbivores and species with diets containing a mixture of animal and vegetable foods. Insectivores are

not included because, while there are medium-sized insectivores in the tropics, there are none at high latitudes.

3.   *Large mammals* weighing more than 20 kg. This class, like the medium-sized mammals, includes herbivores and species with mixed diets, and it contains no insectivores.

3.3A   Small Mammals

The most northerly small herbivores are lemmings *(Dicrostonyx)* (Figure 3.7) which extend beyond the arctic continental coasts to islands even closer to the polar ice cap (Irving 1972). Other herbivorous rodents, such as mice (Cricetidae) and the ground squirrel *(Citellus)*, live as far north as the continental coastline. This also marks the range limit of the most northerly insectivores, shrews (Soricidae) (Burt and Grossenheider 1976).

Rodents live in the desert by avoiding desert conditions (Bartholomew and Dawson 1968), and small herbivores and insectivores living in the arctic follow a similar strategy. At the beginning of winter, they build nests and a network of burrows under the snow,

Figure 3.7   Collared lemming *(Dicrostonyx hudsonicus)* eating Indian paintbrush in northern Canada. (Animals Animals/© Ted Levin.)

which they do not leave until spring. Snow is a poor conductor of heat and thus protects the underlying soil from excessive chilling; temperatures under the snow can be as much as 15°C warmer than exposed soil (Formozov 1946). Insects, green leaves, stems, nuts, and berries on the soil surface and rhizomes and tubers under it apparently provide a plentiful food supply, since many of these animals continue to reproduce and raise litters during winter months. The only animal known to hibernate at this latitude is the ground squirrel (Hook 1960).

In sum, small herbivores and insectivores in the arctic manage to stay warm and well fed and are entirely dependent on snow cover for their good fortune. The consequences of abandoning this cover are dire: Formozov (1946) watched common voles that were disturbed from their nests by ermine try to flee across the surface of the snow only to freeze to death 3 or 4 m from the spot where they emerged.

Moving south, a broad range of rodents (Rodentia), shrews (Insectivora), bats (Chiroptera), and the pika (Ochotona), smallest member of the Lagomorpha, swell the assemblage of small herbivores and insectivores. Snow cover is less certain in some of these areas, and animals cannot count on it to insulate themselves from the cold as they do further north. Instead, they hibernate, thereby solving many of the problems presented by the cold and food shortages of the northern temperate winter.

Hibernation in small mammals is characterized by (1) a reduction of body temperature, (2) a reduction in oxygen consumption, (3) periods when breathing stops entirely, (4) a torpor more profound than sleep, and (5) intermittent arousal accompanied by activation of the major heat-producing mechanisms (Gunderson 1976). It is also associated with a prior buildup of fat reserves, which serve not only as a source of energy through the long fast but also as an insulative layer. Some animals supplement these reserves with a larder from which they feed during periods of arousal, and the pika goes so far as to cure hay in the late summer sun, storing up to a bushel for the winter (Bourlière 1964; Walker et al. 1964). All small mammals that hibernate compress their annual cycle of growth and reproduction into those months when they are awake and active (Kalabukhov 1960).

With one exception, arctic mammals do not hibernate. Environmental temperatures in the arctic drop so low for so long that a hibernating animal's internal body temperature would drop to a point at which basic life functions would cease. No one yet understands how the ground squirrel (Figure 3.8), lone hibernator of the Arctic Circle, solves this problem. In contrast, winter minimum temperatures in temperate regions are neither so low nor so prolonged. Most hibernators are small, weighing no more than 3 or 4 kg. Hibernation is most common among rodents, but it is also seen in bears, bats, certain members of the Insectivora, and one primate.

Migration is another way of coping with the problems of winter. Migrations are predictable movements, usually measured in tens or hundreds of kilometers, during which

Figure 3.8   A photograph of the beautiful golden-mantled ground squirrel (*Lateralis* sp.) in Colorado. (Animals Animals/© Marty Stouffer.)

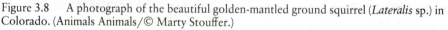

animals change their feeding and drinking areas depending on environmental conditions. Migration does not mean perpetual motion: when suitable conditions are reached, migratory species remain until the habitat deteriorates or better conditions are available elsewhere. Most, but not all, migrations occur annually (Delany and Happold 1979). In general, small mammals are not sufficiently mobile to reach warmer climes when winter sets in; bats are the exception. In fact, most temperate-zone bats migrate or hibernate at the onset of winter (Griffin 1970).

Some mammals show an increase in body size with increasing latitude. The correlation, called *Bergmann's rule,* was long believed to result from the thermoregulatory advantages of large size: large animals need less energy to stay warm because of their low surface-to-volume ratio. In 1955, however, Scholander pointed out that many species do not obey the rule at all or show weight increases so slight as to be physiologically insignificant; moreover, many of the features studied as relevant to the rule actually play no role in heat exchange.

The correlation probably exists for different reasons in different animals. For example, McNab (1971) suggests that large carnivores cannot find large enough prey at high latitudes to survive; their absence permits smaller carnivorous species, which compete with them elsewhere, to increase in size and eat whatever large prey are available. Boyce (1979) argues that large size is advantageous at high latitudes because animals are better able to survive periods of resource shortage in winter.

In summary, small herbivores and insectivores in the middle latitudes commonly adopt an "If you can't beat 'em, join 'em" strategy, by which they live off stored fat and drop their body temperature as outside temperatures decline. Migration is a solution available only to bats, which are capable of covering large distances on the wing. The increase in body size seen in some species is more likely due to competitive release (Subsection 10.2B) or the enhanced survival afforded by a large body than to improved thermoregulation.

3.3B   Medium-Sized Mammals

There are fewer species of medium-sized herbivores in arctic and temperate regions than of small species. Mammals falling into this category include two lagomorphs, the hare and rabbit (Lepidae), and four rodents, the beaver (Castoridae), mountain beaver (Aplodontiidae), porcupine (Erethizontidae), and marmot (Sciuridae) (Figure 3.9). Of these, only the arctic hare *(Lepus)* extends to the arctic islands or even as far north as the arctic continental coasts although the others range into northern Canada.

None of these mammals migrate, and only the marmot hibernates. The primary adaptation of hares, rabbits, and beavers to cold is a thick fur coat that provides excellent insulation (Scholander 1955; Scholander et al. 1950). The porcupine, best known for the impressive set of quills that protects it against predators, is also protected from the elements by another, less obvious covering, a woolly coat (Walker et al. 1964). All these mammals make use of the physical environment as well. Hares and rabbits burrow into snow to make shelters; beavers retreat to dens constructed in water; and porcupines retire to rock piles, caves, and hollowed-out tree trunks.

Of the four causes of food scarcity identified in Section 3.2, snow cover seems the least important for these animals. During the autumn, beavers build up a store of sticks and logs that they anchor under water to provide a winter larder. Ognev (1962:357), for example, notes that "in the Kulikovskii forest (U.S.S.R.) in 1934 each family stored 108 aspens 8 to 35 cm in diameter. . . . The author considers such a store excessive." This store, excessive or not, means that beavers do not have to flounder through drifts of soft

Figure 3.9   Medium-sized herbivores that live in cold regions of the world include hares, rabbits, beavers, porcupines, and marmots (also known as woodchucks or groundhogs).

snow in search of food. Porcupines are likewise reported to store food in their dens, although they are less systematic about it than beavers (Taylor 1935). The problem of mobility is minimal for porcupines; they live in forests and are as at home in the trees as on the ground, although they apparently are clumsy in both locales (Taylor 1935; Bourlière 1964)! Rabbits and hares do not store food, are completely terrestrial, and are thus the most susceptible to the effects of heavy snow. They have limbs and paws specialized for burrowing, however, and can dig through snow to reach buried food plants; only when the snow is particularly deep or compacted must they depend on whatever plant parts protrude above it. Because the feet of lagomorphs are broader and more heavily furred at high latitudes than in warmer regions, they are well insulated and their weight is spread widely, minimizing the risk of breaking through the snow crust. At 1 to 5 kg they are among the lightest mammals in the medium-size weight category and can support themselves on quite thin surface layers.

The greatest problem faced by these medium-sized herbivores in winter lies not in finding food as such but in its quality. Even beneath the snow, much of the available plant material is *browse,* which is higher in fiber, lower in nutrients, and harder to digest than young leaves, flowers, and fruits. In a temperate forest in winter, browse includes such items as bark, the underlying phloem, dormant buds, needles of coniferous trees, and woody parts of hardy shrubs and herbs protruding above the snow or reachable by digging. *Phloem* contains sap, which is rich in easily absorbed sugars and proteins, but its initial collection and processing are time- and energy-consuming for animals not equipped to gnaw their way through tough bark. Beavers and porcupines have strongly developed and continuously growing incisors set in a very robust skull and, hence, just the kind of highly efficient chewing and gnawing apparatus needed. What we know of their diet in the wild suggests that they make good use of it. In wintertime, porcupines are reported to subsist on the bark, phloem, and needle leaves of conifers (Taylor 1935), and the stomach contents of a series of beavers collected in winter contained primarily phloem with small amounts of bark (Morgan 1868).

Specializations of the digestive tract also help these herbivores extract the maximum nutritional benefit from their diets. All lagomorphs, many rodents (including beavers and porcupines), as well as the Perissodactyla, Proboscidea, and a few primates (Subsection 5.4B), are *hindgut fermenters.* Part of the food they ingest is broken down, or digested, in the stomach by enzymes produced by the animals themselves. The products of digestion are then absorbed through the lining of the stomach and the small intestine. Still-intact food particles are broken down when they reach the enlarged hindgut, or large intestine, in a process of fermentation carried out by bacteria and protozoa. Tough foods are thus digested by the double action of enzymes and microbial fermentation.

Absorbing the products of fermentative digestion presents hindgut fermenters with a problem. These products are formed only in the cecum and colon, and in most animals, it is generally too late for much absorption to occur. Some hindgut fermenters, including the lagomorphs, solve this problem by *coprophagy.* They excrete two types of fecal pellet, one moist and one dry; moist pellets are reingested and swallowed with little or no chewing, and food broken down on its first passage through the gut and then excreted is now absorbed into the bloodstream. Other hindgut fermenters, including beavers and porcupines, manage to absorb these products through the walls of the large intestine on the first passage (Parra 1978).

In summary, medium-sized herbivores in arctic and temperate biomes increase their protection against the cold in winter by growing a thick, insulative coat, except for the marmot, which hibernates. Beavers and porcupines have particularly strong incisors, giving them access to the sap-rich phloem of trees. Lagomorphs dig through the snow in

Figure 3.10   Raccoons *(Procyon* spp.), nocturnal mammals, live predominantly in trees. (I. DeVore/Anthro-Photo.)

search of nutritious plant parts and resort to browse when they must. All these nonhibernating species boost their digestive efficiency by hindgut fermentation.

Raccoons *(Procyon* spp.) (Figure 3.10) are members of the order Carnivora, but in the temperate forests of the northern United States they have a mixed summer diet containing a substantial amount of fruit as well as fish, frogs, small mammals, and an assortment of grubs and insects. In winter, they hole up in a den when temperatures drop much below freezing, emerging to forage only on warmer days (Schoonover and Marshall 1951; Sharp and Sharp 1956). The pickings are thin, though. Fecal samples collected from denned animals in February contained dead leaves, wood, bark, and raccoon hair (Stuewer 1943). Since raccoons lack the digestive specialization needed to derive much nutritional value from browse, they feed heavily on fruit during the autumn, accumulating fat to serve as a nutritional reserve in winter. Even so, winter weight losses of up to 2 kg have been recorded in Michigan forests among adult male raccoons initially weighing about 7 kg (Stuewer 1943), and studies in Minnesota have shown that most juveniles, yearlings, and adults lose half their body weight over the winter, with many juveniles dying from starvation or complications caused by parasites (Mech, Barnes, and Tester 1968; Schneider, Mech, and Tester 1971). Raccoon metabolism has not been studied in detail. Although raccoons enter a state of lethargy in cold weather, they do not seem to experience the marked reduction in metabolic rate that characterizes true hibernators.

Figure 3.11   Grazing mountain sheep (*Ovis* spp.). (Etter/Anthro-Photo.)

3.3C   Large Mammals

Most large herbivores belong to one of two orders, the Artiodactyla and Perissodactyla. Members of these orders are together referred to as *ungulates,* meaning hoofed mammals. Large ungulates live as far north as the arctic islands. All these most northerly forms are artiodactyls, although within historic times horses (Perissodactyla) were present in subarctic biomes. Today the northern limit of the geographical range of horses is the margin of the central grasslands of Eurasia. Caribou *(Rangifer,* family Cervidae) and musk oxen *(Ovibos,* Bovidae) are the artiodactyl representatives of the continental arctic, and moose *(Alces,* Cervidae), mountain sheep *(Ovis,* Bovidae),* and bison *(Bison,* Bovidae) range to the northern edges of the New World forests. Moving south into temperate zones, the assemblage of large herbivores increases to include more cervids, such as elk, mule deer, and white-tailed deer; a bovid, the mountain goat; and the pronghorn, lone member of the Antilocapridae. Most of these families are represented in the Old as well as the New World.

Like medium-sized forms, large herbivores grow thick winter coats that provide insulation against the cold: a shorn mountain sheep is as sensitive to low temperatures as a naked person (Geist 1971). Behavioral responses to the cold can also be important. Mountain sheep, for example, choose warm places to forage, concentrate their activities during the warmest part of the day, reduce their energy expenditure by minimizing nonfeeding activity, and rest in sheltered places with their legs bunched under them to decrease heat loss (Geist 1971) (Figure 3.11).

Large size reduces potential problems faced by herbivores in finding food and surviving on poor-quality items. For one thing, the mobility of large animals is less affected by snow cover. Bison, for example, can plough through snow chest deep without apparent difficulty (Soper 1941). Smaller species in this size range are less able to do this; in

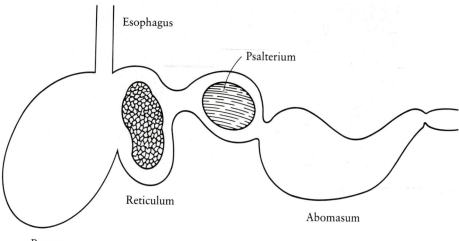

Figure 3.12   Diagram showing the structure of a ruminant's stomach. (Adapted from Chapman and Barker 1972.)

fact, the distribution of certain Eurasian bovids and cervids is correlated with depth of snow cover (Formozov 1946). In general, however, the mobility of large herbivores in snow allows them to travel widely in search of food. A second advantage of large size, an extension of the first, is that it permits migration from one biome to another. Annual migrations are typical of many ungulates at high latitudes. For example, at the onset of winter, caribou travel 200 to 400 km from the tundra north of the timberline to the northern forest fringe (Banfield 1954). A third advantage is that the rate at which a mammal uses up energy per unit of its weight decreases with increasing body size (Subsection 4.3A). In contrast, there is an approximately one-to-one relationship between gut capacity and body size (but see Chivers and Hladik 1980 and Subsection 5.4B). Together, these factors permit a large mammal to keep food particles in its gut for a longer period than a small mammal can and hence to digest them more thoroughly (Demment and Van Soest 1983).

In addition to the general advantages conferred by large size, many artiodactyls benefit from specializations of the digestive system permitting them to derive still greater nutritional value from a poor diet. In contrast to the hindgut fermentation process of the perissodactyls and others, microbial fermentation takes place in a huge, four-chambered stomach, or *foregut* (Figure 3.12) in most artiodactyls.

Foregut fermenters with a four-chambered stomach are called *ruminants*. Food is swallowed rapidly with little chewing and enters the first chamber of the stomach, or *rumen,* where it undergoes a softening process. It is then regurgitated into the mouth and chewed (hence the expression "chewing the cud"), so that it is broken down into smaller particles and mixed further with saliva. The second time the food is swallowed, it goes into the second chamber, or *reticulum;* from there it goes into the third chamber (*psalterium*) and finally into the fourth, the *abomasum*. Microbial fermentation takes place primarily in the second and third chambers, and gastric digestion in the fourth. Consequently, microbial fermentation in a ruminant largely precedes digestion by enzymes produced by the animal itself, rather than following digestion as in a hindgut fermenter. Products of the digestive process are absorbed through the walls of the stomach and the small intestine (Parra 1978).

In sum, ruminants are characterized by a capacity to ingest large quantities of low-grade food in a short period; they then grind up and chemically pretreat this food at leisure and pass it slowly through their voluminous stomachs, where the food is broken down with the help of fermentative action by bacteria. In this way, ruminants transform into assimilable nutrients vegetable matter from which other animals can derive little nutritional value. Microbial fermentation also occurs in the stomachs of a number of marsupials, tree sloths, and leaf-eating monkeys, but none of them have a four-chambered stomach, and the function of fermentation in these medium-sized forms is probably different (Section 3.4).

Pigs belong to the Suidae, one of three nonruminating families in the order Artiodactyla. While lacking the digestive specializations of other temperate-zone ungulates, the European wild hog (*Sus scrofa*) (Figure 3.13) ranges through broad-leaved deciduous and boreal forests in temperate Eurasia. Pigs have a different solution to the problems of surviving in winter. Their distribution in Eurasia is closely related to snow depth, with the northern limits of their range coinciding almost exactly with the 30–40-cm snow line (Formozov 1946). Cover, not cold, is critical: there are pigs in central Russia, where average temperatures in January range from −20°C to −30°C but snowfall is light. Pigs are not heavily furred, and it is unclear how they withstand such cold temperatures; probably fatty tissues under the skin provide them with an insulative layer. Present only in areas with little or no snow, pigs forage over huge distances for food on the surface of the ground or dig for it in the soil with their hooves, tusks, and noses. In winter, they feed primarily on leftovers from the autumn's crop of acorns, beechnuts, and fruits in broad-leaved forests and on pine cones in more northerly boreal forests; they supplement these food items with grass roots, insect larvae, and carrion (Walker et al. 1964; Formozov 1946).

In summary, all large herbivores but the pig combat the cold of northern winters with a thick, insulative coat. Deep snow seems to have little effect on the largest animals in this size class, although it makes travelling and finding food more difficult for smaller forms. Most large herbivores are of necessity browsers at high latitudes, and many spe-

Figure 3.13   Wild boar *(Sus scrofa)* crowd together to drink at the edge of a pond. (Animals Animals/©M. Harikrishnan.)

cies migrate to more favorable biomes at the onset of winter. Large herbivores can survive better than small ones on poor-quality food, and most of them also have specialized digestive tracts that enable them to extract the maximum nutritional value from such food. The pig is an exception to this general picture. It is not a ruminant, its diet is not predominantly browse, and it does not migrate in winter. Rather, it lives in areas where snowfall is light and there it forages over huge areas to secure a relatively high-quality diet of items found under as well as on the ground.

Black bears *(Ursus americanus)* (Figure 3.14) range through the temperate forests of North America. Like raccoons, they are members of the order Carnivora and have a mixed diet that includes plant food as well as meat. Weighing from about 120 to 150 kg, these bears are of course very much bigger than raccoons.

In the springtime, black bears eat more grasses, leaves, roots, and tubers than meat. During the summer, they gorge themselves on ripe berries, forage for small mammals such as mice, voles, and shrews by turning over stones and small logs, and catch fish if they live near water (Walker et al. 1964). In winter, bears remain in their dens for up to five to six months at a stretch. There they hibernate. Hibernating bears differ from hibernating small mammals in several ways (Folk, Larson, and Folk 1976). Bears have a lower surface-to-volume ratio than small hibernators, so they lose heat more slowly.

Figure 3.14   Young black bear *(Ursus americanus)* climbing. (Animals Animals/©Leonard Lee Rue III.)

This permits them to halve their metabolic rate and still maintain a body temperature within 7°C of their normal summer temperature. Although bears have periods of wakefulness most days, they go for months without eating, drinking, urinating, or defecating. Small hibernators, in contrast, rouse themselves every few days, raise their body temperature, move about, and urinate, and some species also eat and defecate. Finally, unlike small hibernators, which are reproductively active only during months when they are not hibernating, bears give birth in the middle of winter. Afterwards, the mother resumes her deep sleep, waking only intermittently to care for her young. The cubs themselves do not hibernate, but suckle and sleep warmly cuddled up against their mother. Weighing less than half a kilogram at birth, by the time a cub leaves the den three months later, its weight has reached 2 to 4 kg. Like many other hibernators, bears fatten up for winter by eating sugar-rich fruits. Even so, most of them lose 15–30% of their weight over the winter, and nursing mothers can lose up to 40%. Animals that fail to put on enough weight in the summer starve during the winter months.

3.4   The Case of the Langurs and Macaques

Both subfamilies of Old World monkeys have temperate-zone representatives. Macaques (Cercopithecinae) and langurs (Colobinae) live in mixed deciduous and needle-leaved evergreen forests covering the southern and eastern slopes of the Himalayan foothills (*Macaca mulatta, M. arctoides, Presbytis entellus,* and *Rhinopithecus roxellanae),* the mountains of northern Japan (*M. fuscata),* and the Atlas Mountains in Morocco (*M. sylvanus).* These primates share with other temperate-zone and high-altitude mammals adaptations that protect them from the cold (Figure 3.15). As winter approaches they grow a heavy insulative coat. Northern macaques and langurs are generally heavier than their southern counterparts (Napier and Napier 1967; Roonwal and Mohnot 1977), and there is evidence, albeit inconclusive, that Japanese and rhesus macaques may be specialized physiologically so that they can maintain their body temperature at lower environmental temperatures than macaques living in the tropics (Tokura et al. 1975). As temperatures drop, monkeys also adopt behavioral mechanisms to conserve heat. For example, Japanese macaques spend much of their time sunning themselves on south-facing slopes and sleep in deciduous trees, which, unlike evergreen conifers, do not accumulate snow that subsequently falls on their heads. (In summer they use evergreens and deciduous trees with equal frequency; in autumn, while deciduous trees are dropping their leaves, Japanese macaques sleep exclusively in evergreens!) They also travel shorter distances each day and move less frequently in winter than in summer, and they huddle alone or together at night and during snow storms (Suzuki 1965). Observations in Pakistan suggest that rhesus monkeys have a similar repertoire of behavioral responses to the cold (Goldstein 1984).

Temperate-zone primates seem better equipped to cope with low temperatures than with scarce and poor-quality food. Consider first the macaques. Like most primates, macaques are vegetarians. Japanese and rhesus macaques are highly terrestrial, and for much of the year they feed heavily on foods on or close to the ground. Since they neither store food nor hole up and hibernate in winter, they must search for food each day.

Figure 3.15   A heavy insulative coat is one of the adaptations to cold shown by primates such as these rhesus macaques *(Macaca mulatta)*, which live in temperate forests in Eurasia. (Courtesy of W. Keene.)

Figure 3.16   Travel on the ground becomes difficult for macaques in deep snow unless the crust is thick enough to hold their weight. (Courtesy of W. Keene.)

Deep, soft snow makes this difficult, because macaques are heavy enough (7–14 kg) to sink through all but the thickest crust (Figure 3.16). Movement through the trees is possible only if their crowns abut and are not encased in snow and ice.

Since macaques are poorly equipped for digging, the depth of snow in winter significantly affects their access to food. This has been systematically studied in Japanese macaques (Suzuki 1965). Table 3.4 shows the number of days of snow cover per year and the maximum snow cover in different parts of the geographical range of these ma-

Table 3.4   Snowfall in Japanese macaque habitat

Habitat	Days of snow annually	Maximum snow cover (cm)
Takasakiyama	12	10
Haranomachi	15	30
Shimokita	110	90
Mount Azuma	110	170
Shiga Heights	140	250

Source: Adapted from Suzuki 1965.
Note: Table entries estimated from data recorded at weather stations nearest to each site.

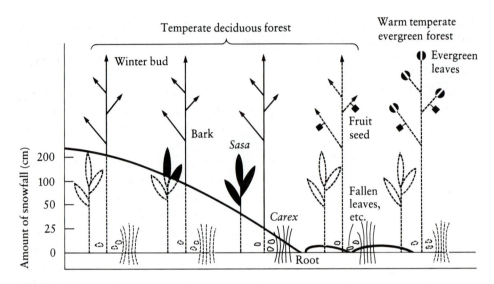

Figure 3.17   A schematic representation of the effect of snow cover on the winter diet of Japanese macaques. Solid lines indicate parts eaten; dashed lines indicate parts not eaten. (Adapted from Suzuki 1965.)

caques. Figure 3.17 shows schematically the effect of snow cover on their diet in winter. Five distinct patterns of winter feeding behavior can be discerned that directly relate to the amount of snow cover as well as to changes in the actual abundance of different food items:

1.   *Green-leaf eating* (Takasakiyama). In this broad-leaved evergreen zone, leaves are available throughout the winter and provide the main dietary component in that season.
2.   *Remnants eating* (Haranomachi, Kinkasan Island). Light snow cover does not prevent animals from eating fruit, seeds, and roots from the forest floor in addition to leaves, fruits, and seeds left on trees.
3.   *Carex eating* (Mount Ryozen, Shimokita Peninsula). Snow cover is still sufficiently light so that the forest floor is exposed in some places, and animals eat substantial quantities of the grasses *(Carex* and *Miscanthus)* growing there.
4.   *Sasa eating* (Mount Azuma). Leaves of *Sasa,* a tall grass, emerge above snow even in mid-winter and are fed on extensively. Ground layer vegetation (including *Carex* and *Miscanthus)* is otherwise completely covered and not eaten.
5.   *Bark eating* (Shiga Heights). Only defoliated trees stand above the snow. Animals eat large amounts of winter buds and bark.

Exposed to harsh winters in the Atlas Mountains, barbary macaques show a seasonal shift to browse like that reported in the Japanese macaques at Shiga Heights. Their diet consists primarily of cedar needles and cones during periods of heavy snow, and they also strip the bark of cedar trees and eat the soft underlying layer. This practice causes considerable damage to the trees, and the animals are shot as pests by the local people (Deag 1977; Taub 1977).

None of the macaques are well suited to a diet high in browse. They lack continuously growing teeth specialized for gnawing, and because their digestive system is without the bolstering effect of foregut or hindgut fermentation, they cannot harvest or digest woody, fibrous foods as efficiently as beavers, porcupines, hares, and rabbits do. We therefore might expect to find evidence of marked winter weight loss among macaques. The only data available are for a population of rhesus macaques *(Macaca mulatta)* that other researchers and I studied in northern Pakistan. Animals were trapped and weighed in November, before the onset of the brunt of winter, and again the following May, in the late spring. All adults except young males showed weight loss of up to 1 kg, and the juveniles showed no weight gain. Still, most temperate-zone mammals lose weight in winter, and this weight loss is minor compared with the losses recorded in some raccoon populations. Moreover, no monkeys in our study population died during the winter months. These findings are hard to interpret, however, because the animals were difficult to find and to follow in winter, and our knowledge of their winter diet is sparse (Goldstein 1984). In January, animals were sighted eating items high in fiber and low in nutritional value, such as the tough leaves of the evergreen oak *(Quercus dilatata)* and pine needles *(Pinus wallichiana)*. On the other hand, they were also seen eating a sweet, sticky exudate of the pine trees as well as roots of ground herbs on slopes where the snow had melted or been blown away. The exudate almost certainly contained easily digested sugars, and roots are often a rich source of several nutrients. In early February, a common shrub *(Vibernum foetens)* began to flower, and over the next month or so animals were repeatedly sighted feeding on the profuse blossoms (Figure 3.18), providing easily digested sugars and likely other nutrients as well (Section 4.2). It is therefore possible that the animals managed to keep the browse content of their diet quite low. It would be particularly interesting to have comparable weight and mortality data for the Shiga Heights Japanese macaques, for which a diet high in browse is better documented.

Hanuman langurs *(Presbytis entellus)* are the best studied of the northern colobines (Figure 3.19). They have been the subjects of three studies in highland Nepal, where they range between about 2500 to 3500 m above sea level (Bishop 1979; Boggess 1976, 1980; Curtin 1975). Grassy meadows are common in this area, making up almost half of the home range of one of the groups studied. When the deciduous trees, which dominate the forests, lose their leaves in autumn, langurs exploit these meadows intensively. They feed on the mature leaves of herbs and shrubs and, in the earlier part of winter, on the fruit of one particular shrub *(Cotoneaster sp.)*. They supplement these items with browse, including the leaves of evergreen oaks, bark, bare branches, roots, and also with

Figure 3.18   In northern Pakistan, flowers of the vibernum shrub are an important food for rhesus macaques in the mid and late winter. (Courtesy of W. Keene.)

the remains of human crops. Although snow falls intermittently in winter, the ground is not continuously or deeply blanketed. During brief periods of heavy snowfall, the animals avoid open habitat and range in forests where they generally forage and travel in the trees (Boggess 1980). The absolute scarcity of food, rather than its inaccessibility under snow, seems to be the major problem these animals face, and they forage daily over greater distances in winter than in summer, eating some foods, like oak leaves, that are available year round but are untouched in other seasons.

Like macaques, langurs lack teeth and jaws specialized for gnawing, but unlike macaques they do have a voluminous stomach in which microbial fermentation takes place (Subsection 5.4B). Yet it is unlikely they could survive for long on a diet of browse, because the energy needs of langurs are higher relative to their gut capacity than those of large herbivores, and they could not digest browse fast enough to meet those needs. Langurs normally eat large quantities of leaves, and foregut fermentation probably evolved in them and in other medium-sized herbivores, like sloths and certain marsu-

Figure 3.19   Hanuman langur *(Presbytis entellus)* in the bare trees of winter at Melemchigaon, Nepal. (Courtesy of J. M. Bishop.)

Figure 3.20   A Japanese macaque *(Macaca fuscata)* contemplates a long winter, its preferred foods almost completely covered by snow. (Courtesy of D. Sprague.)

pials, as a specialization for a diet of leaves rather than browse (Parra 1978; Freeland and Janzen 1974; Demment and Van Soest 1983). It is an efficient way of breaking down cellulose, an otherwise indigestible component of leaves, and it may also neutralize the poisons often present in leaves. In short, langurs are probably little better equipped than macaques to survive wintertime scarcities of nutritious food.

In summary, the few primates living outside the tropics face the same problems posed by a temperate environment as other medium-sized mammals. They respond to these problems in various, though not necessarily the same, ways. Macaques move shorter distances each day in winter, thereby conserving energy; in contrast, langurs move further, visiting widely scattered resources. Neither macaques nor langurs have limbs specialized for digging, and when snow is deep they must seek out patches of ground where it has melted or blown away or else feed on plant parts protruding above it (Figure 3.20). Probably neither extracts much nutritional value from browse. Though the browse content of their diet surely goes up in winter, it is unclear by how much. Some reports indicate that animals continue to find more nutritious resources as well; others suggest a primary role for browse. A better understanding of the digestive physiology of these primates as well as more detailed information about diets, winter weight loss, and mortality in temperate populations in particular are needed before we shall really be able to evaluate the nature and success of their responses to the stresses imposed by temperate winters.

3.5   Discussion

The purpose of this discussion is to integrate the data presented in Part I. Our framework is the set of questions underlying the material in these chapters: how do mammals other than primates survive at high latitudes? What general features distinguish tropical-zone primates from temperate-zone mammals of equivalent size? Do macaques and langurs have special adaptations permitting them to survive in northern areas? Finally, what general conclusions, if any, do the answers to these questions allow us to draw about the ecology and evolutionary success of the primates?

In temperate regions, the most critical problem facing herbivorous mammals in winter seems to be the absence of nutritious foods. Small mammals solve the problem by hibernating or living under the snow and eating grasses, roots, and leftover fruits and nuts on or in the soil. The size and gut specializations of large mammals permit them to survive on a diet high in browse. Medium-sized mammals are unable to move browse through their digestive tract in sufficient bulk and ferment it fast enough to meet their relatively high energy requirements (Parra 1978), and even in winter most of them are selective feeders. Hares, rabbits, porcupines, and beavers pass high-quality food rapidly through their stomachs, extracting most of its value with their own digestive juices. Remaining particles and more fibrous foods not broken down in the stomach are fermented in the hindgut, and the products are either assimilated there or reingested in the feces and absorbed on the second go-round.

The difficulty of extracting nutrients from low-quality vegetable food and consequent intense competition for scarce, high-quality items may help explain why there are few medium-sized herbivores outside the tropics. Their success in surviving the winter in temperate biomes depends on the possession of specializations of the hindgut and on their ability to uncover or have in reserve high-quality foods to supplement the low-quality browse most readily available. From this perspective, raccoons seem particularly ill fitted to live in northern forests, a suggestion that the winter mortality data from Michigan and Minnesota appear to bear out. It is no coincidence, I believe, that raccoons are the most avid wintertime raiders of human garbage cans, a fine source of high-quality food.

Figure 3.21   Three callitrichids from the forests of Amazonia (left to right): saddle-back tamarin *(Saguinus fuscicollis)*, red-handed tamarin *(S. midas)*, and black and red tamarin *(S. nigricollis)*.

What distinguishes primates from other mammals of equivalent size surveyed in this chapter? If small rodents can survive at high latitudes, why not small primates, like the callitrichids (Figure 3.21) and most strepsirhines? The callitrichids are a family of small, New World primates. Though the limits of their range are poorly delineated, we know that they have an exclusively tropical distribution. There are none north of Panama, even though the Central American forests extend into Mexico. The precise southern limits of their distribution are not known (Heltne and Thorington 1976), but there are no reports of them in the temperate montane forests to which tropical forest gives way on the flanks of the Andes. Likewise, most strepsirhines—the suborder of primates to which bush babies and lemurs belong—live in tropical forest. Only in South Africa and southern Madagascar do we find them in woodlands, where temperatures in winter fall close to or below freezing at night.

Even though they too are herbivores and insectivores, the biology of small primates differs from that of the intrepid little herbivores and insectivores discussed in Subsection 3.3A. First, small primates are specialized for life in the trees, they come to the ground rarely or not at all, and they are not equipped to exploit resources under the ground. Lacking continuously growing incisors for gnawing or hands and feet specialized for digging, they are ill suited to survive where winters are long and cold, and where most of the high-quality food is on or below the surface of the ground, often in the form of nuts and seeds with thick protective coats. Second, unlike many other small mammals, primates, with one exception, do not hibernate. The exception is the dwarf lemur *(Cheirogaleus medius)* (Figure 3.22). During the wet season in western Madagascar when food is abundant, these lemurs build up fat reserves in their tails, which double or triple in girth. During the dry season, when temperatures drop and food is scarce, they curl up in deep holes in trees and remain totally inactive, their body temperature dropping close to that of the external environment. Dwarf lemurs hibernate for at least six and in some cases up to eight consecutive months. Mouse lemurs *(Microcebus murinus)* also build up fat reserves in their tails, allow their body temperature to drop, and enter a state of lethargy during the dry season; however, they become active again every few days, their torpor is less profound than the dwarf lemur's, and they cannot be considered true hibernators (Hladik, Charles-Dominique, and Petter 1980; Petter-Rousseaux 1980; Andriantsiferana and Rahandraha 1973; Bourlière, Petter, and Petter-Rousseaux 1965; Russell 1975).

Turning from the smallest to the largest living primates, the Great Apes, we find the range of gorillas and chimpanzees limited to tropical Africa. Highland gorillas include elfinwood and high-altitude grassland in their home ranges but spend most of their time in forests lower on the mountain slopes. Large areas of African rain forest contain no gorillas, so that the question is not only what limits their distribution to rain forest but also what determines their distribution within it. (Hunting by humans provides a partial though probably not complete answer to the latter issue.) Most chimpanzees live in rain forest and seasonal forest, but at the limits of their distribution they are found in woodland and savanna-mosaic. The third Great Ape, the orang-utan (Figure 3.23), today lives in the rain forests of insular Southeast Asia, but in the Pleistocene it was wide-spread in forested environments on the mainland. There is no evidence that any of the largest primates, orang-utans included, ever had a temperate-zone distribution, with the possible exception of a few dryopithecids and, of course, ourselves.

To explain why the ranges of the large primates are limited to the tropics, we need information we can never obtain about their distribution, biology, and competitive re-

Figure 3.22   The dwarf lemur *(Cheriogaleus medius)* of Madagascar is the only primate known to hibernate. Note the thickness of this lemur's tail. (Courtesy of B. Freed, Duke University Primate Center.)

Figure 3.23   The solitary habits of the orang-utan *(Pongo pygmaeus)* distinguish it from the more sociable Great Apes, the chimpanzee *(Pan* spp.) and gorilla *(Gorilla gorilla).* (Courtesy of A. Mitchell.)

lationships in the past. As with the smallest primates, it is nonetheless illuminating to contrast their biology today with temperate-zone herbivores. The main food of chimpanzees and orang-utans is fruit. Gorillas feed mostly on leaves, shoots, and stems, items containing a higher proportion of materials that are difficult to digest. Though they have relatively large stomachs (Hladik 1967), gorillas are not foregut fermenters. In short, none of the Great Apes have the specializations needed for a diet of browse. Why are there no large, ruminating primates? This is an unanswerable question but, following Martin (1981), we may speculate that the energy turnover associated with a diet of browse is too low to meet the needs of animals with the highest brain-size-to-body ratio of any living mammal except ourselves. In other words, even if we equipped the living Great Apes with the stomachs of ruminants, they still could not support themselves on browse.

The only large mammals to survive the long temperate winter without gut specializations are pigs and bears. Limited to areas with little snow, pigs dig for roots, tubers, and fruit overlooked in the autumn and protected from decay by thick casings. This diet requires efficient digging equipment that in the case of pigs, consists of a tough, bony snout, tusks, and hooves. Together these make up a tool kit equaled by no nonhuman primate. Bears hibernate, unlike any other large mammals or medium-sized mammal except the marmot. I know of no discussion of bear biology in relation to that of other mammals that casts light on this anomaly.

Turning last to the medium-size range of mammals, we already noted that the stresses at high latitudes seem to represent particular problems for this group. Too large to live under the snow or hibernate, too small to migrate or make efficient use of browse, medium-sized mammals subsist on a mixed diet of browse and more nutritious items. They are characterized by a range of behaviors, such as storing food, and by morphological specializations, such as limbs suited for digging and hindgut fermentation, that help them walk the ecological tightrope in wintertime. Langurs and macaques, the two medium-sized primates found in temperate forests, exhibit none of these behavioral and morphological specializations. In fact, from what we know of their ecology, I believe that they comform well to the typical primate pattern.

What, then, is this typical primate pattern? In Section 1.4 we reviewed inferences made by Cartmill (1972) and Sussman (in press) about the feeding ecology of the early primates, and a picture emerged of small animals feeding in the terminal branches of trees and shrubs on an array of resources including flowers, fruits, gums, nectars, and insects. The selection and manipulation of small food items at night requires acute powers of visual discriminaton, precise hand–eye coordination, and the grasping hands and feet generally characteristic of terminal-branch feeders. All these features we see in primates today, and regardless of the accuracy of the reconstruction, I believe that selectivity and discrimination are key characteristics of the ecology of modern primates of all sizes. Their distribution is closely constrained by the need for sustained access to readily digestible foods rich in energy and nutrients. The presence of macaques and langurs in temperate forests shows that the supply need not be constantly available or at least not always at the same level. I suspect, however, that the distribution of macaques and langurs is extremely sensitive to factors influencing this supply. These include the proximity of crops planted by people, the species composition of the herb layer in the forest, the depth and duration of snow cover, and the abundance of competitors. In short, for the most part macaques and langurs at high latitudes retain the dietary selectivity that characterizes all primates and do not show the adaptations to temperate-zone life exhibited by other medium-sized mammals.

The foods most attractive to primates are most continuously available in rain forest, and it is here that primates are most diverse or, if you will, successful. They are specialists in seeking out the cream of the plant world's crop. They do it well, and part of doing it well seems to involve having a large brain. However, while the possession of a large, expensive-to-run brain may help ensure the success of primates in the tropics, I believe that it may also be important in limiting their distribution to areas with an abundance of high-quality food. More generally, being a finicky feeder with a big brain carries a host of implications not only for a primate's foraging behavior but also for its pattern of reproduction and social life. Much of the remainder of this book is an exploration of these implications.

PART II

PRIMATES AS INDIVIDUALS

Introduction

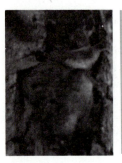

IF a species is to persist, its members must find sufficient food to grow and to survive as adults, and they must reproduce in sufficient numbers to replace themselves as they die. Part II looks at biological and behavioral patterns associated with these activities in primates and at attempts to explain these patterns in a broader evolutionary context.

The environment contains a vast array of plants and animals that differ in size, physical structure, and chemical content. A primate ignores most of this array, selecting as foods just a few items that provide varying amounts of nutrients at varying times. Survival and reproduction depend on an individual's ability to locate and harvest these foods in a combination that will meet its nutritional needs. At least three kinds of information are needed to understand why primates vary in their choice of foods: (1) the nutritional content of food items; (2) the kinds and quantities of nutrients needed by different primates and the consequences of not fulfilling these nutritional requirements; and (3) how primates differ in their ability to extract nutrients from particular food items and, more specifically, the effect of differences in body size, teeth, and digestive tract on this ability. In Chapter 4, we begin with the first issue and look at the constituents of food and how these constituents are distributed among plants and animals. Then we consider the balance of nutrients a primate needs to stay healthy. Finally, we review field data about the effects that variations in quantity and quality of food have on primate survival and reproduction.

In Chapter 5 the emphasis shifts from food as such to the particular combinations of food items comprising the diets of different primates. We also examine the morphological and physiological specializations associated with particular diets. The chapter ends with a discussion of models attempting to predict the optimal diet.

Chapter 6 turns the discussion from survival to reproduction. We have already seen (Section 1.4) that if we know a mammal's weight and its brain-to-body-weight ratio, we can predict its life history pattern fairly accurately. Large mammals live longer than small ones, take longer to mature, produce fewer offspring at longer intervals, and begin to reproduce relatively late. These trends are particularly marked in the primates that not only are quite large but also have a high brain-to-body-weight ratio. However, although reproductive patterns are broadly predictable from body and relative brain size, size does not allow us to predict, for example, which female will give birth most frequently in a particular social group of a particular species. There is no evidence that in this context smaller means quicker. The biological determinants of reproductive patterns are complex and beyond the scope of this book to examine fully. Here we look simply at the components of a female's reproductive cycle, how these components vary in timing and duration, and how this contributes to intraspecific variation as well as interspecific differences that are contrary to those that would be expected from body and relative brain size.

In short, Chapter 6 uncovers some of the variability concealed by the broad approach taken in Section 1.4. In so doing, it illustrates the capacity of primates to change their life history pattern rapidly and the pathways by which these changes can occur. This, in turn, provides us with a biological context in which to place theories about the evolution of life history patterns that will be considered in Chapter 7.

CHAPTER FOUR

Food

4.1 Constituents of Food

 FOOD contains chemicals belonging to one or more of the seven major nutrient categories: proteins, carbohydrates, fats, vitamins, water, minerals, and trace elements. Many plants use secondary compounds to defend themselves against herbivorous animals (Ehrlich and Raven 1965; Janzen 1969; Freeland and Janzen 1974; Rhoades and Cates 1976; however, impressive evidence has accumulated showing that plants use secondary compounds to defend themselves against herbivorous animals (Ehrlich and Raven 1965; Janzen 1969; Freeland and Janzen 1974; Rhoades and Cates 1976; Feeny 1976; Glander 1982). Some insects also contain secondary compounds, obtained from their plant foods and subsequently stored in their own bodies.

Before describing these constituents of food, let us first clearly distinguish between a food's *nutritional content* and its *nutritional value* to a particular animal. A food's nutritional content is usually described in terms of the major nutrient categories. Specifying its nutritional value is more complex. First, variation within nutrient categories can affect a food's value. For example, animals use certain kinds of plant protein more efficiently than others. Second, a food's value to an animal is partly determined by the proportion of nutrients that the animal can extract and digest. Because species vary in their ability to perform these tasks, a food with a particular nutritional content may have a high nutritional value for members of one species and a low value for members of another. In the following description, we shall consider the nutritional content of different food items and how variations in the major categories of nutrients affect their nutritional value.

4.1A Proteins

Proteins are used by animals primarily as building blocks. They are essential for growth, reproduction, and regulating the body's functions, and they provide about 12% of the caloric, or energy, needs of most mammals (Munro 1969).

Table 4.1   Classification of the amino acids

Essential	Nonessential
Lysine	Glycine
Tryptophan	Alanine
Histidine	Serine
Phenylalanine	Norleucine
Leucine	Aspartic acid
Isoleucine	Glutamic acid
Threonine	Proline
Methionine	Citrulline
Valine	Tyrosine
Arginine	Cystine

Proteins are large molecules made up of *amino acids* strung together by *peptide links*. Amino acids are composed of the elements carbon, hydrogen, oxygen, nitrogen, and in some cases sulfur and phosphorus; the links contain carbon, nitrogen, and oxygen. Because protein molecules are too big to be readily absorbed, they are first *digested;* this breaks down the peptide links and frees the constituent amino acids, which can then be absorbed. A series of transport systems assist in absorption and later transfer amino acids from the bloodstream to the body's cells, where they may be built back up into complex proteins.

Several factors cause the nutritional value of protein to vary. Because different amino acids may use the same transport systems, an excess of one may result in deficient transport of another despite adequate dietary intake (Allison 1964; Munro 1964). Even more important are the pattern and composition of amino acids making up proteins. Although there are only about 20 amino acids, each protein molecule contains several hundred of them, and there are thousands of different proteins in living things. The nutritional value of a particular protein to an animal depends in part on how similar the protein is in pattern and composition to the proteins that the animal ultimately synthesizes from it. For example, the pattern and composition of amino acids in milk, eggs, and meat render them a source of high-quality protein, while the pattern of amino acids in vegetables tends to be less similar to the pattern in animal protein, and vegetable proteins often lack adequate amounts of one or more of the essential amino acids. *Essential amino acids* (Table 4.1) cannot be synthesized by the animal and must therefore be supplied in the diet. The so-called nonessential amino acids are just as important to an animal's well-being: *essential* and *nonessential* refer only to *dietary* requirements. Most mammals, as well as many fish, insects, and protozoans, must include eight to ten amino acids in their diet; ruminants are the exceptions here, since protein synthesis in the digestive tract by microbes makes ruminants dietarily independent of the essential amino acids (Schmidt-Nielsen 1975).

4.1B   Carbohydrates

Carbohydrates are a primary source of energy for animals. In some forms they provide roughage, which is ingested food that provides bulk due mainly to its high-fiber content. Although roughage has no direct nutritional value, its presence is important, for it gives the gut something to get a grip on, stimulating the efficient movement of food during digestion.

In their simplest form, carbohydrates are compounds of carbon, hydrogen, and oxygen, although many have other components as well (Sharon 1980). It is convenient to divide carbohydrates into three biochemical classes. *Monosaccharides* are simple, water-soluble sugars. There are over 200 of them, of which the best known are glucose and fructose (fruit sugar). Monosaccharides are the primary products of photosynthesis, and they are the only form of carbohydrate that can cross the lining of the gut. Because monosaccharides can be metabolized immediately after absorption, they are an animal's most quickly available energy source. *Metabolism* is the term used to cover all chemical changes that go on in the body's tissue; energy metabolism refers to those changes whereby fats, carbohydrates, and proteins are broken down and oxidized to release energy or are synthesized into energy-storing compounds.

Monosaccharides rarely occur as single units in nature. Usually they are combined into the larger molecules comprising the other two biochemical classes of carbohydrates, *oligosaccharides,* which contain 2 to 10 monosaccharides, and *polysaccharides,* which are larger still, containing from 11 to as many as 26,000 monosaccharides. Sucrose (cane sugar), maltose (malt sugar), and lactose (milk sugar) are all oligosaccharides. Each is a compound of two simple sugars that can be easily split up during digestion; absorption of the resulting monosaccharides can then take place.

Carbohydrates are the most abundant group of biological compounds in the world, and the most abundant carbohydrate is the polysaccharide *cellulose* (Sharon 1980). Together with hemicellulose and lignin, cellulose is the major structural material of plants, forming primary components of cell walls and giving plants their fibrous texture. In fact, these polysaccharides are often called *structural carbohydrates.* Vertebrates do not produce enzymes to break them down efficiently, and for many mammals plant cell walls are not only a largely inaccessible source of energy but also a barrier to the more easily digested cell contents. For these species, structural carbohydrates provide dietary roughage.

Animals with a major proportion of structural carbohydrates in their diet have microorganisms in their digestive tract that can break down these compounds. How effectively the microorganisms do this depends on their environmental conditions, such as the temperature, alkalinity, and rate of passage of food (Parra 1978). Among mammals, ruminants offer the best accommodations for these microorganisms: consequently, ruminants are probably the most efficient digesters of structural carbohydrates although lignins, which are structurally more complex than cellulose, present digestive problems even to them (Harkin 1973).

Not all polysaccharides play structural roles. Some, such as starch and glycogen, are *nonstructural* compounds. Starch is the principal form in which plants store carbohydrates, and glycogen is the principal storage polysaccharide in animals. Animals use the enzyme *amylase* to break down starch into its component monosaccharides; this process is accelerated if the starch is first cooked — hence, no doubt, the assiduousness with which we bake our bread and potatoes. For animals without ovens, bacteria may help in the initial breakdown (Schmidt-Nielsen 1975).

Chitin is an important constituent of the exoskeleton, or shell, of insects. It makes up a major proportion of the total weight of an invertebrate — about one-third of a shrimp's weight is shell, for example — and it may represent about half the energy value of an invertebrate (Kay and Sheine 1979). Chitin is a polysaccharide with characteristics intermediate between those of the structural and nonstructural polysaccharides of plants; like nonstructural plant components, it is digested by enzymes produced by the animal but, like structural elements such as cellulose and lignin, it is not easily or completely digestible (Kay and Sheine 1979).

4.1C   Fats

Fats and oils (liquid forms of fat) another major source of energy for an animal. Indeed, a given volume of fat supplies about twice as much energy as the same volume of carbohydrate. While these energy sources are largely interchangeable for metabolic purposes, they are not completely so. For example, the human brain uses only carbohydrate as an energy source, and carbohydrates are needed to break down fats and extract their energy.

Fats, like carbohydrates, are basically compounds of carbon, hydrogen, and oxygen, but they contain many more hydrogen atoms than carbohydrates. They are highly insoluble in water; the vertebrate pancreas secretes an enzyme, lipase, which breaks them down into glycerol and fatty acids, both of which can be absorbed across the gut lining.

4.1D   Vitamins

Vitamins, like proteins, fats, and carbohydrates, are *organic compounds,* that is, compounds that contain the element carbon and are combustible. They are formed by plants from carbon dioxide, water, and in some cases nitrogen. Unlike the nutrients considered so far, they are needed only in very small amounts. Four fat-soluble vitamins (A, D, K, and E) and two water-soluble ones (C and the B complex) have been discovered. Vitamin B was first named shortly after the turn of the century, but by 1930 it had become apparent that "vitamin B" was actually a mixture of several vitamins with little in common except that they were all water soluble, all found in large amounts in the liver, and, unlike C, all contained nitrogen (Bogert, Briggs, and Calloway 1973). Since then, this group has come to be known collectively as the "B complex."

Vitamins do not provide energy, as do carbohydrates and fats, or building materials, as do proteins, but small amounts play a critical role in regulating the body's metabolism. In this they resemble some hormones, but while the latter are synthesized from proteins within the body, vitamins must for the most part be supplied from external sources.

The importance of specific vitamins in the diet is best illustrated and best known from deficiency studies. Insufficient vitamin A, for example, causes stunting of growth, an inability to see well in dim light or more serious eye problems, diseases of the skin and of the membranes lining the respiratory and digestive tracts, and abnormalities of the enamel-forming cells of the teeth. Vitamin D is crucial for the metabolism of calcium and phosphorus, and a deficiency causes stunted growth and bone deformities. (For further discussion of the roles of these and other vitamins, see Bogert, Briggs, and Calloway 1973).

4.1E   Water

Water is the largest single component of animals and of most plant parts. In humans, for example, it makes up 45–75% of body weight. It constitutes between 70 and 80% of plant parts except seeds, which contain little water, and organs (e.g., tubers) specially adapted for storing or transporting fluids, which contain more. Because fatty tissue holds much less water than lean, fat animals contain relatively less water than thin members of the same species. However, the water content of a particular individual remains steady over time, a balance achieved by matching water intake to water loss. A few mammals, such as the camel *(Camelus)* (Figure 4.1) and the kangaroo rat *(Dipodomys),* are specialized for desert life and can tolerate major losses of water, but most (including primates) experience extreme discomfort if water loss decreases their body weight by 10–15%, and a 20% drop is likely to be fatal (Schmidt-Nielsen 1975).

4.1F   Minerals and Trace Elements

Minerals and trace elements are crucial for many aspects of body function. Neither are needed in large amounts, and trace elements are necessary only in tiny quantities indeed. Before discussing their roles, some terms must be clarified. Four elements (oxygen, carbon, hydrogen, and nitrogen) make up 96% of an animal's weight, and another 30 or so make up the remaining 4%. Six of these (sodium, potassium, chlorine, calcium, phosphorus, and magnesium) are present in relatively large quantities and are referred to as minerals or inorganic salts. Both usages are loose ones, for these substances are not min-

Figure 4.1   Camels exhibit many adaptations to a hot, dry climate. They tolerate water losses twice those of most other mammals and they have a high internal body temperature that helps them store heat during the day and decrease heat flow from the environment; their thick, insulative fur also serves to reduce heat gain. (© Margaret Thompson/Anthro-Photo.)

erals in the geologic sense, and although they may exist in foods or tissues as inorganic salts, a substantial quantity are found in organic compounds (Bogert, Briggs, and Calloway 1973). A seventh element, sulfur, is present at levels comparable to the minerals, but since sulfur is always acquired from organic rather than inorganic compounds, it is classified separately. The other 23 or so elements in the body (which actually qualify equally well—or poorly—for the label *mineral*) together comprise no more than 0.01% of body weight and for this reason are generally called trace elements.

The part played by minerals and trace elements in the body's metabolism can be compared with that played by vitamins. Sodium, for example, regulates the quantity and distribution of fluid in the body; potassium is necessary for normal cell function, and a deficit can result in muscle disorders; magnesium and calcium are both involved in regulating the neuromuscular system. Among the trace elements, iron and copper are important for synthesizing hemoglobin, zinc helps regulate growth processes, and iodine is necessary for producing thyroid hormone.

4.1G   Secondary Compounds

A discussion of *secondary compounds* takes us away from what is of value in an animal's food to what is potentially of harm, although it is important to point out that the term *secondary compound* should not be used interchangeably with *toxin* or *poison*. Toxicity is not an inherent characteristic of any molecule, and a particular compound can be highly toxic to members of one species while having no effect on another (Glander 1982; Janzen 1979). Moreover, the detrimental effects of secondary compounds are not always caused by their toxicity to an animal (see below).

Approximately 12,000 secondary compounds have now been identified in the plant world, and each year about a thousand previously undescribed molecules are added to the list (Levin 1976). Swain (1977) speculates that there may be as many as 400,000 in all. Most are found in the seeds, roots, shoots, leaves, flowers, and fruits of the *angiosperms*, or flowering plants. They belong to many different biochemical groups, of which we shall consider just two, *alkaloids* (comprising over 4,000 of the 12,000 compounds known) and *tannins*. These groups illustrate the two major kinds of threat presented to herbivores by secondary compounds. Certain alkaloids act as *toxins* in certain species: they leave the animal's stomach and enter cells where they disrupt metabolism. Tannins, in contrast, are digestibility-reducing agents that act in a herbivore's gut to reduce the availability of plant nutrients. Tannins do this by binding with food proteins; certain of the other digestibility-reducing agents do it by binding with digestive proteins (i.e., enzymes) (Rhoades and Cates 1976).

Alkaloids are small molecules that are metabolically inexpensive for a plant to produce (i.e., a plant expends little energy in producing them). They are usually present in small amounts, for even in tiny concentrations they pass through the gut wall of many herbivores and may interfere readily, sometimes drastically, with ongoing metabolism. Some herbivorous insect species have evolved a physiological response to a particular alkaloid, so that individuals not only are unharmed by it but actually key into it, feeding exclusively on plants that contain it (Price 1975; Ehrlich and Raven 1965).

Tannins and other digestibility-reducing agents do not have to pass through the lining of the gut, and they are usually larger molecules than the toxic secondary compounds. They are metabolically more expensive for the plant to produce and are generally present in larger concentrations. Again unlike toxins, they have dosage-dependent effects. Eaten in small amounts, they have little or no impact, but in larger quantities they have adverse consequences whose severity corresponds to the dosage. For example, as the tannin concentration in the food of insects increases, their growth rate is progressively reduced (Feeny 1976). Tannins function in such a way that animals cannot evolve physiological responses to neutralize their effects as they can for some alkaloids.

4.2   How Foods Vary

Foods vary in nutritional and secondary compound content and also in patterns of distribution through space and time. In Subsection 4.2A we look at such variations among different classes of food items. Note that in generalizing to food types, we are overlooking considerable variation within each category. For example, the protein content of young leaves of different tree species on Barro Colorado Island, Panama, varies from a low of 7% of dry weight to a high of 20%. *On average*, however, young leaves contain more protein than mature ones (Milton 1979). We shall consider the following types of food: animal, fruit, seed, flower, leaf and stem, woody stem, sap and gum, and underground plant parts.

Trace elements and secondary compounds are excluded from the description. Animal foods usually contain all essential trace elements. Plants, in contrast, contain some of them but rarely all, and the distribution of these chemicals in different plant parts is extremely unpredictable (Boyd and Goodyear 1971; Casimir 1975). The concentration of some elements is influenced by climate, temperature, and soil type, and so it makes little sense to describe the typical trace element content of plant foods (Bogert, Briggs, and Calloway 1973). Secondary compounds are considered separately in Subsection 4.2B, because the complex and variable distribution of these substances is more appropriately considered in a general discussion of plant defense mechanisms.

4.2A   Variation in Content and Distribution

Animals

As a food, *animals* provide a plentiful supply of high-quality protein, fats, and fluid. Of the essential minerals, only magnesium is not present in significant amounts. Organs such as the liver and kidney are good sources of vitamins except vitamin C. Insects with exoskeletons are a rich source of carbohydrates for species able to digest chitin. Carbohydrate is the one major nutritional component in short supply in organisms without exoskeletons. Although all carbohydrates are converted to monosaccharides for absorption, transport, and metabolism, monosaccharides form little of the composition of body tissue. The storage carbohydrate glycogen occurs mainly in liver cells, and like the simple sugars, it contributes little to total body weight.

Insects are the animal food most commonly eaten by primates. From a primate's perspective, insects are small and often highly mobile, and their harvest demands considerable skill (Subsection 5.2A). They live at all levels in the forest, from the soil to the treetops. They tend to be scattered quite widely, although there are exceptions. One of the most striking exceptions is the termite mound: chimpanzees take advantage of the concentration of prey concealed within these mounds by spending long periods fishing

Figure 4.2   Adult female chimpanzee *(Pan troglodytes)* uses a grass stem to catch ants while her infant watches. (Jim Moore/Anthro-Photo.)

for them with specially prepared grass stems or strips of bark (Goodall 1965) (Figure 4.2).

The foraging patterns of primates show that although insects may be a scattered food, they do not occur randomly. For example, mangabeys *(Cercocebus albigena)* in Kibale Forest, Uganda, focus their foraging activities on dead wood, bark, and the insides of hollow sticks, vines, and lianas (Figure 4.3); blue monkeys *(Cercopithecus mitis)* and red-tailed monkeys *(C. ascanius),* in contrast, search the surfaces of leaves and bare, live branches and trunks (Struhsaker 1978a). In the forests of Colombia, squirrel monkeys *(Saimiri sciureus)* often follow groups of capuchins *(Cebus capuchinus),* and while the capuchins poke in nooks and crannies for prey, the squirrel monkeys pounce on insects the capuchins stir up but overlook (Thorington 1967). Ecologists have provided direct evidence of the patterned spatial distribution of insects and of marked temporal population cycles (Price 1975; Leigh and Smythe 1978).

Some primates, notably baboons *(Papio cynocephalus),* macaques *(Macaca* spp.), and chimpanzees *(Pan* spp.), also eat vertebrates ranging in size from lizards and nestling birds to young antelope (Butynski 1982) (Figure 4.4). Moreover, primates themselves occasionally fall prey to other primates. For example, baboons sometimes prey on vervet monkeys *(Cercopithecus aethiops)* (Altmann and Altmann 1970; Harding 1973, 1975; Hausfater 1976; Rose 1977a; Struhsaker 1967a; Strum 1981), and chimpanzees occasionally kill baboons, red-tailed monkeys *(Cercopithecus ascanius),* blue monkeys

Figure 4.3   Mangabeys *(Cercocebus albigena)* in the Kibale Forest, Uganda, energetically investigate cracks and crevices in trees and epiphytes in search of insect prey. (Courtesy of L. Leland.)

Figure 4.4   Adult male baboon *(Papio c. cynocephalus)* eating a young gazelle at Amboseli, Kenya. (Courtesy of C. Saunders.)

(C. mitis), and red colobus monkeys *(Colobus badius)* (J. Goodall 1968; McGrew 1979; Teleki 1973; Wrangham 1974). Large vertebrates tend to be difficult to catch and are often difficult to subdue, however, and the contribution of vertebrate prey, primate or otherwise, to the diets of nonhuman primates is in each case very small.

Fruits

Fruits vary markedly in composition depending on whether the plant relies on animals or some other mechanism to disperse its seeds. The fruits of animal-dispersed species (edible fruits) are composed primarily of a succulent flesh that provides a rich source of simple sugar (fructose) and water but very little protein or fat and few structural carbohydrates or vitamins. Some fruits contain significant amounts of potassium, but for the most part fruits are mineral poor (Bogert, Briggs, and Calloway 1973). Plants that in effect encourage animals to eat their fruit by making them nutritious at the same time defend their seeds. This is done either by providing their seeds with a thick, protective coat so they pass undigested through the animal's gut or by filling them with secondary compounds so the animal will spit them out before swallowing the flesh of the fruit. The fruits of species that do not depend on animals to disperse their seeds tend to be small, inconspicuous, unappetizing, or plain poisonous.

How is edible fruit distributed in space and time? The answer depends on whether we consider all the plant species in a forest together or one species at a time and, if the latter, which particular species. It also depends on how big a patch of habitat we consider and in which biome. In tropical rain forest, fruit is produced in small amounts by most tree species and in abundance by a few widely scattered trees (Charles-Dominique 1975). Seasonal changes in climate are slight and do not confine the cycles of growth and reproduction among plants to certain times of the year, and synchrony among and within species is low. Since plants produce fruit at different times of the year, fruit of some kind is usually available year-round in a large patch of forest. In a small patch, such as many primate social groups occupy, there are likely to be periods when few or even none of the trees are fruiting.

Spatial and temporal patterns of availability of fruit belonging to a particular species are complex, and a primate in search of a particular fruit faces several problems. First, it must find a member of that species. Plants vary in their spatial distribution (Figure 4.5). Individuals of some species occur in widely separated pockets; this is a clumped distribution. Others are evenly dispersed throughout the forest. In yet a third pattern, individuals are randomly scattered. To harvest the fruit of a particular species efficiently, then, a primate needs to know the location of individual plants. Second, it must know approximately when that species and, in some instances, when each member produces fruit. In some species, all individuals produce fruit at the same time each year or at regular intervals of years. In others, all individuals produce fruit at the same time of year, but production occurs irregularly from year to year. In both cases, species members are

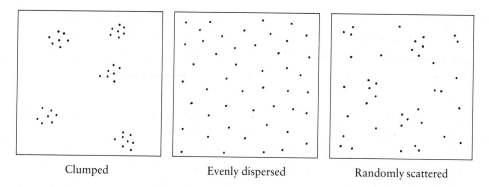

| Clumped | Evenly dispersed | Randomly scattered |

Figure 4.5   A schematic representation of different ways in which resources may be distributed across habitats.

closely synchronized. However, there are also species whose members do not produce fruit at the same time so that in a given month one individual is producing fruit while another 200 m away has yet to produce flowers.

This variation is often considered as a spectrum from highly predictable species, all of whose members produce fruit at the same time each year, to highly unpredictable ones, whose members produce fruit at different times and at irregular intervals. I believe these terms are best avoided. To say that a resource is predictable implies that the animals have a mental map (Altmann and Altmann 1970) of its distribution in space and time, giving them foreknowledge of the resource's availability. This is possible but largely undocumented, and animals may be using other methods to find food, such as monitoring potential resources at frequent intervals or simply wandering at random through the forest.

In seasonal forest, fruiting tends to be highly synchronous among and within species and so there is an abundance of fruit for a short period in the same season each year and little or no fruit at any other time. Fruit also tends to be available only at certain times of year in tropical savanna, where there are clear wet and dry seasons, and plants synchronize their cycles of production with the cycles of rainfall. Fruit-bearing trees and shrubs are usually distributed in widely scattered clumps.

Seeds

Seeds are high in fats and starch (Janzen 1971), and some provide a significant source of protein or particular vitamins. They contain little water, few minerals, and few or no simple sugars. Structural carbohydrates are present in the heavy protective coat that surrounds many seeds, which prevents most herbivores from extracting a seed's contents mechanically (with their teeth) or biochemically (with digestive enzymes). Such seeds are usually swallowed by animals whose primary interest is in the sugar-rich flesh surrounding them, and they pass through the digestive tract unharmed, emerging in the feces ready to germinate. The distribution of seeds in space and time is like that of fruit.

Flowers

Flowers contain simple sugars in their nectar, but less than half as much as in fruit. Moreover, flowers are usually much smaller than the fruits to which they give rise, although they are often more abundant. Pollen contains protein but is difficult to digest (Howell 1974). The water content of flowers is high, and the structural carbohydrate, vitamin, and mineral content very low. The distribution of flowers is like that of fruit and seeds, although plants often produce more flowers than fruit.

Leaves and Stems

Leaves and stems vary in content according to age. In general, the older the leaf or stem, the thicker and more inflexible are the walls of its constituent cells. This means that the cell contents of *young* leaves and stems not only are more accessible to a herbivore but also tend to make up a greater proportion of the total cell mass. Thus these young parts provide a good source of protein and water together with some roughage in the form of structural carbohydrates. They contain little or no fat and are low in simple sugars. *Mature* foliage contains a much higher percentage of structural carbohydrates, and proteins within the cell walls are largely inaccessible to animals lacking the digestive specialization necessary to break down these walls. Of all plant parts, leaves are generally the richest source of vitamins, and they also provide a good supply of some minerals, in particular calcium, phosphorus, and magnesium. Sodium and chlorine, it should be noted, are usually either absent or present only in minute amounts in leaves (or any other plant part for that matter). Only plants specialized to grow in salty conditions and some freshwater plants contain high sodium levels (Botkin et al. 1973; Oates 1977a).

In tropical rain forests, trees keep their leaves throughout the year, and in the plant community as a whole, mature leaves are the single most abundant, evenly distributed, and consistently available item. While less abundant and less evenly distributed than their mature counterparts, leaves at earlier stages of development are also generally available at any time in a large patch of rain forest. However, a primate intent on eating the leaves of a particular species faces more problems. First, it must locate a member of that species. Unlike fruit, mature leaves are present year-round. Many primates prefer immature leaves over mature ones, however. Leaf production in most plants is cyclical and may not be synchronized among individuals. As with fruit, then, animals must either learn the temporal patterns of leaf production for their food species or else monitor members of those species at frequent intervals.

In seasonal biomes, many trees lose their leaves for part of the year. Those that keep them have leathery or needle leaves for protection from cold and drought. This means that edible, mature leaves are not constantly available, and immature leaves are present only during the short, annual growth period.

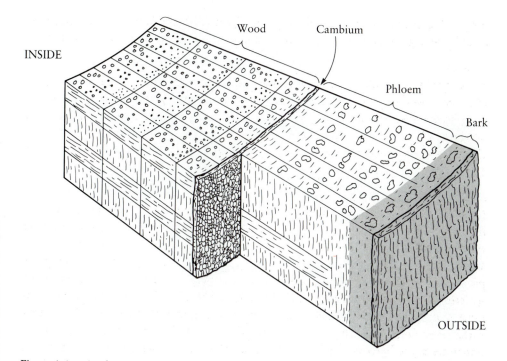

Figure 4.6 A schematic representation of a cross section through a woody stem. The most nutritious layers are the phloem and cambium.

On the tropical savanna, grass blades and leaves of other ground herbs are widely and evenly distributed. However, their nutritional content changes with age as does that of tree leaves, and the cycle of production is limited by the seasonal rainfall. As a result, this resource is a nutritious food only during certain months of the year.

Woody stems

Many plants have *woody stems*. These stems are made up of layers with different functions, including outer protection (bark), transportation of nutrients and waste products *(phloem)*, and structural support *(wood)* (Figure 4.6). The thin layer of *cambium* is the site of new growth, making wood toward the inside and phloem toward the outside. Woody stems are high in fiber, composed largely of cells with thick and, for many animals, indigestible walls. Wood and bark are often of little value even to ruminants because they tend to be high in the structural carbohydrate lignin, which is not broken down by microbial fermentation (Harkin 1973). The most nutritious components of a woody stem are its thin layer of cambium and the wider layer of phloem. Both have thin-walled cells, and if the bark is pried off, they provide a rich source of protein.

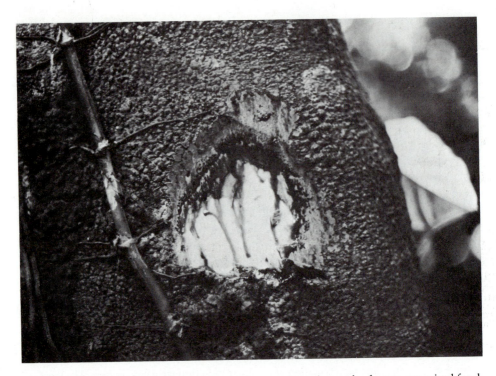

Figure 4.7   Gums like this one, exuded to seal a wound on the trunk of a tree, are prized food for some primates. (Courtesy of M. Plotkin.)

Sap and Gum

Sap and gum are found in plant stems. *Sap* refers to the juices, or circulating fluids, present in all plants. It contains water, minerals, and the products of photosynthesis. This last component makes it a good source of simple sugars for animals able to tap the vascular system of plant stems, through which sap flows up and down. *Gum* is produced in response to injury and infection by special cells present in many species of tree and shrub (Figure 4.7). It initially occurs as a soft, sticky material, but in contact with air it hardens to form a solid, transparent or translucent mass that prevents the spread of infection by blocking the plant's vascular system. Gum is composed primarily of polysaccharides, with small amounts of protein, fiber, and minerals (particularly calcium, potassium, and magnesium). Although its chemical structure presents unique problems of digestion, the gum of some species is a valuable source of carbohydrates for animals able to solve these problems (Hausfater and Bearce 1976; Bearder and Martin 1980). Gum is not be be confused with *resin*, another exudate which has similar functions but is of no nutritional value.

Figure 4.8   Apart from ourselves, baboons (*Papio* spp.) are one of the few primates that exploit the underground parts of plants. Some baboon populations, such as these animals at Gilgil, Kenya, feed heavily on grass corms in the dry season. (Photography by A. Richard.)

Gum is produced in response to the activities of a variety of woodboring beetles and moths as well as a few primates. When an insect- or primate-inflicted wound has been plugged, gum production stops at that site, but it starts up again there or elsewhere as soon as new damage is done. Thus, although rarely produced in large quantities at a single location, gum is usually available in small amounts throughout the year. It is most common on tree trunks, which is where beetles, moths, and primates are most active. As a result, not only the digestion but also the harvest of this food requires special adaptations (Subsection 5.2B).

Underground Plant Parts

Underground plant parts, including roots, tubers, rhizomes, corms, and bulbs, all serve as storage sites for plants, particularly in areas where water availability fluctuates seasonally. Such storage organs contain water, carbohydrates in the form of simple sugars and starches, and protein (Noy-Meir 1973; Hatley and Kappelman 1980). Because it is underground, this rich source of nutrients is rarely used by most primates, who are ill equipped to dig for their food. Baboons, however, depend on it heavily: out on the African grasslands, they spend long hours extracting and eating grass *corms,* the swollen underground bases of stems (Figure 4.8).

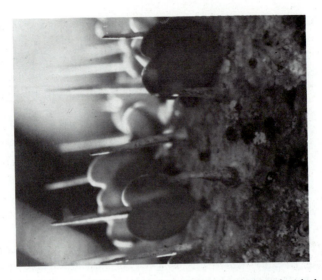

Figure 4.9   Many trees and shrubs in southern Madagascar are covered with sharp thorns such as these. Note that this tree (*Allauadia ascendens*) has leaves that are smaller than its thorns. No one understands how the sifaka (*Propithecus verreauxi*) manages to leap between thorn-covered tree trunks without injury. (Courtesy of R. Dewar.)

4.2B   Variation in Defense Mechanisms

Secondary compounds can be divided into two functional categories, as already noted: toxins and digestibility-reducing agents. Under what circumstances do plants produce each type of compound? There is no simple answer to this question because plants use different mechanisms in varying combinations to defend themselves against herbivores. Some grow physical barricades such as hairs and thorns (Figure 4.9), which make it difficult for an animal to chew or even approach the protected plant part. Others have a pattern of spatial dispersion or a growth cycle that makes it hard for a herbivore to know where and when the plants will be available as food; this is called spatial or temporal escape. In still others, the cycles of growth and reproduction are closely synchronized among individuals so that more food of a particular sort is available for a short time than the predator population can consume. As a result, a single plant is unlikely to have all its new leaves, flowers, or fruit devoured. This is called predator satiation. Finally, there are chemical defenses provided by secondary compounds.

Although we are far from understanding how plants deploy these various defenses, we can make a few general statements about the distribution of secondary compounds (Feeny and Bostock 1968; Levin 1971, 1976; Janzen 1971; Cates and Orians 1975; McKey 1974). Plants under strong selective pressure for rapid growth, early maturation, or the production of readily dispersed seeds tend to have the less expensive toxins as a protective system. In the tropics, toxin-producing plant species tend to be those that colonize newly available or recently disturbed areas. Plants that develop slowly and have longer reproductive periods tend to have more elaborate chemical defense systems. These plants are usually found in environments that have been undisturbed for long periods of time.

Secondary compounds are more common among tropical than temperate species: plants at high latitudes have an extremely short growing season, and many are annuals that must reach maturity and reproduce in a single season. Under these conditions, even toxins may be too expensive metabolically to be worth producing, and plants tend to rely on the brevity of their life cycle to make their detection by herbivores unlikely.

Different parts of particular plants tend to show variations in secondary compound content. Leaves, for example, contain high levels of secondary compounds compared with flowers and are more likely to contain digestibility-reducing compounds. This is because leaves usually last the entire growing season in deciduous species and for much longer than that in evergreens; flowers, in contrast, last only a short time.

Many animals likely to fall prey to primates defend themselves by flight or by hiding. However, some insects store plant secondary compounds in their bodies for defensive purposes, and it is well known that predatory birds can learn to avoid such species. It has yet to be shown that these compounds are harmful to insect-eating primates, but there are many field observations of primates eating only part of an insect and discarding the rest. It is possible that animals learn which part of the insect contains the sequestered chemicals and discard it (Glander 1982).

4.3   Nutrients: How Much Is Enough?

If we want to know what mix of nutrients an animal must obtain from its food to secure an adequate diet, we must first decide what we mean by *adequate*. This presents a problem. The simplest definition of an adequate diet is that it does not produce weight loss or signs of specific deficits. According to this definition, many different diets can be adequate. Yet closer scrutiny may show that these diets do vary subtly in their effects: for example, one may slow down growth rates slightly, another may speed them up, a third may prolong the interval between births, while a fourth may shorten it. Such variations can have a significant effect on an individual's lifetime reproduction. We might therefore want to distinguish between more and less adequate diets or perhaps between an adequate and an optimal diet, the latter being one that maximizes survival and reproduction. In practice, however, nutritional science is still in its infancy, and only the simplest distinction between adequate and inadequate is possible. We need to increase our understanding of the precise influences of age, sex, size, reproductive state, and activity pattern on an animal's nutritional needs before we can measure the subtler effects of variations in its dietary intake.

This section first reviews a few general considerations about energy and protein requirements in animals of different size, age, and reproductive state and then looks at the handful of nutrient analyses of the diets of primates in the wild. This will demonstrate not only the general range of primate requirements but also the general range of problems to be solved in this area of research.

4.3A   General Energy and Protein Requirements

An animal's energy requirements are met by the carbohydrates, fats, and to a lesser extent proteins in its diet (although, energy stored in the body can be drawn upon as long as such stores last). Energy is exchanged, or used up, in basic life processes such as breathing and circulating blood and also in all the activities necessary for survival. In the course of these exchanges, some energy is dissipated directly in the form of heat. Surplus energy is stored in the body.

The rate at which energy is exchanged in basic life processes is called the *basal metabolic rate* (BMR). It is most commonly measured by estimating the volume of oxygen an animal consumes when at complete rest and in a fasting state. It is usually expressed in kilocalories per gram or kilogram of body weight, and it decreases with increasing weight. In other words, large animals use less energy per unit of body weight than do small ones (Table 4.2).

Table 4.2   Observed rates of oxygen consumption in mammals of various body sizes

Animal	Body mass (g)	Total O_2 consumption (ml O_2/h)	O_2 Consumption per gram (ml O_2/g/h)
Shrew	4.8	35.5	7.40
Harvest mouse	9.0	22.5	2.50
Kangaroo mouse	15.2	27.3	1.80
Mouse	25	41	1.65
Ground squirrel	96	98.8	1.03
Rat	290	250	0.87
Cat	2,500	1,700	0.68
Dog	11,700	3,870	0.33
Sheep	42,700	9,590	0.22
Man	70,000	14,760	0.21
Horse	650,000	71,100	0.11
Elephant	3,833,000	268,000	0.07

Source: Adapted from Schmidt-Nielson 1975.

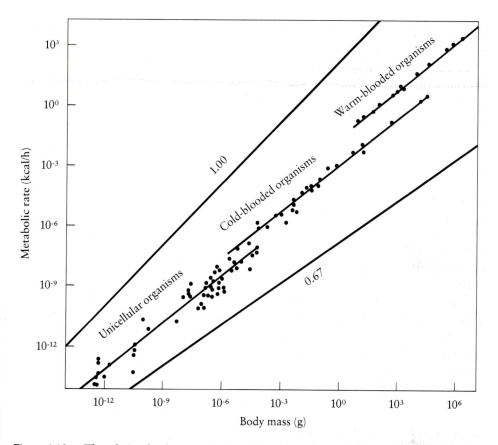

Figure 4.10   The relationship between body weight and basal metabolic rate among organisms of widely varying size and taxonomic affinity. (Adapted from Schmidt-Nielsen 1975.)

When the weights and BMRs of a range of organisms are expressed as logarithms and plotted against each other on a logarithmic scale, the points fall more or less along a straight line. The slope of this line is described by the equation

$$V_{O_2}/W = 3.8 \times W^{-0.25} \qquad (1)$$

where V_{O_2} is the volume of oxygen consumed in milliliters per hour and W is the animal's weight in grams (Schmidt-Nielsen 1975). This relationship between BMR and weight allows us to make general predictions about an animal's energy needs for basal metabolism from knowledge of its weight alone.

If the energy expended on basal metabolism by individual animals is calculated and the data are again plotted logarithmically against their weight, a straight line with a positive slope of 0.75 is obtained (Figure 4.10). This means that although a rabbit, for example, uses more energy in total than a mouse, it does not do so in direct proportion to its bigger body. Although a rabbit weighs about 100 times more than a mouse, it only uses about 32 times as much energy to support basic life processes. The equation in this case is derived from (1) by multiplying both sides of (1) by body weight W:

$$V_{0_2} = 3.8 \times W^{0.75} \tag{2}$$

Calculating an animal's energy requirements for basal metabolism is a useful first step in estimating its total energy needs, but two points should be emphasized. First, the relationship between size and BMR shown in Figure 4.10 is for organisms of widely varying weight. If the relationship is plotted for organisms from a narrower weight spectrum, much more variation in BMR becomes apparent (Eisenberg 1981; McNab 1980). The BMR of some bats, for example, is far lower than we would predict (Figure 4.11). These findings suggest that it is better to determine the BMR of a particular species empirically rather than to infer it from weight.

In addition, even within a species, where animals can be assumed to have similar BMRs, the total energy needs of different individuals may vary widely (Coelho 1974). A very active animal has high total energy needs compared with its energy needs to maintain basal metabolism alone, while one that sits around most of the time needs little energy beyond that required for basal metabolism. The energy expended in activities like walking, running, and climbing is difficult to specify. Nevertheless, there again seems to be an economy of scale, and the amount increases only as a fractional power of body weight: other things being equal, a big animal expends relatively less energy performing a particular activity than does a small one (Tucker 1970; Taylor, Schmidt-Nielsen, and Raab 1970).

Because protein is used primarily to build the body's structures, it is particularly important during growth phases (although it remains a necessary dietary item even after maturity). The body's proteins are continuously broken down and resynthesized during adult life; each time resynthesis occurs, about 20% of the amino acids must be supplied from food. Adult human males, for example, degrade and synthesize about 400 g of protein every day, an amount far in excess of their intake in food, which is likely to be 100 to 150 g/day (in the United States, at least) (Schmidt-Nielsen 1975). Protein is also continuously lost in the form of cells shed from the lower intestines and from the skin, nails or claws, and hair. In humans, this drain amounts to about 28 g of protein a day, a loss that must be compensated for if the individual is not to waste away.

Protein also provides many mammals with about 12% of their energy needs, and it follows that protein requirements per unit of body weight diminish with increasing total body weight; indeed, with increasing body weight, the rate of protein turnover decreases in a manner similar to that of energy metabolism (Munro 1969). Still, protein requirements are very hard to specify. One gram of protein per kilogram of body weight

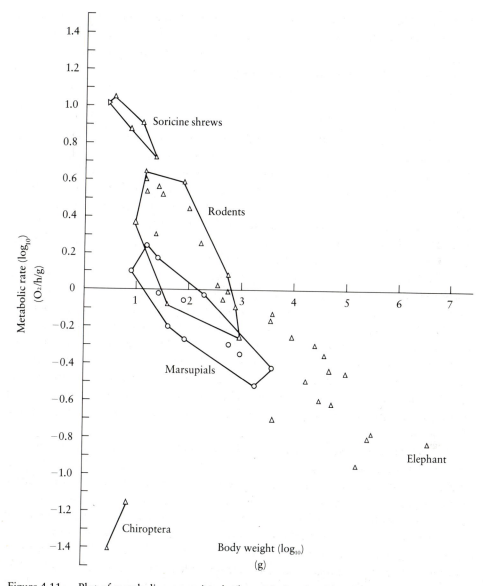

Figure 4.11   Plot of metabolic rate against body weight in selected rodents (Rodentia), bats (Chiroptera), shrews (Insectivora), and marsupials. (Adapted from Eisenberg 1981.)

per day is the standard amount recommended by international health organizations for adult humans, but in practice people remain healthy on a tremendous range of intakes (Bogert, Briggs, and Calloway 1973); the possible subtle effects of these huge differences in intake have yet to be systematically investigated.

The essential amino acids are one aspect of protein requirements that can be specified. If any are omitted from the diet, normal body processes are disrupted, and for humans we even know the minimum amounts needed of each. Such precise information is not available for other primates, but we do know that they, too, require 8 to 10 amino acids in their diet.

With these general observations in mind, and assuming that other things (like activity levels) are equal, how might we expect animals differing in size, age, or reproductive state to vary in their energy and protein intakes? First, consider immature animals. They are growing, and being smaller than adults, they need less total energy; however, they metabolize it faster. Their diet should therefore contain a high proportion of protein and easily metabolized food compared with that of adults. Second, consider species in which members of the two sexes differ in size. The smaller of the two, usually the females, need less total energy but metabolize it faster; accordingly, we might expect the smaller sex to include a higher proportion of easily metabolized food in its diet than does the larger sex. A similar argument can be applied to the diets of species of different body sizes (Section 5.3). Predictions about differences in the diets of males and females are complicated by changes in the reproductive state of the latter and associated shifts in nutritional requirements. We are still some way from being able to determine the relative effects of body size and reproductive state upon female and male dietary requirements.

4.3B   Nutrient content of natural diets

The study of nutrition in the wild is fraught with practical problems. Elusive subjects can be shot and their stomach contents inspected, but this technique shows only what was eaten during the preceding few hours and has the added disadvantage of eliminating those particular animals from the population and from further study. Alternatively, dietary data can be collected by direct observation if the animals themselves and their environment permit it. With time and patience, many primates can be habituated to the presence of an observer (Figure 4.12), and their generally large size, slow movements, and individual distinctiveness make them good subjects for intensive observation. Yet even under the best of conditions, vegetation may block an individual primate from view for minutes or hours at a time, and the gaps in data that inevitably result mean that fine distinctions among individuals are likely to be lost.

Estimates of time spent feeding on different food items are available for many primates (e.g., Clutton-Brock 1977b, 1977c; Clutton-Brock and Harvey 1977b). To estimate a diet's nutritional content, however, we must also know the *feeding rate*, or quantity of food ingested per unit time. Although time spent feeding on an item may accurately reflect that item's contribution to the total quantity of food eaten, it does not necessarily do so. This is a major problem. Hladik (1977b) compared feeding rates and

Figure 4.12   My field assistant, Mr. K. Augustin, standing beneath a habituated sifaka *(Propithecus verreauxi)* in southern Madagascar. (Photography by A. Richard.)

Figure 4.13   The Hanuman or gray langur *(Presbytis entellus)* (right) and the purple-faced langur *(P. senex)* (left) are found in many of the same forests in Sri Lanka. The Hanuman langur's range includes much of South Asia, but the purple-faced langurs are found only on this island south of India.

time spent feeding on different foods by two langur species in Polonnaruwa Forest, Sri Lanka (Figure 4.13). In the purple-faced langur *(Presbytis senex)*, the percentage of time spent eating an item was fairly closely correlated with the percent contribution by weight of that item to the diet, but in the gray langur *(P. entellus)* there was a considerable discrepancy between the two values. For example, the gray langurs spent 28% of feeding time eating large figs, but these composed 77% of the total fresh weight that the animals ingested that day. In contrast, all other fruits together accounted for 17% of fresh weight ingested and 47% of feeding time. The discrepancy was even more striking for flowers and leaf buds, where the proportion of time spent feeding was 10 times higher than the weight of food ingested (Table 4.3).

Table 4.3   Time spent feeding and food intake for two langurs (*Presbytis senex* and *P. entellus*) at Polonnaruwa, Sri Lanka

Sample eaten	Time spent feeding		Fresh-weight intake		Dry-weight intake	
	Minutes	Percent	Grams	Percent	Grams	Percent
	Presbytis senex					
Mature leaves of *Adina cordifolia*	57	30	165	30	56	30
Mature leaves of *Schleichera oleosa*	40	21	150	27	87	46
	51% of feeding time spent on leaves		57% of the fresh weight ingested from leaves		76% of the dry weight ingested from leaves	
Young leaves and shoots of *Schleichera oleosa*	52	27	115	21	23	12
Young leaves of *Garcinia spicata*	27	14	70	13	7	4
Leaf flushes (and some flowers) of *Walsura piscidia*	9	5	30	6	9	5
	46% on young leaves and shoots		39% from young leaves and shoots		21% from young leaves and shoots	
Green fruits of *Adina cordifolia*	5	3	20	4	6	3

Source: Adapted from Hladik 1977b.
Note: Presbytis senex observed 8 April 1969 from 6:32 to 18:30h; *Presbytis entellus* on 15 June 1969 from 5:30 to 18:30h.

Continued on page 152

Sample eaten	Time spent feeding		Fresh-weight intake		Dry-weight intake	
	Minutes	Percent	Grams	Percent	Grams	Percent
			Presbytis entellus			
Fruits of *Ficus benghalensis*	36	28	1200	77	251	71
Fruits of *Drypetes sepiaria* (stones not included in weight)	19	15	126	8	32	9
Fruits of *Schleichera oleosa*	20	16	68	4	22	6
Fruits of *Walsura piscidia* (with seeds eaten)	11	9	24	2	6	2
Fruits of *Alseodaphne semecarpifolia*	10	8	40	3	10	3
Young leaves of *Streblus asper*	5	4	4	0.3	0.7	0.2
Young leaves of *Tamarindus indica*	10	8	13	0.8	3	0.8
Flowers of *Tamarindus indica*	2	2	3	0.2	0.8	0.2
Leaves of *Alangium salvifolium*	6	5	60	4	23	7
Leaves of *Mimosa pudica*	9	7	14	0.9	4	1

Notes (brace, fruits group):
- Time spent feeding: 28% of feeding time spent on *Ficus benghalensis*; 47% on other fruits
- Fresh-weight intake: 77% of the fresh weight ingested from *Ficus benghalensis*; 17% from other fruits
- Dry-weight intake: 71% of the dry weight ingested from *Ficus benghalensis*; 20% from other fruits

Notes (brace, leaves/flowers group):
- Time spent feeding: 13% on young leaves and flowers; 12% on mature leaves
- Fresh-weight intake: 1.3% from young leaves and flowers; 4.9% from mature leaves
- Dry-weight intake: 1.2% from young leaves and flowers; 8% from mature leaves

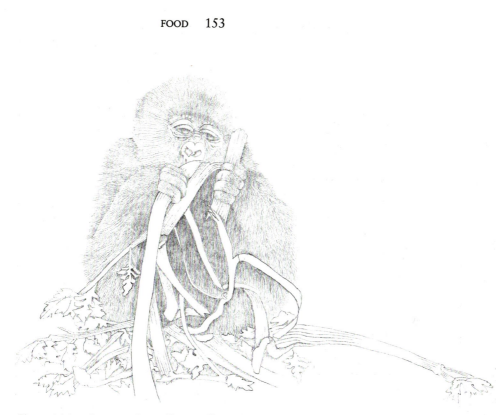

Figure 4.14   A mountain gorilla *(Gorilla gorilla)* feeding on wild celery *(Peucedanum linderi)*. (Drawn from a photograph by the National Geographic Society in *Gorillas in the Mist* by Dian Fossey.)

Another important distinction that bears repeating is between a food's nutritional content and its value to the animal eating it. The value of a food item depends not only on its contents but also on the form of those contents and the efficiency with which an animal can digest them. Other things being equal, the digestibility of vegetable food decreases as its structural carbohydrate content increases, but in reality crude analyses of a diet's total nutrient content are unlikely to give a very accurate picture of its value.

A number of researchers have analyzed the nutrient content of items eaten or avoided by their subjects (e.g., Coelho, Bramblett, and Quick 1976; Casimir 1975; Glander 1981; Hladik 1977b; Milton 1979, 1980; Gaulin and Gaulin 1982), but only a few studies provide estimates of the total nutritional content and value of the diet. Let us briefly consider four such studies.

Mountain Gorillas
Mountain gorillas (Gorilla gorilla beringei) were studied in the Kahuzi region of Zaire, west of Lake Kivu, for a period of seven months during which 273 hours of data were collected (A. G. Goodall 1977) (Figure 4.14). Because of the shyness of the subjects

Table 4.4   Analysis of diet of gorillas

Food item	Wet-weight intake (kg)	Dry-weight intake (kg)	Available water (l)	Net available energy (kcal)	Net available protein (g)
		Male (200 kg)			
Basella (leaves)	10.0	1.0	9.0	3,328	300
Urera (leaves)	10.0	1.5	8.5	3,976	180
Urera (bark)	10.0	3.0	7.0	3,589	90
Total	30.0	5.5	24.5	10,892	570
Bamboo (shoots)	30.0	3.0	27.0	10,579	1104
		Female (100 kg)			
Basella (leaves)	6.0	0.6	5.4	1,997	180
Urera (leaves)	6.0	0.9	5.1	2,386	108
Urera (bark)	6.0	1.8	4.2	2,153	108
Total	18.0	3.3	14.7	6,535	396
Bamboo (shoots)	18.0	1.8	16.2	6,304	662

Source: Adapted from A. G. Goodall 1977.
Notes: Predicted basal metabolic energy requirement for a 200-kg animal is 3,722 kcal; for a 100-kg animal, 2,213 kcal. Gorillas eat bamboo shoots almost exclusively when they are in season.

throughout the study and the denseness of the vegetation, it was impossible to keep an animal in sight for long. Consequently, the data on diet are not complete.

Gorillas are highly dimorphic. Males weigh between 140 and 180 kg, and females about half that, 70 to 110 kg (Napier and Napier 1967). From data on fecal output, a daily wet-weight intake of 30 kg of food was calculated for adult males and 18 kg for adult females. These estimates were corroborated by direct observations of feeding. Since detailed data on every food eaten in the course of a day were not collected, the energy and protein value for two simplified diets were worked out, one composed of three food items commonly eaten throughout the year and the other of bamboo shoots, which were eaten almost exclusively when they were in season (Table 4.4). The first of these diets would yield about 10,900 kcal and 570 g of protein for a male and 6500 kcal and 396 g of protein for a female. A similar quantity of bamboo yields slightly less energy (10,579 kcal for males and 6304 kcal for females) but twice the protein (1104 g and 662 g for males and females, respectively). Table 4.4 suggests why gorillas are so rarely seen drinking free-standing water: their diet supplies males and females with 24–27 l and 15–16 l of water, respectively.

Figure 4.15   Once a hilltop, Barro Colorado became an island when Lake Gatun was created during the construction of the Panama Canal. Howler and cebus monkeys were present on the island from the outset, but spider monkeys were only introduced, in small numbers, in the 1960s.

Goodall compared his estimates of energy intake for the gorillas studied with estimates of their basal metabolic requirements as inferred from body size (Subsection 4.3A) and found that the actual energy intake was almost three times higher. Likewise, his estimates of the protein available to them were between three and six times greater than the commonly touted figure of 1 g of protein per kilogram of body weight.

Since Goodall attempted to distinguish the nutritional content of food items from their nutritional value, his figures represent estimated values rather than content. Lacking direct information on gorillas, he used data from the pig, another nonruminant of about the same body weight, to make this distinction. He concluded that gorillas digest and assimilate as much as 84% by weight of certain items (e.g., bamboo shoots) and as little as 38% of others (e.g., certain rarely eaten leaves).

Howler, Spider, and Cebus Monkeys

Howler, spider, and cebus monkeys (Alouatta palliata, Ateles geoffroyi, and *Cebus capuchinus)* were studied for about one year on Barro Colorado Island, Panama (Hladik and Hladik 1969) (Figure 4.15). For each species, the quantity of different food items eaten

Table 4.5   Percentages of nutrients in the diets of six primate species

Nutrient	Howler monkey (Alouatta palliata)	Spider monkey (Ateles geoffroyi)	Cebus monkey (Cebus capuchinus)	Purple-Faced langur (Presbytis senex)	Gray langur (Presbytis entellus)	Sportive lemur (Lepilemur mustelinus)
Protein	9.6	7.4	14.4	11.5	14	13.6
Soluble sugar	21.7	33.7	26.3	10	10	4.9
Fat	3.2	4.9	15.8	4	7	1.8
Cellulose	13.6	11	7.6	74.5	69	15.1
Complementary fraction	51.9	43	36	—	—	64.6

was estimated on between 26 and 34 days scattered throughout the year, and samples of similar food items were then analyzed for their nutritional content. From these two data sets, the total nutritional content of each species' diet was estimated.

The results of this study cannot be compared with those on the mountain gorilla. The intake of specific nutrients was published only as a percentage of total intake, and quantities of carbohydrates and fats were measured rather than the energy content of these nutrients. The distinction between the nutritional content of foods and their value was inferred rather than investigated experimentally. However, the data for the three subjects of this study can be compared with each other and with data obtained for two Old World monkeys and one lemur species whose diets were analyzed similarly by Hladik and his colleagues (see below).

Table 4.5 shows the nutritional content of the diets of howler, spider, and capuchin monkeys (as well as of the species considered below), broken down according to the percentage of dry weight made up by protein, soluble sugar, fat, cellulose, and the *complementary fraction*. This last component includes minerals and all insoluble sugars except cellulose; in other words, it is made up predominantly of the least digestible elements in the diet. Some variation is apparent in the dietary constituents of the howler, spider, and cebus monkeys. For example, the capuchin diet contains more protein and fat than either of the others, while soluble sugars occur at higher levels in the spider monkey's diet than in the howler monkey's. The biggest difference is in the complementary fraction, which ranges from a high of 52% in the howler monkey to a low of 36% in the capuchin.

Purple-Faced and Gray Langurs

Purple-faced and gray langurs (Presbytis senex and *P. entellus)* were studied at Polonnaruwa, Sri Lanka, for a period of about one year (Hladik 1977b). In this time, quantitative data on diet were gathered during 220 hours of observation (Table 4.5). Although

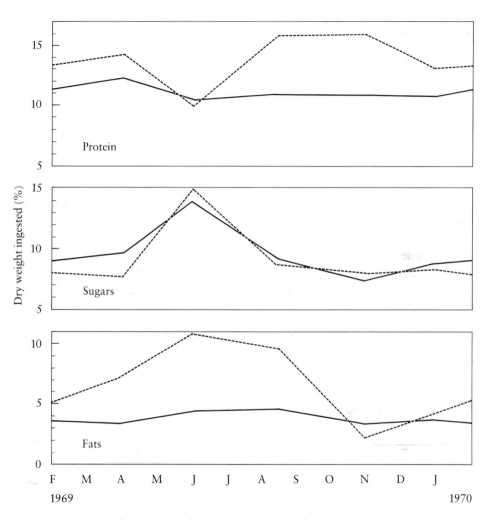

Figure 4.16   Seasonal variation in the content of the diet of the purple-faced langur *(Presbytis senex)* (solid line) and the gray langur *(Presbytis entellus)* (dashed line). (Adapted from Hladik 1977b.)

the analysis is less detailed than in the preceding study, the techniques used were similar and the results show another aspect of nutritional variation: the percentages of protein and fat in the purple-faced langur's diet were essentially consistent over time, but these figures varied considerably from month to month in the gray langur's diet (Figure 4.16). Note that although both langurs fall within the range of values given for protein intake in the three New World monkeys, they had much lower intakes of soluble sugars and fats.

Figure 4.17   The sportive lemur *(Lepilemur mustelinus)* is found in forests all over Madagascar except in the extreme north. (Photography by A. Richard.)

Sportive Lemur

The sportive lemur (Lepilemur mustelinus) was studied intensively by Charles-Dominique and Hladik (1971) for six weeks at the height of the dry season in Madagascar (Figure 4.17). Despite its limited scope, this study is included here because it suggests a dietary pattern that is unusual among primates. The sportive lemur lives in a wide range of forest environments; these data come from a population living in the xerophytic forest that covers large tracts of southern Madagascar. Members of this lemur species are nocturnal; the subjects studied slept most of the day and passed the night in brief bouts of feeding and moving interspersed with long periods of immobility. At the height of the dry season, their diet consisted mainly of the tough leaves and flowers of just three tree species. These items contained a high proportion of cellulose and other hard-to-digest or completely indigestible materials and a very low level of soluble sugars and fats.

Charles-Dominique and Hladik estimated that the energetic contribution of readily assimilable nutritive elements (proteins, fats, and soluble sugars) was only about 13 kcal every 24 hours and that for basal metabolism alone 23 to 30 kcal were needed daily. The authors suggest that animals make up the difference by extracting energy from cellulose in a process also found in the gorilla, lagomorphs, and some rodents. Cellulose is degraded by bacterial action in the voluminous large intestine, and the products are absorbed during a second passage of food through the gut, achieved by *coprophagy,* or feces eating. According to Charles-Dominique and Hladik, sportive lemurs derive an estimated additional 10 to 15 kcal per 24 hours in this way. Russell (1977) has challenged this idea on the basis of a four-month study of the same population during which he did not witness a single episode of coprophagy. Whoever proves correct, here we have a primate that has such a low activity level, at least at certain times of year, that its total energy requirement barely exceeds the requirement for basal metabolism above.

4.4   Food, Survival, and Reproduction

It is a difficult task to determine precisely an animal's nutritional requirements, but the importance of meeting those requirements is amply documented by a growing literature concerning the effects of food deprivation. In principle, too little food, too much food, or the wrong food (food containing toxins or digestibility-reducing compounds) can all have adverse effects. In practice, the first of these problems is probably most common, and the second least so: obesity is widespread only among human populations in the Western world and among artificially provisioned animals (most of them in the Western world, too!). Regarding the third possibility, animals rarely die of poisoning in the wild as far as we know, and avoiding the wrong foods undoubtedly plays an important part in food choice. In this review, "too little food" includes both overall nutrient deficiency *(undernourishment),* and the deficiency of a specific nutrient *(malnourishment).* The distinction between undernourishment and malnourishment is important, particularly in trying to define the constituents of an adequate diet, but in the wild it is a distinction that can rarely be made.

First, what evidence is there that food scarcity increases mortality rate? It is known to be important in controlling the adult populations of many carnivores (Schaller 1972; Kruuk 1972; Jordan, Shelton, and Allen 1967). Even when food is abundant, starvation is a common cause of death among the young, for they reach a stage of development when they are too large to be nourished adequately by their mother's milk but likely to be driven away from kills made by adults; at this time they are often still too inept to kill enough on their own to feed themselves.

For herbivores, the importance of food as a limiting factor has been a matter of more debate. It has been argued that the plant world presents them with more than enough food and that other factors, such as predator pressure, limit population size (Hairston, Smith, and Slobodkin 1960; Slobodkin, Smith, and Hairston 1967). This argument is circular, however, since the population size of predators itself depends closely upon that of their prey. Besides, all that is green is not food, and herbivores do on occasion starve.

In the East African savanna, for example, large numbers of ungulates have been found dying at the height of the dry season in some years; although the specific cause of death has rarely been identified, starvation is a likely possibility during this annual period of food scarcity (Kruuk 1972). Schaller (1972) estimated that between 14% and 21% of the meat scavenged by lions in the Serengeti National Park came from animals that had died from disease or malnutrition. On the island of Saint Kilda in the Outer Hebrides, sheep mortality rates reach a peak in late winter and early spring, when forage is still abundant but of very low digestibility. Gwynne and Boyd (1970) suggest that deaths may be due to the sheep's inability to take in enough digestible material to meet their energetic needs.

In medium- and large-sized mammals, the mortality rate tends to be high among juveniles (Section 7.5). Data on the pattern of mortality among primates are scarce, but the evidence so far suggests that it is similar. For example, during a food shortage on Sri Lanka, deaths among toque macaques *(Macaca sinica)* were concentrated in younger age groups (Dittus 1975, 1977a, 1979). Struhsaker (1976) attributes the high mortality rate of young vervet monkeys *(Cercopithecus aethiops)* to their inability to cope with a seasonally low-quality food supply.

Infant mortality rates are strongly, if indirectly, affected by food availability, since the production of an adequate supply of milk depends upon the nursing mother's ability to secure ample food herself. Although her needs have yet to be well specified for primates other than humans, we know that they are high (Kerr 1972). In her study of baboon *(Papio cynocephalus)* mothers, Jeanne Altmann (1980:56) comments, "Even with fairly conservative estimates of energetic demands, a mother could not provide all caloric requirements for herself and her infant beyond six to eight months of infant age and probably could do so up to that age only with difficulty and major restructuring of other aspects of her life." Reduced milk production due to nutritional stress on the mother has been proposed as a major factor contributing to infant deaths in this population (Altmann et al. 1977). Weight data for the Japanese macaque *(Macaca fuscata)* provide further evidence of the stress experienced by nursing mothers (Hazama 1964). Females weigh most just prior to giving birth; thereafter their weight declines steadily until the infant is weaned; females who miscarry regain their preconception weight fastest, followed by those with stillbirths and finally, with a considerable lag, by those who raise their infants successfully.

Experimental approaches have been used to study the relationship between infant survival and the mother's food intake and milk production. In one experiment, gestating and lactating female white-tailed deer *(Odocoileus virginianus)* were fed diets containing varying amounts of protein. No fawns of females on a high-protein diet died of malnutrition; in contrast, 4 out of 15 in an intermediate-protein group and 8 out of 19 in a low-protein group were born dead or died as a result of lactational failure (the drying up of milk) (Murphy and Coates 1966). Growth rates are lower in baboon infants with mothers fed a low-protein diet; the effect is probably due to lowered milk production rather than to lower protein content of the milk, since low-protein milk formulas fed in larger quantities to infants do not affect their growth rates (Buss and Reed 1970).

Food scarcity reduces the reproductive success of mothers by increasing the infant mortality rate. It directly and indirectly curtails female reproductive rates in other ways as well (Sadleir 1969a, 1969b; Gaulin and Konner 1977; Altmann et al. 1977). A reduction in the food supply is known to reduce growth rates in many mammals; this, in turn, delays the onset of sexual maturity and thus the *age at first reproduction*. For example, unprovisioned red deer *(Cervus elaphus)* do not reproduce until the third breeding season after birth (i.e., at about 29 months), whereas deer reared on a highly nutritious diet give birth for the first time in their second year, at about 17 months (Youngson 1970). Female gray squirrels give birth at the age of 7 months during years when the nut crop is particularly large, although in average years the first litter is born at about 12 months (Smith and Barkalow 1967). Among primates, differences have been recorded of two years or more in the ages at first reproduction between captive or provisioned and wild populations of macaques (C. Burns in prep.), baboons (Altmann et al. 1977), and chimpanzees (Tutin and McGinnis 1981).

The *conception rate* is affected by changes in ovulation rate and by the number, motility, and longevity of sperm. Except under experimental conditions, it is usually impossible to distinguish these various effects. Most reports of differences in ovulation rates are of North American cervids. White-tailed deer, for example, have reduced ovulation rates under poor range conditions (Ransom 1967). In domesticated cattle, sperm counts decrease with reduced total caloric intake (Moustegaard 1977).

Captive female primates are known to have sexual cycles without releasing an egg from the ovaries, but the causes of these anovulatory cycles are not known. In the wild, the nutritional status of females probably affects their ability to conceive and carry a fetus, if not to ovulate, but there is no direct evidence of this. Among humans, it is known that malnutrition can inhibit ovulation (Short 1976).

Pregnancy probably increases the nutritional requirements of females, but neither the specific level of increase nor the consequences of failure to reach this level have yet been studied in detail.

The *interbirth interval* varies inversely with infant mortality rate. In other words, if an infant dies, its mother is likely to give birth sooner than if the infant survives. Taking into account variation due to differences in infant survival, the time taken to conceive again after a birth is longer if food is scarce.

Female longevity affects reproductive success in that, other things being equal, reduced longevity decreases the number of offspring a female can bear during her life. Since mortality rates increase in all age groups when food is scarce, longevity is reduced on average.

4.5   Summary

Even though we know something about the distribution of nutrients among different types of food items and about the general effects on animals of general differences in the quality and quantity of food, we still know very little about precisely how much and what kinds of food a primate needs to stay alive, reproduce, and raise its offspring. Two approaches to the latter problem were presented in this chapter. The first is to predict dietary requirements from general principles of animal physiology. The second is to discover them by analyzing the content of the diets of wild animals, with the assumption that these animals are meeting their requirements. Each approach has drawbacks and advantages, and above all, each generates more questions than it answers: therein, perhaps, lies the greatest value of each.

Principles of animal physiology have been most useful in calculating a mammal's minimal energy requirements. Specifically, energy needs for basal metabolism can be approximated from a knowledge of body weight, and small animals have higher metabolic rates per unit body weight than large ones. However, basal energy needs vary sufficiently among similarly sized species as to require direct, species-by-species measurement if nutritional requirements are to be specified with precision. In addition, an animal's basal energy requirements may be a small fraction of the total energy it requires to find food, escape predators, and so forth.

Data from the wild indicate that species vary widely in their intake of particular nutrients. Among the six species for which data are presented in Table 4.5, the percent dry weight of the diet made up of protein ranged from 7% to 14%; of soluble sugars, from 5% to 34%; of fats, from 2% to 16%; and of the complementary fraction, or least digestible portion of the diet, from 43% to 65%. How is it that howler monkeys, for example, thrive spending 48% of feeding time on leaves, a rich source of protein, while in the same forest spider monkeys appear to do equally well spending only 22% of their time on leaves and little or no time on any other protein-rich food (Milton 1981b)?

Part of the variation we see is surely due to differences in how the data were collected and to assorted biases in the data (Section 5.2). Much of it, however, represents real differences among species. To understand the causes of this variation, we must take into account several factors, including (1) the varying physical and chemical properties of food items with broadly similar nutritional contents, (2) differences in the nutritional requirements of species due to differences in their BMRs and activity patterns, and (3) differences in the ability of species to digest and assimilate particular types of food and their speed in doing so (Soest 1981).

CHAPTER FIVE

Primate Diets: Patterns and Principles

5.1   Introduction

IN this chapter we survey primate diets (Section 5.2) and look at explanations for dietary variations among species. An animal's size and food-processing apparatus (i.e., its teeth and gut) constrain its food choices, and dietary variations among species can often be related to morphological and physiological differences. We consider relationships between body size, food-processing apparatus, and diet in Sections 5.3 and 5.4. Optimal foraging theory attempts to explain how natural selection shapes feeding patterns. It is evaluated by testing predictions about short-term changes in the feeding behavior of animals responding to shifts in the quality and abundance of their food. We consider the value of optional foraging theory for primate research in Section 5.5.

5.2   Dietary Patterns

The dietary patterns of primates are classified in Table 5.1 according to the type of food that the animals most commonly eat, including insects and a variety of plant parts (Figure 5.1). Plant foods are first classified according to growth form. Nonwoody plants, or herbs, include grasses and other species that have a low-growth habit and die back each year. Woody plants include trees and shrubs. Although they may lose their leaves seasonally, their woody parts persist from year to year. A second set of divisions has been made according to plant part (leaves, fruit, etc.), because plant parts have characteristic nutritional properties that influence the food choices of primates, and this chapter is primarily an exploration of factors governing these choices.

Because most primates spend much more time eating one type of food than any other (Table 5.1), the major categories in our dietary classification are defined according to the type of food most commonly eaten. "Most commonly eaten" refers to time spent feeding for some species and to quantity eaten for others. Because these two ways of assessing an animal's diet may give markedly different results, data in Table 5.1 should

Table 5.1 Composition of diets of selected primate species

Species by Dietary Category	Insects	Gum and Sap	Fruit	Seeds	Flowers	Leaves[a]	Herbs	Reference
Insectivores								
Arctocebus calabarensis[b]	85	—	14	—	—	—	—	Charles-Dominique 1977
Galago demidovii	70	10	19	—	—	—	—	Charles-Dominique 1977
Mirza coquereli[b]	50	20	30	—	—	—	—	Hladik 1979
Gummivores								
Cebuella pygmaea	33	67	—	—	—	—	—	Ramirez, Freese, and Revilla 1978
Galago crassicaudatus	5	62	21	—	—	—	—	Doyle and Bearder 1977
G. elegantulus[b]	20	75	5	—	—	—	—	Charles-Dominique 1977
Phaner furcifer[b]	15	70	15	—	—	—	—	Petter, Schilling, and Pariente 1971 Hladik 1979
Frugivores eating insects								
Callicebus torquatus	14	—	67	—	—	13	—	Kinzey 1977
Cebus capuchinus[b]	20	—	65	—	—	15	—	Hladik and Hladik 1969
Cercocebus albigena	26	—	59	—	3	5/0	—	Waser 1977a
Cercopithecus aethiops	17	—	48	—	23	12	—	Struhsaker 1967a
C. ascanius	22	—	44	—	15	11/4	—	Struhsaker 1978a
C. pogonias[b]	14	—	84	—	—	2	—	Gautier-Hion 1978
C. cephus[b]	10	—	79	—	—	8	—	Gautier-Hion 1978
Galago alleni[b]	25	—	73	—	—	—	—	Charles-Dominique 1977
Miopithecus talapoin[b]	36	—	43	—	2	2	—	Gautier-Hion 1978
Perodicticus potto[b]	10	21	65	—	—	—	—	Charles-Dominique 1977
Saguinus geoffroyi[b]	30	—	60	—	—	10	—	Hladik and Hladik 1969
Frugivores eating leaves								
Ateles geoffroyi[b]	1	—	80	—	—	20	—	Hladik and Hladik 1969
A. belzebuth	—	—	83	—	—	7	—	Klein and Klein 1977
Cercocebus galeritus[b]	3	—	73	—	1	14	—	Gautier-Hion 1978
Cercopithecus mitis	20	—	45	—	11	8/12	—	Struhsaker 1978a
C. neglectus[b]	5	—	74	—	3	9	—	Gautier-Hion 1978
C. nictitans[b]	8	—	61	—	1	28	—	Gautier-Hion 1978
Hylobates lar	4	—	60	—	1	36	—	MacKinnon and MacKinnon 1978
Lemur catta[c]	—	—	60	—	6	24	6	Sussman 1974
Macaca fascicularis	1	—	87	—	3	1	—	Wheatley 1978
M. sinica	2	—	95	—	6	0/6	—	Dittus (Hladik and Hladik 1972)
Pan troglodytes[b]	4	—	68	—	—	28	—	Hladik 1977b

Table 5.1   Composition of diets of selected primate species (continued)

Species by Dietary Category	Insects	Gum and Sap	Fruit	Seeds	Flowers	Leaves[a]	Herbs	Reference
Frugivores eating leaves continued								
Papio cynocephalus ursinus	—	—	72	—	—	28	—	Hamilton, Buskirk, and Buskirk 1978
P. c. anubis[c]	—	—	54	—	—	—	—	Dunbar and Dunbar 1974a
Pongo pygmaeus	1	—	50	—	2	29	—	Rodman 1977
Presbytis melalophos	—	—	60	—	12	38	—	MacKinnon and MacKinnon 1978
P. obscura[c]	—	—	46	—	6	38	—	MacKinnon and MacKinnon 1978
Propithecus verreauxi[c]	—	—	—	—	—	33	—	Richard 1978a
Gramnivores								
Colobus satanas	—	—	—	58	5	19/18	—	McKey 1978
Folivores								
Alouatta palliata	—	—	12	—	18	20/44	—	Glander 1978
A. seniculus	—	—	28/14[d]	—	5	7/45	—	Gaulin and Gaulin 1982
Colobus badius	3	—	6	—	16	29/46	—	Struhsaker 1975, 1978a
C. guereza	—	—	13	—	2	14/60	—	Oates 1977a
Gorilla gorilla	—	—	2	—	2	86	—	Fossey and Harcourt 1977
Hylobates syndactylus	6	—	40	—	5	0/50	3[e]	Chivers 1977
Indri indri	—	—	25	—	5	1/34	—	Pollock 1977
Lemur catta[c]	—	—	34	—	8	44	15	Sussman 1977
L. fulvus	—	—	25	—	4	71	—	Sussman 1977
Lepilemur mustelinus[b]	—	—	—	—	48	51	—	Charles-Dominique and Hladik 1971
Presbytis senex[b]	—	—	28	—	12	60	—	Hladik and Hladik 1972
P. entellus[b]	—	—	45	—	7	48	—	Hladik and Hladik 1972
P. obscura[c]	—	—	35	—	7	24/32	—	Curtin 1976, 1980
Propithecus verreauxi[c]	—	—	33	—	9	20/26	—	Richard 1978a
Herb eaters								
Macaca mulatta	1	1	9	—	4	19	66[f]	Goldstein 1984
Papio cynocephalus anubis[c]	2	—	12	—	—	—	80	Harding 1976
P. c. cynocephalus	—	15	27	1	5	0/4	42	Post 1982
Theropithecus gelada	—	—	—	—	—	—	90	Dunbar 1977a

Note: Species discussed in text are not listed here if quantitative data on their feeding habits are unavailable.

[a] Where entries contain two figures, figure to left of the bar represents mature leaves; to right, immature leaves.
[b] Figures refer to percentage of total dry weight of stomach contents. (Otherwise, figures refer to percent of total time spent feeding.)
[c] Species assigned to two categories.
[d] Ripe/unripe.
[e] Roots.
[f] Roots and herbs.

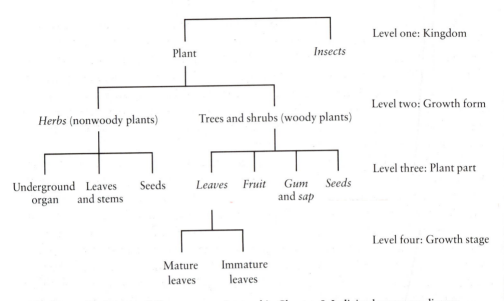

Figure 5.1   Derivation of dietary categories used in Chapter 5. Italicized terms are dietary categories found in Table 5.1. The term *herbivore* refers to animals that eat plants; these plants may be nonwoody (i.e., herbs) or woody (i.e., trees and shrubs). In other words, all herb eaters are herbivores, but not all herbivores are herb eaters.

be compared only when they were collected in the same way.

The dietary classification scheme contains several inconsistencies, which are made clear in Figure 5.1. For instance, there is no subdivision by type of insect. Some researchers have made or proposed subdivisions according to the microhabitat preferences of insects (e.g., Struhsaker 1978a; Thorington 1967; Terborgh 1983), their patterns of movement and predator avoidance (e.g., Charles-Dominique 1977), and the presence or absence of noxious substances in their bodies (Charles-Dominique 1977; Glander 1982), but these distinctions have yet to be applied widely. Figure 5.1 subdivides leaves but not fruit according to growth stage, even though primates may be highly selective about fruit, eating some and discarding others from a single plant (Gaulin and Gaulin 1982; Chivers 1977; Waser 1977a). However, few observers have collected systematic data on fruit at different stages of ripeness. This is partly because it is often difficult to distinguish these stages when they are not indicated by color changes. I suspect it is also partly because the possible nutritional and ecological importance of the distinction has only recently begun to receive much attention. Other inconsistencies in Figure 5.1 exist for similar reasons.

These comments bring us to a more general note of warning about classifications (see also Section 2.1 and Hinde 1974). Natural phenomena do not necessarily, and probably rarely, fall into clear-cut classes. When data are sorted, the researcher's boundaries between classes often do not reflect boundaries in the real world. For example, a few species in Table 5.1 are assigned to one class rather than another on the basis of a percentage-point difference, and others are assigned to two classes because they have markedly different diets in different parts of their geographic range. We shall discuss these cases individually as we come to them.

Classifications of all but the simplest phenomena conceal forms of variation they are not designed to take into account. For instance, Table 5.1 does not distinguish between species that feed on a wide range of food parts or food species (or both) and those that eat a few parts or species (or both). This distinction, often referred to as one between *generalists* and *specialists,* has a variety of important ecological implications. In addition, Table 5.1 is based on yearly averages and makes no distinction between species with diets that change little from month to month and those with diets that vary in composition. For example, a species that spends 75% of its time eating leaves every month is indistinguishable in the table from one that eats leaves 100% of the time for six months and only 50% of the time for the other six months of the year. Animals that make rapid and dramatic changes in their diet are often called *opportunists,* although the term is rarely defined rigorously.

In short, if we used a variable such as the number of food species in the diet, we would probably group the data presented in Table 5.1 differently. Thus, it is important to remember that our classification is designed to show only the aspect of dietary variation that is particularly relevant to the relationship between diet and species biology.

Most primates spend from 40% to 80% of annual feeding time on one of six types of food: insects, gums and saps, fruit, seeds, leaves, and ground herbs. Table 5.1 groups 48 species according to the food that predominates in their diet. *Insectivores* eat primarily insects; *gummivores,* gums and in some cases saps; *frugivores,* fruit; *gramnivores,* seeds; and *folivores,* leaves. No distinction is made between the leaves, flowers, fruits, seeds, and roots of herbs because animals often eat whole plants, roots and all, and when they are more selective, it is frequently difficult to see which part of the plant is being eaten and which discarded. Frugivores are further subdivided in Table 5.1 according to the second most common food in their diet, since most species supplement fruit with either insects or leaves.

Animals that eat substantial quantities of both animal and plant food are often labeled *omnivores,* although it is rarely clear when a quantity qualifies as substantial. Indeed, omnivory is almost never defined carefully. Thus, while often described as omnivores (Gaulin 1979; Hamilton, Buskirk, and Buskirk 1978; Altmann and Altmann

1970), baboons actually spend little time overall eating animal food. Because of this vagueness, the term is not used here.

5.2A   Insectivory

The primary component of an insectivore's diet is insect prey, although most insectivorous primates include other small animals in their diet as well as a substantial amount of plant material. Insectivorous lorisids in Gabon, for example, eat beetles, caterpillars, moths, grasshoppers, locusts, ants, centipedes, millipedes, spiders, slugs, snails, and frogs in addition to fruit and gum (Charles-Dominique 1977).

All the primates listed as insectivores in Table 5.1 are strepsirhines. Two are lorisids, members of the family present in Africa and South and Southeast Asia. The third is *Mirza coquereli,* Coquerel's mouse lemur, a diminutive cheirogaleid from Madagascar. The aye-aye *(Daubentonia madagascariensis)* may also belong in this category, but there is not enough information to assign it with confidence (Petter 1977). Tarsiers *(Tarsius* spp.) are excluded from Table 5.1 for the same reason, although they are actually the only primates that have not been observed eating any vegetable food at all. Preliminary observations and the examination of a few stomach contents indicate that the diet of this member of the haplorhine suborder is made up exclusively of insects and other small animals, particularly lizards (Fogden 1974; Niemitz 1979; MacKinnon and MacKinnon 1980a).

Members of the subfamilies Galaginae and Lorisinae capture prey in different ways. By running and leaping about, galagines stir up insects concealed in the vegetation around them and then use their excellent vision and hearing to spot and grab these insects in flight. Prey are always captured by hand. Lorisines, in contrast, locate their prey by sniffing around as they slowly climb through the trees. Actual capture is accomplished by grabbing the prey with one hand. The bulk of lorisine prey consists of slow-moving, often strong-smelling insects that are easy to find and catch, a category of food generally ignored by galagines (Charles-Dominique 1977).

The aye-aye and the tarsier each have their own unique set of hunting techniques. The aye-aye spends much of the night searching for insect larvae in dead branches, slowly moving along and carefully inspecting every inch. It is unclear whether prey are detected by sound or smell or, most likely, by a combination of both. The areas of the brain associated with smell are well developed in the aye-aye, and it also sports a pair of bat-like ears unusually large for a primate. Once larvae are detected, the aye-aye strips off the overlying bark with its large and powerful incisors and probes the hole so made with its strange, spindly middle finger (Figure 5.2). By repeatedly stirring the larvae around, the aye-aye reduces them to a puree, and it is this, rather than individual larvae, that the aye-aye finally extracts with its finger and eats. In sharp contrast to the insecti-

Figure 5.2   The elusive and highly endangered aye-aye *(Daubentonia madagascariensis)* of Madagascar. The large ears and acute hearing of the aye-aye help it to locate larvae hidden under the bark of trees, and it then uses its continuously growing incisors and spindly middle finger to uncover and extract the prey. (Adapted from a photograph by Jean-Jacques Petter in *The Primates,* Life Nature Library.)

vores of Africa and Asia, the aye-aye apparently ignores adult insects. In fact, in captivity an aye-aye weighing almost 3 kg has been seen running away from insects (Petter 1977)!

Tarsiers have been observed preying on insects and other small animals that feed on fallen fruit (Fogden 1974). The tarsier scans the forest floor for up to 10 minutes at a time from a perch a meter or so above the ground (Figure 5.3), leaping as far as 2 m onto a prey and killing it with a bite. Fogden suggests that hearing as well as sight plays a part in the location of prey, and he describes hunting episodes in which tarsiers grope in and sniff at the leaf litter on the ground. Although tarsiers use touch and smell at this stage in the hunt, their actions also cause cryptic insects to move so that they can be seen, and tarsiers then pounce on them in the usual way.

All insectivorous primates (except the tarsier) supplement their diet with fruit or gums or both. Fruit is the most frequent supplement, perhaps because few species have surmounted the problem of digesting gums. The Senegalese bush baby *(Galago senegalensis)* (Figure 5.4) is one of the exceptions. Although the bulk of its diet is composed of insects, it eats gum throughout the year; in the cold of winter, when insects are scarce, almost half its feeding time is devoted to gum collection (Bearder and Martin 1980).

Figure 5.3   The tarsier (*Tarsius* spp.) has long been recognized to be morphologically distinct from all other primates. Yet recent studies have begun to reveal unique behavioral features too, such as its dietary habits: it has never been seen eating anything except insects and small vertebrates. (Courtesy of D. Haring, Duke University Primate Center.)

Figure 5.4   The Senegalese bush baby *(Galago senegalensis)* from southern Africa is one of the few Old World primates that regularly includes gum in its diet. (© M. P. Kahl/NAS Collection-Photo Researchers.)

Figure 5.5   Coquerel's mouse lemur *(Mirza coquereli)* eats the secretions of larvae during the cold, dry months of the austral winter in Madagascar. (Courtesy of B. Freed, Duke University Primate Center.)

Before leaving the insectivores, Coquerel's mouse lemur (Figure 5.5) deserves special mention as the species that fits least well into this category. During the wet season, this cheirogaleid has a varied diet of flowers, fruits, gums, and insects. With the onset of the dry season, food becomes scarce, and animals devote about 60% of feeding time to licking the branches of trees used by colonies of a particular larva. When disturbed, this larva secretes a sweet liquid and releases a few drops onto its support; it is this liquid that Coquerel's mouse lemurs lick up with such enthusiasm (Petter, Schilling, and Pariente 1971, 1975). The rest of their dry season diet is composed of insects and probably small vertebrates as well, but only by including time spent feeding on the secretions of larvae can we describe them as insectivores.

5.2B   Gummivory

Primates that feed primarily on gums can be divided into two groups according to how they obtain this food: (1) species that gouge holes in trees to stimulate exudate flow and that probably eat some sap as well as gum and (2) species that concentrate on gums exuding onto the bark of trees in response to insect damage (Bearder and Martin 1980). The first group includes the pygmy marmoset *(Cebuella pygmaea)* and the fork-crowned lemur *(Phaner furcifer)* (Coimbra-Filho and Mittermeier 1978; Ramirez, Freese, and Revilla 1978; Petter, Schilling, and Pariente 1971, 1975; Charles-Dominique and Petter 1980), and the second includes the thick-tailed bush baby *(Galago crassicaudatus)* and the needle-clawed bush baby *(G. elegantulus)* (Charles-Dominique 1977). This distinction can also be applied to species that rely secondarily on exudates. Thus marmosets *(Callithrix* spp.) can gouge holes when foraging for gum (Figure 5.6), while mouse lemurs *(Microcebus* spp.), bush babies *(Galago* spp.), pottos *(Perodicticus potto),* and tamarins *(Saguinus* spp.) rely on the activities of insects (Martin 1972; Charles-Dominique 1977; Izawa 1975; Garber 1980).

Let us briefly consider the pygmy marmoset as an example of the first group. Recent research has done much to explore the ecology of this primate (Izawa 1975; Hernandez-Camacho and Cooper 1976; Kinzey, Rosenberger, and Ramirez 1975; Moynihan 1976; Ramirez, Freese, and Revilla 1978; Coimbra-Filho and Mittermeier 1978), al-

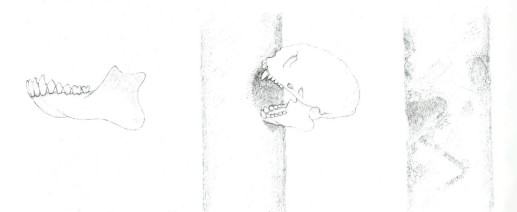

Figure 5.6   The gouges in the dry, wooden cage fixtures (right) were made by captive marmosets *(Callithrix jacchus).* The center drawing shows a marmoset skull placed in a gouging position. The animal braces itself by keeping the upper incisors dug firmly into the bark; it uses the lower incisors and incisorlike canines to scrape and gouge (left). (Adapted from photographs by R. Mittermeier.)

though longer studies are still needed to document a complete annual cycle. Pygmy marmosets devote much of their feeding time to collecting and consuming exudates (Table 5.1), using their lower incisors to gouge holes on the bottom of tree trunks. It is unclear why most holes are made close to the ground: perhaps the flow of exudates is stronger there than further up the tree.

Marmosets show a number of anatomical specializations for gummivory (Fleagle 1984; Sussman and Kinzey 1984). For example, they have claws to help them cling when moving up and down tree trunks, instead of nails like other primates. The size and shape of holes they make vary with tree species. Holes in some trees are consistently small and shallow (12–15 mm in diameter and 4–10 mm deep), while in others they are large pits 20 mm across and 15 mm deep. Rates of exudate flow are also variable; some trees produce a glob the size of a marble in less than half an hour, others a small drop. A five-month study in northeastern Peru (Ramirez, Freese, and Revilla 1978) showed that trees exude gum for at least that long, although they probably do not do so indefinitely. In any case, holes are regularly excavated further to prevent scar tissue from forming, until they finally become too deep to be dug into easily. Animals concentrate their activities on eight or nine trees at a time, moving rapidly from hole to hole over the riddled trunk of each.

The ecology of needle-clawed bush babies suggests that the second group of gummivores may differ from the first in more ways than one. Like pygmy marmosets, galagines visit the same exudate sources repeatedly, but while the former move among 8 or 9 trees, the latter collect tiny quantities of gum at 500 to 1000 sites in one night (Charles-Dominique 1977). Gum droplets are licked or scooped up with the tooth scraper (Subsection 5.4A) from nooks and crannies in the bark of tree trunks and branches and also from the surface of some lianas. Gum is produced but briefly at one spot on lianas, and bush babies locate new outlets each night by smell. Like pygmy marmosets, needle-clawed bush babies show a number of specializations for gum feeding on vertical supports, and although they do not (their common name notwithstanding) in fact have claws, each nail bears a keel that ends in a short point, which helps them negotiate large trunks and branches inaccessible to other strepsirhines.

Gummivores supplement their diet with fruit or insects or a combination of both. The pygmy marmoset, for example, begins its day feeding on gum and ends it excavating holes to produce more gum, but it turns its attention to insect foraging in the late morning. This activity involves moving slowly along branches, looking carefully under limbs, and turning over leaves in search of ants, moths, and spiders. Once located, prey are captured with a quick pounce. The slow progression during insect foraging is in sharp contrast to the fast spurts with which these animals move at other times (Ramirez, Freese, and Revilla 1978).

While fruit is a rare item in the pygmy marmoset's diet, it plays an important role in that of the thick-tailed bush baby (Galago crassicaudatus). Indeed, substantial supplies of fruit in some seasons seem to be essential, because this species lives only in areas where fruit is abundant during at least half the year (Doyle and Bearder 1977). In other words, its range may be determined by the distribution not of its most commonly eaten food but of a supplementary item.

5.2C   Frugivory

Many primates are frugivores. Whereas insectivory is found only among strepsirhines and one haplorhine, the tarsier, and gummivory only among strepsirhines and one family of New World monkeys, there is at least one frugivore in every primate family. Some frugivores spend a substantial amount of time looking for and eating insects, while others supplement their diet with leaves.

Primates are highly selective about the stage of ripeness of their fruit (Chivers 1977; Waser 1977a). Before a fruit ripens, it may contain high levels of toxins. Once ripe, it is likely to be consumed by one of the many fruit eaters of the animal world. In other words, eat it too soon and it may be poisonous; wait too long and it will be gone. Consequently, the best strategy for a primate may be to select fruit that is just barely ripe (cf. Diamond 1975). The field researcher is often made painfully aware of this strategy when standing under a tree where primates are feeding, for overhead a continual process of inspecting, feeling, sniffing, and biting into fruit goes on, and rejects are dropped. Once viewed as a symptom of clumsiness if not downright wastefulness, this rain of fruit is now seen as a sign of extreme selectivity.

Few frugivores derive nutritional benefit from seeds, which tend to be physically or chemically well protected. A notable group of exceptions are the so-called nutcracker primates (Kay 1981), all of which have particularly thick tooth enamel. They crack open hard-coated fruits, such as oil palm nuts, and eat their contents. Species that may be included in this group are the mangabey (Cercocebus albigena) (Waser 1977a), orangutan (Pongo pygmaeus) (MacKinnon 1974), and brown capuchin monkey (Cebus apella) (Terborgh 1983). The coats of the palm nuts upon which capuchins feed are so hard that a human needs a hammer or heavy stone to open them! Apart from this handful of species, frugivorous primates generally pass seeds through their digestive tract and excrete them intact (Hladik and Hladik 1967; Waser 1977a; Subsection 11.4B) or else extract them before eating the fruit's pulpy flesh.

The first species to present a classificatory problem is the ring-tailed lemur (Lemur catta). It is listed as both a frugivore and a folivore in Table 5.1. Intraspecific variation in dietary pattern is probably greatest in species that occupy several biomes and are presented with a different array of resources in each. Unfortunately, such variation has received little systematic attention, and for several reasons it is often hard to tell whether the differences are in fact as great as they appear to be.

Figure 5.7   Ring-tailed lemurs *(Lemur catta)* feed at all levels in the forest in southern Madagascar, but they usually travel on the ground. In the past, there were other terrestrial primates on the island, but today, with one possible exception *(L. coronatus)*, the ring-tailed lemur is the only species commonly found on the ground. (Photography by A. Richard.)

The ring-tailed lemur (Figure 5.7) lives in a range of vegetation in southern and south-western Madagascar, from lush gallery forests along the banks of rivers to the dry brush and scrub vegetation that takes over as distance from water increases. Sussman (1974, 1977) studied the feeding ecology of this species at two locations, Berenty and Antser-ananomby. At the first site, animals lived exclusively in gallery forest, whereas at Ant-serananomby they spent most of each day foraging through dry brush and scrub. The data suggest that at Berenty the ring-tailed lemur is frugivorous and at Antserananomby folivorous. However, when the time of year that the observations were made is taken into account, another possibility emerges. One population (at Berenty) was studied after the onset of the annual rainy season, while the other was studied in the middle of the dry season. Primates living in seasonal environments often show marked *seasonal* variations in diet (Clutton-Brock 1977c), and in this instance it is impossible to tell whether the variation reflects an overall difference in dietary pattern between the two populations or a seasonal difference in diet common to both.

5.2D   Gramnivory

The only primate known to spend enough time eating seeds to qualify as a gramnivore is the black colobus monkey *(Colobus satanas),* member of a subfamily generally noted for its leafy diet. Preliminary evidence suggests that saki monkeys *(Pithecia* and *Chiropotes* spp.) may also belong in this category (Mittermeier and van Roosmalen 1981). In a study of black colobus monkeys in Cameroonian rain forest, McKey (1978) found that the animals spent over half their feeding time on seeds. Seeds of all but one species were partially or completely digested, for they did not emerge intact in the feces. Moreover, they were eaten for their own sake rather than as an incidental component of a fruit; many lacked surrounding flesh, and encased seeds were removed before being eaten, the flesh discarded untasted. The reasons for this unique dietary specialization are discussed in Subsection 5.4C. It has yet to be established whether black colobus monkeys are also thick-enameled nutcrackers, but McKey's descriptions suggest that they are.

5.2E   Folivory

Leaves of woody plants are the most important item in the diet of primate folivores. Folivory is found in a handful of strepsirhines, one New World monkey, two apes, and several colobines. Table 5.1 indicates two major trends among folivores. First, they supplement their diet with fruit and, to a lesser extent, flowers. Animal prey plays little part in the diet of any folivore. Second, in every case where the observer has made such a distinction, animals eat more immature than mature leaves. Indeed, two species *(Indri indri* and *Hylobates syndactylus)* feed on immature leaves to the complete or almost complete exclusion of mature ones.

Three species assigned to the folivore category deserve further comment. The feeding ecology of mantled howler monkeys *(Alouatta palliata)* (Figure 5.8) has been studied in Costa Rica in evergreen gallery forest merging into upland deciduous vegetation (Glander 1975, 1978, 1981). Animals were observed for 2071 hours over a 14-month period, and the figures in Table 5.1 are from this work. An equally intensive study of mantled howler monkeys on Barro Colorado Island produced a higher figure for fruit, but the animals were still folivorous, spending 39% of feeding time on immature leaves, 9% on mature leaves, 42% on fruit, and 10% on flowers (Milton 1978, 1979, 1980). The results of a second study of the Barro Colorado Island population, comprising 407 hours of observation spread over 6 months, are in close agreement: 49% leaves, 39% fruit, and 6% flowers (Smith 1977).

The diet of red howlers *(A. seniculus)* in montane forest in Colombia is quite similar to that of mantled howlers as judged by time spent feeding on different items. However, if the quantity of different plant parts eaten by red howlers is used instead, a different picture emerges. Almost three-quarters of their diet by weight is fruit (Table 5.2). We do not know whether a similar discrepancy exists between these two measures of food

Figure 5.8   Ecology of the mantled howler monkey *(Alouatta palliata)* has been studied in Costa Rica and on Barro Colorado Island, Panama. In both places, animals spend more than a third of feeding time eating young leaves. (Courtesy of K. Glander.)

consumption in mantled howlers, although limited data collected by the Hladiks (1969) suggest that it does.

Dusky leaf monkeys *(Presbytis obscura)* are classified in Table 5.1 as folivores according to Curtin's data and as frugivores according to the MacKinnons'. There are at least three ways of accounting for the differing results of these two studies. First, a primate's diet may vary considerably from year to year depending on local fluctuations in plant growth and reproduction. Fluctuations of this kind have been documented for the very forest in which Curtin and the MacKinnons studied dusky leaf monkeys. In the first year of a five-year study, siamangs *(Hylobates syndactylus)* (Figure 5.9) spent 32% of feeding time on fruit, 58% on leaves, 9% on flowers, and 2% on insects. Yet over all five years, fruit made up a much higher percentage of their diet. The probable explanation for the discrepancy is that fruit was unusually scarce in the first year, following a superabundance the preceding year (Chivers 1974, 1977; Chivers, Raemaekers, and Aldrich-Blake 1975). The dusky leaf monkey was studied by Curtin in 1970–71 and by the MacKinnons in 1973. Variation in the abundance of resources between these years may have resulted in dietary differences.

Table 5.2   Measures of the diet of the red howler monkey (*Alouatta seniculus*)

Measure	Mature leaf	Young leaf	Ripe fruit	Unripe fruit	Flower and inflorescence	Petiole
% feeding time	7.5	44.5	28.4	13.9	5.4	0.1
% dry weight eaten	7.4	17.1	55.7	17.2	2.4	0.2
% protein eaten	12.8	21.7	40.7	19.5	5.3	0.1
% dry weight eaten/ % feeding time	0.99	0.38	1.96	1.24	0.44	2.00
% protein eaten/ % feeding time	1.71	0.49	1.43	1.40	0.98	1.00

Source: Adapted from Gaulin and Gaulin 1982.

A second reason for the differing results may be that the data were collected by different methods. All sampling methods create some bias or distortion in the data, the nature and degree varying according to the method used. Since Curtin and the MacKinnons used different methods, the apparent differences found in the diet of dusky leaf monkeys may result from biases introduced by these methods.

Finally, small samples may give an unrepresentative picture. Samples of behavior observed on a few days scattered over seasons or even years may produce a distorted picture if behavior varies much from day to day. Extensive daily variation in diet has been found in baboons (Post 1982) and probably occurs in other species too. Curtin collected 356 hours of data in about 120 days spread over a year. The MacKinnons studied five species over a period of six months and do not list the observation time for each separately, but necessarily data collection on any one of the five must have been intermittent. If the feeding behavior of the dusky leaf monkey does vary from day to day, then researchers collecting a limited number of samples on different days from one another could observe differences in diet even if they use the same sampling techniques.

The Malagasy sifaka (*Propithecus verreauxi*) is assigned to both the folivore and the frugivore category in Table 5.1. The sifaka (Figure 5.10) lives in a variety of forests in Madagascar, and its ecology and social behavior have been studied in two of them (Richard 1978a, 1978b). One was a *xerophytic* (drought-adapted) spiny forest in the arid south of the island, and the other a much more luxuriant forest in the moist northwest. The same sampling techniques were used in each forest, and in each animals were observed during wet and dry seasons of the same year. In the north animals spent more time feeding on leaves than on fruit (46% to 33%); in the south the situation was reversed (48% fruit to 39% leaves). In addition, more than half the leaves eaten in the north were mature, whereas they were mostly immature in the south. It is difficult to see

Figure 5.9   Like their close relatives the gibbons, siamangs *(Hylobates syndactylus)* often swing by their long arms when traveling in a mode of locomotion called brachiation. This female is supporting part of her weight with one arm while she feeds. (Courtesy of D. Chivers.)

Figure 5.10   The sifaka *(Propithecus verreauxi)* habitually travels by leaping from one vertical support to another. The animal starts with its back toward its destination and then twists in midair so that it lands facing forward. Its powerful legs are used both to provide momentum for the leap and to cushion the landing. (Photography by A. Richard.)

how these differences could be obtained as a consequence of the sampling design. They seem to be real and are almost certainly related to differences in the distribution and abundance of food items in the two forests (Richard 1978a).

5.2F   Herb Eating

Herbs are the most important item in the diet of gelada baboons *(Theropithecus gelada),* yellow baboons *(Papio cynocephalus cynocephalus),* and some olive baboons *(P. c. anubis).* In parts of their geographic range, olive baboons are frugivorous. Rhesus macaques *(Macaca mulatta)* feed most heavily on herbs at the site where their diet has been investigated in greatest detail, although there is evidence suggesting that, like olive baboons, they are frugivorous in some habitats.

Figure 5.11   Gelada baboons *(Theropithecus gelada)* feeding in a highland meadow. Note how most of them are sitting and using both hands to harvest grass. (Courtesy of R. I. M. Dunbar.)

Gelada baboons live at 1500 to 4500 m on the Simēn Plateau in Ethiopia. Their environment is an almost treeless grassland, and more than 90% of their diet consists of grass parts. They are highly selective, avoiding dried-up blades and coarse, reedy species with low nutritional value. During the wet season, geladas eat almost nothing but fresh grass blades (93% of feeding time) that grow in abundance. By the onset of the dry season, grasses have gone to seed and these seeds become the major (70%) item in their diet. Later in the dry season, they feed heavily on grass corms (67%), eating young blades whenever they find them.

Geladas are *bottom-shufflers*. They feed in a sitting position (Figure 5.11), harvesting items directly in front of them and shuffling forward on their haunches as they exhaust supplies immediately within reach. Grass blades are picked with the thumb and index finger, gathered into the palm of the hand, and transferred to the mouth once a good handful has been gleaned. Seeds are harvested by pulling stems through the teeth one at a time, or stripping standing stems with the thumb and index finger. Corms and root systems are dug up with rapid, alternating movements of the hands, which are used almost like little pickaxes. All these harvesting activities are performed with great speed and efficiency (Dunbar 1977a, 1977b). Nonetheless, since the processing time for corms and roots is almost certainly longer than for other items, the proportion of time spent eating them is probably considerably greater than their proportionate contribution by weight to the diet (Harding 1976; Post 1982).

Yellow baboons at Amboseli, Kenya, spend almost a third of total annual feeding time eating grass corms, and for only one month do corms not form the major food item in their diet (Post 1982). Fruit is the second most commonly eaten item, and gum and sap from the fever tree *(Acacia zanthophloea)* the third. This last item is eaten in considerable quantities by some males (Hausfater and Bearce 1976).

Olive baboons have been studied at several sites in East Africa (e.g., DeVore and Hall 1965; Rowell 1966, 1969; Packer 1975, 1977, 1979a, 1979b), but their dietary habits have been systematically investigated at only two. Table 5.1 includes the results of both these studies. One was carried out in open grasslands at Gilgil, Kenya (Harding 1976). One thousand hours of observation spread over a year showed the overwhelming importance of grass in the diet of these baboons. Over half their feeding time is spent on grass blades and seeds, and another quarter on grass corms; fruit accounts for only a small percentage of the diet. Like gelada baboons, the Gilgil olive baboons often sit while feeding, bottom-shuffling as they go (55% of all feeding is done this way). Otherwise, they stand tripedally (i.e., on three legs) and pick food items with their free hand. Again like geladas, they use their hands in pickax fashion to extract grass corms (Rose 1977b).

The second study, carried out in the Bole Valley, Ethiopia (Dunbar and Dunbar 1974a), leads us to classify olive baboons as frugivores as well as herb eaters. Gallery forest along the riverbed in the Bole Valley gives way to scrub vegetation, which in turn is replaced by grassland and scrub on higher slopes and the plateau above. The baboons range up and down the valley, and 137 hours of observations spread over six months showed that almost half their diet consists of fruit. Grass blades and seeds and clover together make up much of the rest, but unlike their conspecifics at Gilgil, the Bole Valley baboons almost never eat grass corms.

The Dunbars (1974a) stress how much less efficient the Bole Valley olive baboons are than geladas at handling grass parts. Instead of transferring blades to the mouth with both hands while bottom-shuffling along, they are more likely to stand tripedally and harvest blades with one hand. On the rare occasions when they dig for corms, they make a poor showing compared with geladas, scraping ineffectually at the earth with one hand and then pulling the pieces out of the ground with the other or with their teeth.

Although the Dunbars' study did not span a full year, it did sample feeding activities in both wet and dry seasons in Ethiopia. Thus, we may tentatively conclude that the apparent dietary difference between the Kenyan and Ethiopian olive baboons is real and not a result of methodological differences between researchers. The difference in diet seems to be associated with different food-harvesting techniques.

Figure 5.12   Young female rhesus monkey *(Macaca mulatta)* in Pakistan sits while gathering grass and clover with both hands. (Courtesy of W. Keene.)

Rhesus monkeys are widely distributed throughout South and Southeast Asia, and they have been studied at a number of sites (e.g., Neville 1968; Lindburg 1971, 1977; Southwick and Siddiqi 1977; Southwick, Beg, and Siddiqi 1965; Goldstein 1984). In northern Pakistan, rhesus live in the mixed coniferous and deciduous forests that blanket the Himalayan foothills at about 3000 m. The climate is highly seasonal: a dry, sunny spring gives way to a long, wet summer monsoon; after another clear period in the autumn, winter sets in with repeated snowstorms, which dump up to 7 m of snow in January, February, and March. The most commonly eaten food in this forest is clover *(Trifolium),* a herb found mostly where human or natural disturbances give rise to patches of open ground. Such patches are few and far between in the forest, but the monkeys seek them out. They sit to pick the clover using both hands (Figure 5.12), or they stand tripedally and use one, but they do not bottom-shuffle (Goldstein 1984).

Other studies provide mixed support for the idea of rhesus monkeys as disturbed-area feeders. During summer months, rhesus monkeys in Katmandu, Nepal, eat large quantities of the herb *Oxalis* (B. Marriott, pers. com. 1979). *Oxalis,* with a growth form similar to that of clover, is also typical of disturbed habitats. Rhesus monkeys in India commonly live close to or even in villages and cities (Southwick, Beg, and Siddiqi 1965), but the diet of these urban monkeys has yet to be studied. Lindburg's (1977) primarily descriptive account of the diet of rhesus monkeys in a forest in northern India contradicts the findings reported so far. His subjects were apparently more like other macaques listed in Table 5.1, eating a preponderance of fruit and few herbs and grasses. Thus, although more data are needed to be sure, the rhesus monkey is probably more of a dietary jack-of-all-trades than Table 5.1 suggests.

5.2G   Summary

Three points emerge from this survey of primate diets.

First, all primates eat one class of foods that contains high levels of protein and one that is rich in carbohydrate. For most strepsirhines, this means insects (for protein and for carbohydrate in the form of chitin) and gums or fruit (for carbohydrate). For haplorhines, it means insects or leaves (for protein and varying amounts of carbohydrate) and fruit (for carbohydrate). The main dietary distinction among haplorhines lies in the extent to which they exploit leaves. In Section 5.4 we shall see how certain species are specially adapted to do this.

Second, although all primates exhibit some flexibility of diet as they respond to temporal shifts in the availability of different foods, so far as we know most species in the long run consistently combine the same kinds of food in their diet. For example, wherever it has been studied, the chimpanzee eats primarily fruit, with a leafy supplement and a few insects thrown in for good measure. This conservatism is reflected in the fact that we assigned only 4 out of 50 species or subspecies to more than one dietary category: the ring-tailed lemur, the dusky leaf monkey, the sifaka, and the olive baboon. The evidence of dietary diversity is strong for only two of these four, the sifaka and the olive baboon. Problems were raised about the classification of three other species, namely the rhesus monkey and two species of howler monkey. Preliminary findings suggest that rhesus macaques may belong with the sifaka and olive baboon as a wide-ranging species with exceptional dietary flexibility. Doubts about the assignment of howler monkeys centered on different measures of dietary composition and apply equally to all the assignments made in Table 5.1.

Finally, conclusions about dietary variation must be drawn with caution, since apparent variation can sometimes be a product of differing sampling procedures. Thus, when data from different species or different populations of a single species are compared, it is essential to check precisely what is being compared. In a rigorous comparison, behaviors are sampled in similar ways, observations span all seasons of the year in the same year or years, and enough days are sampled to eliminate the effects of daily variations.

5.3   Importance of Size in
Understanding Dietary Variation

We have shown that primates display a wide range of dietary patterns. How are we to account for this variation? A mammal's size has major implications for what it does and does not eat. In this section, we explain the reasons for this, look at the association between size and diet among the primates, and discuss the points at which this association breaks down.

Several researchers have demonstrated the importance of size as a predictor of diet among mammals, particularly ungulates (Jarman 1974; Bell 1970, 1971; Geist 1974a, 1974b) and primates (Gaulin and Konner 1977; Gaulin 1979; Kay and Hylander 1978; Hladik 1975; Hladik and Chivers 1978; Chivers and Hladik 1980; Clutton-Brock and Harvey 1977a, 1977b). Insectivorous primates are small, the largest reaching only about 250 g. Most folivores, in contrast, weigh at least 1400 g, and the majority are well over 3000 g. Primates with a high fruit component in their diet tend to fall between these two extremes (Figure 5.13).

The trend can be explained by referring back to points already made. Small mammals have high energy requirements per unit body weight, although their total requirements are small. This means they must eat easily digested foods that can be processed fast; however, these foods do not have to be abundant, since they are not needed in large quantities. Large mammals have lower energy requirements per unit body weight than small ones and can afford to process food more slowly, but their total food requirements are great. This means that their food items must be abundant but not necessarily easy to digest.

Almost all primates, regardless of size, meet part of their energy requirements with fruit, which provides a ready source of simple sugars. It is in how they make up the difference in energy and how they meet their protein requirements that body size is most critical. The biology of small primates dictates that they eat quickly digested foods providing a ready source of protein as well as energy, and animal food fits this description. Lacking specializations for the capture of large prey, small primates derive their protein from insects. Most large primates find insects attractive, too, and few turn up their noses when presented with an unsuspecting vertebrate (Hamilton, Buskirk, and Buskirk 1978; Hamilton and Busse 1978; Butynski 1982). However, animal food is rarely sufficiently abundant and can rarely be captured fast enough by a large primate to meet its dietary needs. In contrast, leaves are among the most abundant edible plant parts in tropical biomes. They also contain high levels of hard-to-digest or completely indigestible carbohydrates. A small primate is unable to process leaves fast enough to extract sufficient energy and nutrients, but a large primate has a lower metabolic rate and can process leaves in great bulk, meeting its protein needs and part of its energy needs by slowly digesting and absorbing the contents of large quantities of this low-quality food.

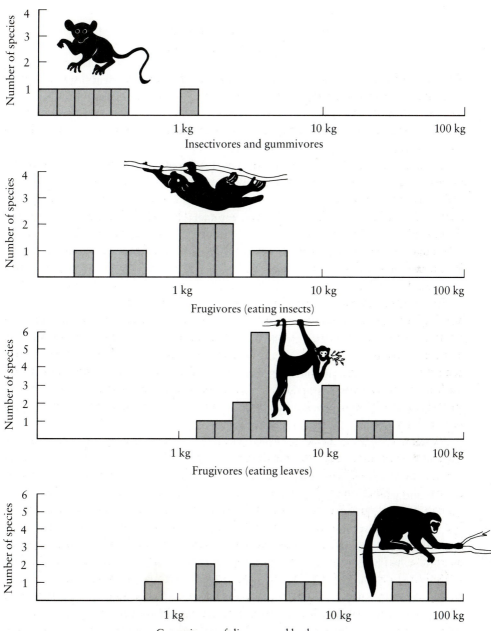

Figure 5.13   This figure shows the frequency distribution of weights of the species listed in Table 5.1, combined into four broader dietary categories. Frugivores that supplement their diet with insects tend to be lighter than those that eat leaves. The lightest primates are insectivores and gummivores, and the heaviest are folivores. Note that the horizontal axis has a log scale.

Frugivores were divided into two groups in Table 5.1, one containing species that supplement the fruit in their diet with insects, the other primarily with leaves. This distinction reflects differences in body weight and is predictable from the argument above. Small frugivores supplement their diet with insects, and large ones with leaves.

Within broad limits, then, body weight predicts the general nature of a primate's diet. Still there are exceptions to the rule, and body weight is not always a good predictor even of general trends. For instance, a small animal can fulfill its total nutrient requirements from low-quality foods if it can extract nutrients fast enough. One way of doing this is to develop specializations of the digestive tract (Gaulin 1979). This may be the case with the sportive lemur *(Lepilemur mustelinus)*, an exceptionally small folivore. It has a copious large intestine and may couple hindgut fermentation with coprophagy (Charles-Dominique and Hladik 1971). By eating its own feces, this lemur may extract more of a food's value on the second passage through the gut and so transform an energy-poor diet into an energy-rich one. As noted earlier, however, coprophagy has not been observed in subsequent studies of this species (Russell 1977). The aye-aye *(Daubentonia madagascariensis)* is another anomaly, for it is much larger than other insectivores; in this instance, size constraints are apparently overcome by a set of highly effective specializations that enable the aye-aye to locate and extract insect prey (Subsection 5.2A). Chimpanzees, among the largest of all primates, seasonally include an unexpectedly large quantity of insects in their diet. This is because they prey heavily upon termites in termite mounds for a few weeks each year. In a sense, a termite mound has the properties of a moderately large carcass rather than of insect prey, and it is this in combination with chimpanzees' own skills in exploiting these mounds that allows them to include such a large insect component in their diet.

There are exceptions to the rule relating body weight to diet, thus, and when finer comparisons are made among species the rule loses its value altogether. Specifically, similarly sized species often have quite different diets. Take red colobus monkeys *(Colobus badius)* and guerezas *(C. guereza)*, which are sympatric in many forests. Both are folivores (Table 5.1), both have similar specializations of the teeth and gut (see Section 5.4), and both are about the same weight. Yet field research shows clear-cut differences in their diets (Struhsaker and Oates 1975; Struhsaker 1978a). Red colobus feed on immature leaves, leaf buds, and flowers from a wide spectrum of species and eat relatively little fruit (Figure 5.14). Guerezas eat many more leaves, mature as well as immature, much more fruit, and almost no flowers, and their foods come from a much narrower range of species (Figure 5.15). The body weight rule fails to account not only for variations such as these but also for differences within a single species.

What causes the rule to break down so completely? There are several reasons. First, it takes no account of the effects of competition. If members of two species compete for food, there are two possible outcomes in the long run. Either one species will outcom-

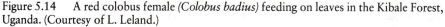

Figure 5.14   A red colobus female *(Colobus badius)* feeding on leaves in the Kibale Forest, Uganda. (Courtesy of L. Leland.)

pete the other and exclude it from its range or even drive it to extinction, or one or both species will modify their eating habits to reduce or avoid further competition (see Chapter 10). Species of about the same weight living in different areas may have different diets because one or both have modified their eating habits in response to competition with other species in their respective environments.

Second, it is important to remember that body weight indicates only a mammal's basal metabolic requirements (Subsection 4.3A). Anyone who has watched a spider monkey racing through the trees all day long and a howler monkey, of much the same size, ambling slowly about the forest with frequent time-outs for prolonged siestas will recognize that an animal's total requirements can vary enormously depending on its activity level. Third, primates have morphological, physiological, and behavioral specializations that constrain and guide their choice of food regardless of their size. Let us turn to these specializations.

Figure 5.15   A guereza female *(Colobus guereza)* and infant.

5.4   Specializations Related to
Harvesting and Processing Food

As we have seen, most primates consistently select particular kinds of food, and different foods have different physical, chemical and, hence, nutritional qualities. Therefore, we might expect primates with similar diets to share traits particularly suited to exploiting certain foods. This section considers features of primate teeth, guts, and behavior believed to be evolutionary specializations relating to particular diets.

A study of dietary specializations in primates would at best assemble information about three factors: (1) the structure and other attributes of the specialized characteristic itself (for example, a tooth's dimensions and the rate and manner in which it wears down with use), (2) the diet of primates with that characteristic, and (3) the actual manner in which that characteristic helps to process dietary items (for example, how finely a particular kind of tooth grinds up a particular food). For instance, an investigator's goal might be to show, first, that species with large incisors all eat fruit and, second, that large incisors help process fruit more efficiently than small ones.

In practice, most studies have gathered information on the first two factors and inferred information about the third from correlations between morphology and diet. Such inferences are beset by the problem of *confounding variables*. This problem occurs when a correlation between two factors is actually due to them both being correlated with a third, unrecognized factor. For example, it has long been argued that large molar teeth are a specialization for leaf eating, since folivores tend to have larger molars than frugivores. However, folivores tend to be bigger than frugivores. Is the large size of a folivore's molars a dietary specialization or a reflection of its large body? This is a difficult question to answer, but probably the second hypothesis is correct (Goldstein, Post, and Melnick 1978; Post, Goldstein, and Melnick 1978).

Setting aside confounding variables, a second problem with the study of dietary adaptations is deciding which aspect of the diet has acted as a selective pressure on the characteristic being studied. We may fail to find a correlation between a particular characteristic and a certain type of diet because we are describing diets inappropriately. At a general level, food texture is evolutionarily important in shaping tooth structure, and food chemistry is important for the gut. But is the strongest selective pressure exerted by the most commonly eaten food or by a rarely eaten food that is critical for survival at certain times? We do not know. Most primate characteristics are almost certainly shaped by several selective pressures at the same time and therefore represent compromise solutions to a series of problems rather than the best solution to a single problem (Gould and Lewontin 1979).

Finally, a third problem in studying evolutionary specializations is the working assumption that the natural world does not change over long periods of time. Most of the studies we shall discuss assume that what a primate is eating today is what it has been eating for the last several hundred thousand, if not several million, years. But environments do change, sometimes quite rapidly, and an animal is not necessarily specialized to eat the foods we see it feeding on now: a species can change its diet (within limits, at least) faster than it can change its teeth.

Despite the many problems, there is sufficient evidence to document a number of specializations associated with the dietary patterns described in Section 5.2. Indeed, given the problems, it is perhaps surprising that morphological variations correspond as well as they do with the dietary patterns of primates in the wild.

5.4A   Teeth

In this subsection we describe several associations between tooth morphology and diet that have been reported. One observation is that gummivores have strong, elongated lower incisors (Coimbra-Filho and Mittermeier 1978; Kay and Hylander 1978). The function of the long, *procumbent* (forward-projecting) lower incisors of strepsirhines has been debated for many years. A combination of behavioral observation (e.g., Buettner-Janusch and Andrew 1962; Richard 1978a; Petter, Schilling, and Pariente 1971) and morphological study (Kay and Hylander 1978) now indicates that the *tooth scraper* formed by these lower teeth plays different roles in different species (Figure 5.16). In some, it is quite a fragile structure that animals use to groom their own and others' fur. Species that eat large quantities of gum have more robust tooth scrapers and vigorously gouge holes in tree trunks with them, thereby inducing a flow of exudates that they then scrape up. The importance of incisors for gummivory is further emphasized by studies of New World primates. Callitrichids can be divided into two groups according to the length of their lower incisors (Coimbra-Filho and Mittermeier 1978; Napier and Napier 1967). Tamarins (*Saguinus* and *Leontopithecus*) show the usual primate condition, with canines much longer than incisors. Marmosets (*Callithrix* and *Cebuella*), in contrast, have incisors as long as their canines. This distinction is explained when we remember that tamarins are frugivores while marmosets are gummivores.

Another observation is that among the higher primates, frugivores have larger, wider incisors than folivores (Post, Goldstein, and Melnick 1978; Hylander 1975). Although the hypothesis has yet to be experimentally demonstrated or systematically studied in the wild, frugivores probably make much greater use of their incisors in preparing their food before chewing it. Such preparation is not necessary for leaves, berries, grasses, buds, flowers, and so forth, but to eat large, tough-skinned fruits a primate has to bite through the outer layer and slice off hunks of the inner flesh. Because the incisors probably wear down rapidly from this use, a frugivore must start life with large ones if it is not to end up with teeth too worn to be useful while its other body parts are still in good working order (Kay and Hylander 1978).

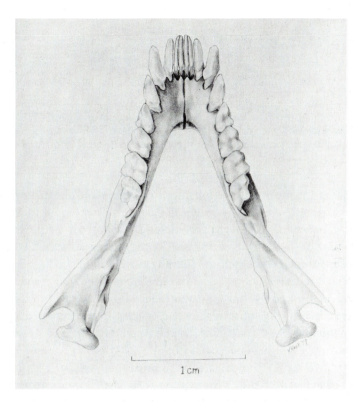

1 cm

Figure 5.16   The sifaka *(Propithecus verreauxi)* uses its tooth scraper, shown here, for grooming and feeding. I have watched animals strip and then eat bark, scrape at hard, dead tree trunks (probably in search of larvae), and gouge deep holes in trees that store water in their trunks. Fragile as it may appear, the sifaka's tooth scraper serves as a robust and useful tool in each of these activities. (Drawing by Virge Kask.)

A third observation, made on Old World monkeys, is that the molar teeth of folivores have higher cusps, more sharply angled and longer shearing blades, and larger crushing basins than those of frugivores (Kay 1975; Kay and Hylander 1978). All cercopithecids have molars with four well-developed main cusps separated by a groove running from side to side. A ridge connects the two front cusps to one another. The back cusps are similarly connected, and the molars are *bilophodont* (having two ridges). However, the cusps of colobines are higher relative to the central groove and more pointed than those of the cercopithecines, and they have longer shearing blades (Figure 5.17). When the upper and lower molars come together, they act like scissors. After food has been cut up by this scissor action, it is crushed and ground in the molar basins. Cercopithecine molars, in contrast, come together with a mortar-and-pestle effect. This dis-

Figure 5.17   Comparison of the lower second premolar and three molar teeth of frugivorous (right) and folivorous (left) cercopithecids. Note the sharper cusps, more sharply defined shearing blades, and larger crushing surfaces of the folivore. (Adapted from Kay and Hylander 1978.)

tinction between the molars of the folivorous colobines and the mainly frugivorous cercopithecines has been interpreted to mean that the molar teeth of folivores are specialized to cut up and grind tough, fibrous foods. Folivorous strepsirhines and the howler monkey also have higher cusps and longer shearing blades than more frugivorous members of their taxonomic groups (Kay and Hylander 1978).

Two recent research developments add weight to the idea of molar specialization for processing leaves. First, folivorous primates reduce leaves to smaller fragments when chewing them than do frugivores (Walker and Murray 1975); second, many mammals digest small particles of food containing structural carbohydrates more completely than they do large particles (Kay and Sheine 1979).

Walker and Murray (1975) compared the stomach contents of gibbons, colobines, and macaques that had been shot in the wild. They found that food fragments were largest in gibbons and smallest in colobines, with macaques falling in between. One possible explanation for this difference is that colobines chew their food longer. To test this, captive macaques and colobines (gibbons were apparently unavailable) were fed escarole lettuce and apples, and the number of times they chewed each mouthful was recorded. Colobines chewed the lettuce fewer times than macaques, suggesting that they do not reduce leaves to smaller fragments by chewing them longer. (All animals chewed the apple much less than the lettuce.) Closer examination of fragments from the wild specimens further indicated that tooth morphology rather than chewing time gave colobines the edge over macaques. Leaf fragments from macaque stomachs were covered with cuts, punctures, and crushed areas corresponding to the morphology of the molar cusps: the teeth apparently lacked sufficient shearing action to cut all the way through the leaf body with every bite. Fragments from colobines' stomachs, in contrast, rarely bore cusp marks on their surface; rather, they had been sheared clean along the edges into confetti-like pieces.

Although similar trends in molar morphology distinguish folivorous and frugivorous apes (Kay and Hylander 1978), they are less marked. Tooth wear proceeds exceptionally fast among apes, and they spend most of their adult lives with quite flat molars. While it is possible that most adults therefore have functionally inferior teeth, there is another explanation. Simply the size of the teeth and power of the chewing muscles of adult apes may yield the same results as highly differentiated cusps do in younger and smaller animals. Molar cusps may thus be functionally important only for the young, helping to process food and acting as guides to assure efficient chewing (Walker and Murray 1975; Gordon 1980).

Four species were assigned to more than one category in Table 5.1. What is the evidence of their molar morphology? The molars of ring-tailed lemurs and olive baboons suggest dietary adaptability, as we would predict. Those of the lemur are almost midway between the frugivores and folivores with which they were compared. While clearly distinct from the teeth of African arboreal folivores, the olive baboon's cheek teeth have higher cusps and longer shearing blades than those of most other cercopithecines (Figure 5.18). Note in Figure 5.18 that the other species that are clustered with the olive

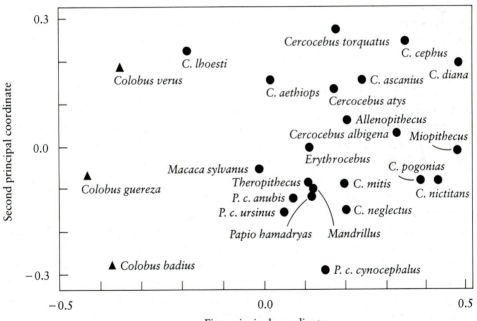

Figure 5.18   Analysis of molar tooth shape among African cercopithecids. The horizontal axis separates folivores (to the left) from frugivores (to the right). The triangles are colobines, the circles cercopithecines. C. = *Cercopithecus*; *P.c.* = *Papio cynocephalus*. Species names are included only when genera contain more than one species. (Adapted from Kay and Hylander 1978.)

baboon are also open-country forms known or likely to eat large amounts of high-fiber food. The molars of dusky leaf monkeys and sifakas show features typical of folivores. The rhesus monkey was assigned to the herb-eating category, but field data indicate that this species has versatile dietary habits, eating primarily on herbs in some areas and fruit in others. The evidence of their molar teeth places them firmly among the frugivores. In summary, features of molar tooth morphology indicate broad dietary habits strongly but not perfectly.

Moving on in our review, we find that the molar teeth of insectivores are more like those of folivores than frugivores (Kay 1975; Sheine and Kay 1977). Kay (1975) noted a major resemblance between the molar teeth of insectivores and folivores, and he inferred that in both cases this morphological pattern somehow increased the efficiency with which members of these dietary classes processed their respective foods. Recent experiments have shed light on the adaptiveness of high-crowned, shearing molars for insectivory (Sheine and Kay 1977; Kay and Sheine 1979). A large proportion of the wet weight and the potential energy value of insects is in the form of chitin, a polysaccharide with characteristics between those of the structural and nonstructural polysaccharides of plants (Subsection 4.2A). Kay and Sheine fed different-sized fragments of chitin to the insectivorous Senegalese bush baby and, by inspecting its feces, compared the completeness with which fragments of different size were digested. At every trial, animals digested significantly more of the smaller fragments; indeed, they refused to eat the largest particles they were offered. Like cellulose and other plant structural carbohydrates, therefore, chitin digestion is improved if it is first finely ground. This suggests why insectivores, like folivores, have high-cusped shearing teeth capable of slicing food into tiny fragments.

5.4B   Gut and Cheek Pouches

A general correlation has long been recognized in mammals between broad dietary categories and certain distinguishing characteristics of the digestive tract (Hladik 1967). Grazers and browsers such as the horse and ruminating ungulates have a long digestive tract (10–30 times their body length) with voluminous reservoirs. Carnivores have a much shorter digestive tract (only 4–6 times body length) with a relatively undeveloped stomach and large intestine, while mammals with mixed diets have digestive tract features between these two extremes.

Preliminary studies of the primates indicated that there was a relationship between diet and gastrointestinal tract morphology within the order as well as between primates and members of other orders (Hladik 1967; Amerasinghe, Van Cuylenberg, and Hladik 1971; Fooden 1964; Jones 1970). Recently a detailed analysis of this relationship was carried out on 48 species of primates, 24 species of wild mammals, and 6 species of domesticated mammals (Chivers and Hladik 1980). Species that feed primarily on animal matter (called faunivores by the authors) have a simple stomach and large intestine and a long small intestine. In folivores, either the stomach or the large intestine, and sometimes both, is voluminous. The gastrointestinal tract of frugivores is intermediate in morphology, varying according to whether the diet is supplemented with insects or leaves. Several studies have concluded that gut size increases linearly (that is, in a one-to-one relationship) with body size in herbivorous mammals (Demment and Van Soest 1983; Parra 1978; Subsection 3.3B), but Chivers and Hladik argue that the situation is more complicated than this. Briefly, they found that in frugivores and hindgut fermenting folivores, the volume of the stomach and large intestine is greater in relation to body size in large species than in small ones with similar diets. In contrast, as foregut fermenting folivores increase in body size, they show a reduction of gut capacity in relation to body size; in faunivores the volume of the stomach and large intestine is linearly related to body size. The authors relate these scaling effects to differences in the diets of folivores, frugivores, and faunivores, and suggest that other studies may not have uncovered these effects because they did not distinguish among herbivores with markedly different diets.

The study of scaling relations of the gastrointestinal tract is fraught with problems, not the least of which are practical difficulties of measurement (Demment 1982). Moreover, it is not altogether clear *what* to measure. The studies discussed here focused on the dietary significance of differences in the surface area of the gut, but others have suggested that surface area has less importance for digestion than the retention time of food in the gut (Hofmann 1973, cited in Demment 1982). In short, before firm conclusions can be drawn, further research is needed, not just into scaling relations but also into digestive physiology.

Digestive activity has been investigated more intensively in colobines than in any other primates. Most studies report high levels of fermentative activity (e.g., Bauchop and Martucci 1968), although one found no evidence of bacterial fermentation and cellulose digestion (Ohwaki et al. 1974). Dietary differences in the experimental subjects may provide the key to conflicting results here: fermentation would not be expected if monkeys were fed little but fruit (Oates 1977a).

Bauchop and Martucci's (1968) work with two species of langur provides conclusive evidence of bacterial fermentation in the stomach, a digestive process not found in any primates except colobines. The colobine stomach is divided into two relatively separate sacs, and one sac is kept at a neutral acid balance (a pH of 5 to 7), probably with the aid of copious saliva secreted from highly developed salivary glands. It is here that fermentation takes place. (A neutral acid balance is necessary to maintain bacterial activity.) The stomach is also voluminous, so that colobines can process food in greater bulk than other primates can.

The products of fermentation are fatty acids that are absorbed and used and a gas mixture that is disposed of by belching. In the two langurs studied, fatty acids were in

Figure 5.19   The cheek pouches of this young rhesus monkey *(Macaca mulatta)* were filled with berries gathered from a vibernum bush. (Courtesy of W. Keene.)

concentrations similar to those found in ruminating ungulates. Microbial fermentation and fatty acid production were continuous and took place at high rates compared with production in domestic ruminants. How important are the products of fermentation in meeting the total energy requirements of these primates? At an average fermentation rate, fatty acids would provide 283 kcal/day. A 4.5-kg langur needs 218 kcal/day for basal metabolism, and so gastric fermentation may play a major role in the colobine energy budget. How important is bacterial fermentation for detoxifying secondary compounds ingested in food? There is as yet no answer to this question despite the suggestion that this, rather than the breakdown of structural carbohydrates, is the primary function of foregut fermentation among medium-sized herbivores (Parra 1978; Freeland and Janzen 1974).

Members of the subfamily Cercopithecinae are distinguished from the Colobinae by the possession of *cheek pouches* (Figure 5.19) as well as by the absence of the gastrointestinal specializations described above. A cheek pouch is an oblong sac that protrudes when full from the lower part of the cheek. Food is moved between it and the mouth through a slitlike opening. Cheek pouches apparently have many functions. They are used in a retrieve-and-retreat pattern of feeding (Murray 1975), either when danger from predators is high or competition with conspecifics for food is intense (Napier and Napier 1967; Gautier-Hion 1971). On these occasions, monkeys quickly cram their cheek pouches with food and then retire to a safer or less crowded spot to chew it. Sometimes the use of cheek pouches may simply increase foraging efficiency. For example, vervet monkeys *(Cercopithecus aethiops)* at Naivasha in Kenya would spend a half-hour on the ground gathering the hard-shelled seeds of fever trees and storing them in their cheek pouches, and then retire to a nearby branch and spend another half-hour cracking them between their molars and eating them (M.D. Rose, pers. com., 1983).

More pertinent to this discussion of digestive specializations, however, is the predigestive role played by the cheek pouches (Murray 1975; Rahaman, Srihari, and Krishnamoorthy 1975). Amylase, the enzyme that breaks down starch, has been found in saliva in the cheek pouches of two cercopithecine species tested (Jacobsen 1970), and 10 bonnet macaques *(Macaca radiata)* took less than five minutes to digest into glucose about 50% of a quantity of starch put into their pouches. It is unlikely that these pouches are very important in cercopithecine digestive processes, but they do seem to provide a helping hand.

5.4C   Choosing the Right Foods

Specializations of the teeth and gut make particular primates especially well suited to exploit particular kinds of food (and, conversely, reduce their efficiency in exploiting others). But as we have seen, frugivores do not eat only fruit or every fruit, any more than folivores eat only leaves or every kind of leaf in the forest. This leads us to a new question: how do primates select the foods they eat? Why does a folivore ignore the leaves of one tree species and spend hours every day eating the leaves of another, closely related species?

Food choice can be studied from at least two perspectives. First, we can look at the immediate chemical and physical signals influencing an animal's decision to eat or avoid a particular food. Second, we can establish criteria that *we* think an animal should use if the animal is *nutritionally wise,* and we can then see whether indeed these criteria distinguish items eaten from those not eaten. An animal shows nutritional wisdom if, regardless of the immediate factors guiding its choices, it selects what we consider to be an adequate diet (Section 4.3). Obviously, the two sets of criteria, ours and the animal's, must produce approximately the same result if the animal is to survive. However, the distinction is important. An animal can select nutritionally valuable foods only insofar as those qualities are reflected in chemical or physical cues that it can detect. There is a real possibility, therefore, that an animal will sometimes make mistakes. Chemical signals are received mainly at receptor sites for taste and smell, and the only chemical components of a food that can influence an animal's selection are those making contact with these receptors (Arnold and Hill 1972; Westoby 1974). Unfortunately, the complexity of such signals has been little investigated, even in mammals as well studied as the domesticated ruminants.

Physical factors may forestall or modify responses to chemical signals. This has been well documented among invertebrates (e.g., Grime, Blythe, and Thornton 1970), and evidence of similar processes in mammals is beginning to emerge (Arnold and Hill 1972). We have already considered bush babies, who reject the largest particles of chitin although they feed enthusiastically on smaller fragments. Howler monkeys (*Alouatta seniculus*) living in Andean cloud forest consistently select the least tough leaves and fruits, and when feeding in a particular tree, they pick large fruits in preference to small ones (Gaulin and Gaulin 1982).

Both of these examples illustrate the convergence between our criteria for selection and those of the primate: the bush baby that rejects large chunks of chitin, probably because they are too big to put comfortably into its mouth, is also rejecting a food that needs prolonged chewing to reduce it to fragments small enough to be digested completely. The howler monkey that picks the least tough leaves, probably because they are easiest to chew, is picking those with the lowest fiber content and hence the highest nutritional value (Harkin 1973; Gaulin and Gaulin 1982).

A food's color may also guide an animal's decision to eat or to reject it (Snodderly 1978, 1979). The few New World monkeys that have been tested, namely squirrel monkeys (*Saimiri*), capuchins (*Cebus*), and woolly monkeys (*Lagothrix*), discriminate best among the shorter wavelengths and colors like green, blue, and violet, and very poorly among the reds, oranges, and yellows at longer wavelengths. In contrast, the Old World primates tested (macaques, baboons, and chimpanzees) have discriminative abilities across the whole spectrum that are comparable to those of humans.

A study of fruits eaten by a New World rain forest dweller, the titi monkey (*Callicebus torquatus*), showed that most were green, either because they were not yet ripe or because they were cryptically colored, presumably to minimize their chance of being spotted by a predator. Snodderly (1978) suggests that the acute discriminative abilities of New World monkeys at shorter wavelengths reflect the importance of color in locating their green food against the green backdrop of the tropical rain forest. The Old World primates tested, he argues, tend to live in drier environments where discrimination among shades of red and brown (i.e., the longer wavelengths) is more important. While it is hard to accept such a simple distinction between New World primates eating green food in a green world and Old World primates eating multihued foods in a rainbow world, the differences in discriminative abilities found by Snodderly are intriguing, and the importance of color in guiding food choices deserves further investigation.

Nutritional wisdom has been most closely studied in laboratory rats (Rodgers and Harper 1970; Rozin 1969). Some rats, it seems, are wiser than others. In one experiment, rats were allowed to choose between food pellets containing different amounts of protein. Individuals spontaneously developed different dietary habits, and those choosing high-protein pellets early in life and lower-protein pellets as adults consistently outlived others (Ross and Bras 1975).

This variability notwithstanding, some degree of nutritional wisdom must be built-in, or else animals would starve in the wild much more often than they do. Physiological research is beginning to uncover the mechanisms underlying such wisdom. For example, a feedback system has been discovered among rats that monitors the physiological effects of food (Garcia and Ervin 1968). Laboratory rats were given taste, visual, and auditory signals followed by an electric shock or nausea (induced by drugs or X rays). They quickly learned to avoid some of these signal-plus-unpleasant-sensation combinations but not others. Specifically, the taste of foods was rapidly associated with subsequent nausea, and these foods were later avoided. Auditory and visual signals were readily associated with electric shocks. Yet rats did not learn to avoid foods with a particular taste even when they were accompanied by a strong electric shock, nor did they learn to avoid audiovisual signals paired with nausea. Moreover, while they slowly learned to associate pain with audiovisual signals when the two were separated by over a minute or so, they quickly associated taste and nausea even when several hours separated the two experiences. Clearly, rats have a highly efficient monitoring system that helps them avoid dangerous foods after only one tasting.

Another laboratory study shows that rhesus monkeys have physiological mechanisms for monitoring the nature and volume of nutrients in their stomach and modify their subsequent behavior accordingly (McHugh, Moran, and Barton 1975). Subjects were fed a variable nutrient preload through a tube inserted directly into their stomachs, and they were then allowed to eat in the normal way. A definite dose–response relationship to the caloric content of the preload was seen in the subsequent normal feeding behavior of the monkeys, regardless of the volume, caloric concentration, or nature of the nutrient preload. Specifically, monkeys ate less when they had been given a nutrient preload, and in almost every case they reduced their caloric intake through the mouth by about the amount previously inserted into their stomach by tube. In other words, these monkeys appear to have quick and accurate physiological means of knowing when they have had enough of a good thing!

Nutritional wisdom is studied differently in the wild. Lacking the elaborate experimental control possible in the laboratory, field investigators try instead to find characteristics that distinguish foods from nonfoods. For ruminants and some birds, there is ample evidence that individuals select high-quality foods from the range available (e.g., Klein 1970; Gardarsson and Moss 1970; Bell 1970; Swift 1948), and the search for similar evidence is now a focus of much primate field research (e.g., Casimir 1975; Hladik 1977a, 1977b; Oates 1977a, 1978; Oates, Swain, and Zantovska 1977; Oates, Watermann, and Choo 1980; McKey et al. 1978; Wrangham and Waterman 1981; Glander 1981). The most extensive data available are for the howler monkey (*Alouatta*), so let us briefly consider this primate's nutritional wisdom (or lack of it).

Gaulin and Gaulin (1982) found that Colombian howler monkeys generally select items with the highest protein content from a class of plant parts. Milton (1979) collected immature and mature leaves from a number of tree species on Barro Colorado Island, including those ignored as well as those eaten by howler monkeys, and analyzed them for total protein and nonstructural carbohydrate content, cell wall constituents, and certain secondary compounds. The analysis suggested that the most important factors determining leaf choice are protein and fiber content, with perhaps some influence from secondary compounds. Howler monkeys choose leaves high in protein and low in fiber.

Finally, Glander's (1981) study of howler monkeys in Costa Rica extends the picture still further. The selectivity of his subjects was initially suggested by the fact that they did not obtain their food from the most common tree species in their range but rather from rare ones, and they spent 79% of feeding time in only 5% of the trees. They discriminated among members of single species as well as among different species, eating the leaves of certain trees while avoiding those of adjacent individuals of the same species. Nutritional analyses revealed a degree of discrimination equal to that found among the Barro Colorado Island howlers. The nutrient and secondary compound content of plant parts varied according to species, season, and the identity of individual plants. Monkeys selected leaves with a high-protein content and low fiber and secondary compound content and ate a sufficient range of species to ensure an adequate intake of essential amino acids (see also Hladik et al. 1971). They maintained this balance of nutrients in their diet by continually adjusting their intake as plant parts changed in abundance and composition. Flowers were plentiful in the dry season and fruit in the wet, and together they provided animals with a year-round supply of carbohydrates. New leaves in some months, and mature leaves whenever new leaves were not available, provided a steady source of amino acids (cf. Smith 1977).

5.5   Theory of Optimal Diet

The approach taken so far in this chapter has been descriptive and interpretative. However, a growing body of ecological literature attempts to predict aspects of an animal's foraging behavior from a knowledge of the quality, abundance, and distribution of its food resources. Ideas about how an animal ought to feed under different conditions are collectively referred to as optimal foraging theory (e.g., MacArthur and Pianka 1966;

Schoener 1971; Westoby 1974; Pyke et al. 1977; Kamil and Sargent 1981). The broad goal of the theory is to investigate how natural selection shapes feeding behavior, although its more immediate aim is to predict changes in the behavior of individual animals over short periods of time. The theory of optimal diets is one subset of optimal foraging theory. Developed primarily with small predatory species in mind, the theory presents many problems when applied to large herbivores. However, these very problems provide useful starting points for exploring the many significant questions about primate diets that still await answers. It is on this questioning note that we end this chapter.

What do we mean by saying an animal ought to feed in a certain fashion? It is simply a brief way of stating that this is how the animal will feed if natural selection has operated to produce the best, or optimal, feeding pattern. (Note that the optimum, or best, is not always the maximum or most.) This, in turn, implies the following: (1) differences in feeding behavior are under genetic control and therefore heritable; (2) natural selection favors individuals that contribute the most genetically to subsequent generations and hence favors the feeding behavior most likely to help bring this about; and (3) natural selection operates more rapidly than the environment changes; in other words, the morphology and behavior of animals can be modified fast enough by natural selection to keep up with new problems presented by the environment. If we accept these three assumptions, then it follows that the feeding behavior of animals will be very close to the optimum for the particular environment in which they live.

The optimal diet is determined in three steps (Schoener 1971; MacArthur 1972; Pyke et al. 1977):

1.   Every potential food item is described in terms of its net energy content and the time it takes to acquire. Thus,

$$\text{value of a food} = \frac{\text{energy content} - \text{energy involved in obtaining it}}{\text{time involved in obtaining t}} = \frac{e}{t}$$

2.   All potential foods in the animals' environment are ranked according to their value. The food with the highest value is the one having the highest energy content and requiring the least time to obtain (i.e., to pursue, handle, and eat); the food with the lowest value is the one having the lowest energy content and taking the most time to obtain.
3.   The optimal diet is determined by starting with the highest-ranked food and then adding food types to the diet in decreasing order of rank. Foods are added so long as the ratio of the net energy value e to the acquisition time t for each new item is greater than the net rate of food intake for the diet without the addition.

These steps, especially the third, are best explained by an example. Let us consider a simple example involving a hypothetical population of trout. Trout feed primarily on

Table 5.3   Example of calculating the optimal diet

Ranked food item	1 Energy content	2 Energy used in capture	3 Pursuit and handling time (t)	4 Value of food item (e^a/t)	5 Encounter rate (r)	Net rate of food intake for the diet $\Sigma (e \times r)$
			Spring			
Emerging mayflies	20	1	1	19.0	2.0	38.0
Nymphs	20	2	2	9.0	1.9	——
Minnows	100	50	7	7.1	0.5	——
			Summer			
Grasshoppers	100	10	5	18.0	0.05	4.5
Nymphs	20	2	2	9.0	0.1	6.3
Minnows	100	50	7	7.1	0.5	31.3

Note: Units of time and energy used in this example are arbitrary.
^{a} e = column 1 — column 2; t = column 3.

small fish and a wide range of insects in, on, and above the water. In early spring we find our trout population lurking just beneath the surface of pools formed by natural dams across streams. Mayflies emerge from the nymphal stage in huge numbers at this time of year, and many are caught in the surface film of streams as they struggle to shed their nymphal skins and dry their wings. As the trout wait, barely moving, trapped mayflies float by above their heads, yet-to-hatch nymphs swim about feeding, and the occasional minnow darts by. By summer, the mayflies have all emerged, mated, laid eggs, and died. Nymphs and minnows are still to be found, and a new source of food, drowned grasshoppers, float by above.

Which of these potential food items should the trout eat in each season and which, if any, should they ignore? From a trout's-eye view, the choicest morsels in springtime are mayflies. Although the energy content of minnows is much higher, these little fish move almost as fast as the trout themselves and take so much time and energy to catch that their value as a food is low. Mayflies, in contrast, cannot escape, and the trout has only to open its mouth and swallow as the current carries them by. Nymphs, which have the same energy content as the mayflies, have to be chased briefly, and their food value is between those of mayflies and minnows. Table 5.3 makes this point numerically (see columns 1 through 4).

Should the trout feed exclusively on mayflies the first-ranked item in springtime? The answer is yes. Mayflies are so abundant then that there is no reason to spend time and energy catching less valuable prey: why bother with a nymph or a minnow when there are endless mayflies for the taking? Indeed, the total amount of food the trout could eat in a given period would actually decline if they ate these other items. By summer, the range of foods available has changed. Grasshoppers now rank highest, followed by nymphs and minnows. Should the trout eat only grasshoppers, as they did mayflies? The answer is no. The ratio of net energy value to acquisition time (column 4 in Table 5.3) for nymphs and minnows is greater than the net rate of food intake for the diet without these items (column 6). Grasshoppers will be eaten whenever they float by, but rarely does this happen. A trout that waited for grasshoppers all day would rapidly be a very hungry trout. Eating nymphs and minnows increases its net rate of food intake.

At least one general point emerges from this illustration: the more common the food with the highest value, the fewer lower-ranked foods the animal needs in its diet. From this it also follows that the scarcer food is, the more diverse the diet will be. A number of studies of insects, fish, and birds support both these predictions (see Pyke, Pulliam, and Charnov 1977; Kamil and Sargent 1981).

How useful is the optimal theory of diet for predicting primate dietary patterns? Certain primates are known to increase the breadth of their diet at times of the year when or in areas where food is scarce (e.g., Richard 1977; Sussman 1977). Yet there are several reasons to question the theory's value in its present simple form when applying it to primates in particular and herbivores in general (Westoby 1974; Freeland and Janzen 1974; Milton 1979; Glander 1981; Pyke, Pulliam, and Charnov 1977).

First, the theory assigns a value to a food according to its net energy content and the handling time associated with feeding upon it. Yet we have seen that plant foods vary in composition, and most lack one or several of the nutrients essential to survival. Primates meet their nutritional requirements by eating foods of differing compositions so that they achieve an adequate *mix of nutrients* rather than simply maximizing their energy intake. Further, plant foods often contain toxins, and in most cases animals must either avoid these toxins completely or at least minimize their intake. Regardless of its energy content, therefore, a plant food that is high in toxins is likely to be of little or no value to a primate. In short, although the theory's ranking of foods may be reasonable for *animal* foods, which for the most part lack toxins and contain almost the whole complement of nutrients necessary for life, it is much too simple for *plant* foods.

Second, the theory measures the cost of acquiring food in terms of the time and energy taken to pursue and subdue the prey. It assumes that all foods are digested at the same rate. While appropriate for animal food, this approach is ill suited for plant foods that obviously do not need to be pursued or subdued but do sometimes present digestive problems. For instance, foods high in structural carbohydrates have to be processed slowly through the gut if their full nutritional value is to be extracted. The rate at which a herbivore can feed, therefore, is probably limited by the rate at which the food can be digested rather than by the rate at which it can be acquired (Westoby 1974). Thus, the value of plant foods should also include a measure of the speed with which they can be passed through the gut with maximum nutrient extraction.

Third, while growth causes changes in the size of animal foods, their nutritional composition stays essentially the same. In contrast, the composition of plant foods changes almost continuously: as young leaves mature, both their fiber and secondary compound content tends to increase; as fruits ripen, the concentration of simple sugars often increases while the level of secondary compounds drops.

These changes present a *tracking problem* for herbivores. Glander (1981) suggests that howler monkeys solve it by constantly sampling potential food items in their environment; he watched single animals gradually lengthen the time they spent eating a new food, and as they did so, others gradually joined that individual until all members of the social group were making use of the new food source. He hypothesizes that when the physiological effects of briefly feeding on a new food are negative, the sampling animal does not return and that food remains unused until sampled later by another animal. In sum, herbivores must continually monitor the status of potential food items, and it is improbable that they can acquire this information without actually sampling items. The assumption of the optimal theory of diet that animals eat only foods with the highest possible value is thus unlikely to hold good for herbivores.

Fourth, in the hypothetical case of the trout, we implicitly assumed that these fish had a fixed number of hours in the day to acquire food and that their goal was to eat as much as possible during that time. In other words, we assumed that they were *energy maximizers*. Some animals, however, may have fixed energy requirements, and their goal may be to meet those requirements in as little time as possible. They are called *time minimizers*. Under what circumstances do animals minimize the time they spend feeding? They may do so if they must expose themselves to predators in order to acquire food; under such conditions, they should try to meet their needs as quickly as possible and retreat to a safer area.

Are primates energy or nutrient maximizers or time minimizers? We do not know the answer, which probably differs from species to species and also from season to season. For example, male baboons spend little time eating when females are in estrus (Post 1982). Competition to mate with estrous females is intense (Hausfater 1975), and to monopolize a female the male has to follow her closely, continuously on the alert for advances from rivals, ever ready to ward them off. At this time, male baboons are probably time minimizers. The distinction between energy maximizers and time minimizers is recognized in the optimal theory of diet (e.g., Schoener 1971), but models and predictions in practice are based on the assumption that animals are energy maximizers (Pyke, Pulliam, and Charnov 1977).

Given these and other problems, is the theory of optimal diets any use at all in studying primates and other herbivores? I believe that it is, if only because it provides a starting point from which to elaborate more complex and appropriate models, a process that is already under way. The studies reported in this chapter that attempt to identify the criteria by which primates select their food are a first step toward establishing a more sophisticated understanding of the value of different foods. Altmann and Wagner (1978) have presented a general model specifying the optimal diet that *does* take into account the variable composition of primate foods. It predicts how much of each food available to an animal it should consume so as to obtain at least the minimum requirement of every nutrient and less than the maximum tolerable amount of toxins; at the same time, either some cost associated with the diet is minimized, such as the time spent foraging or exposure to predators while feeding, or else some benefit obtained from feeding is maximized, such as protein or energy intake. This model has yet to be tested, because there is no population for which enough information has been published about the animals' nutritional requirements and response to toxins on the one hand and the nutritional and toxin content of all potential food items on the other.

In conclusion, the current theory of optimal diets is based on several assumptions about the nature of food and how it is acquired that do not hold up for herbivores. Nevertheless, it is useful. By pointing out how the theory is inappropriate or inadequate, we simultaneously point out the kinds of information needed to deepen our understanding of the factors governing the food choices of herbivores in general and primates in particular.

CHAPTER SIX

Reproduction

6.1   Female Reproductive Cycle

For most of her adult life, a normal female primate is either pregnant or nursing young (Martin 1975). The other phase in her reproductive life (Figure 6.1) is the *sexual cycle,* which includes up to three closely synchronized processes (Rowell 1972a): (1) *ovulation,* the release of eggs from the ovaries; (2) *menstruation,* shedding the newly built-up lining of the uterus; and (3) *estrus,* a set of behavioral changes, specifically the intensification of behaviors associated with mating. Most female primates go into estrus around the time they ovulate, and menstruation follows a predictable number of days later. In some species, however, estrus and menstruation are hard to see or not detectable at all. Ovulation cannot be documented directly from behavior or external changes in morphology, and in the laboratory it is usually pinpointed by daily monitoring of hormonal levels. Captive females sometimes exhibit estrous behavior and menstruate without ovulating; it is not known how often this happens in the wild because ovulation has not yet been studied directly; indeed, it is hard to see how it could be studied without causing great disruption. In short, the sexual cycle is a particularly intractable subject for research under natural conditions. At several points in this chapter, therefore, the term sexual cycle has to be used loosely to refer to a cycle of behavioral and external morphological changes during which ovulation is presumed to occur.

Males of some species are known to have physiological and behavioral cycles related to reproduction (Van Horn 1980). These males produce sperm and show sexual behavior only in certain seasons. Many species have yet to be studied, however, and much remains to be learned about the nature of these cycles. For example, are male cycles directly precipitated by female cycles, or is their timing triggered by other factors? Natural selection will ensure that they are closely synchronized with the female cycle, even if not actually activated by it. Our primary interest here, however, is in the production of offspring, and in mammals the role of the female is more important in this process and her biology places more significant constraints upon the rate at which it can occur. The

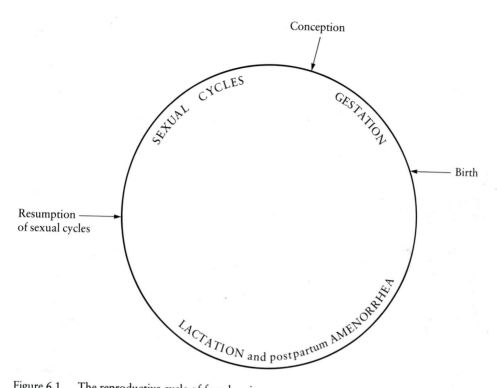

Figure 6.1   The reproductive cycle of female primates.

following discussion focuses upon the reproductive life of female primates, therefore, and we shall simply assume that males provide willing and able partners when called upon.

6.1A   Sexual Cycle

Ovulation

When primate females reach sexual maturity, their ovaries contain a limited number of immature eggs, or *oocytes;* menopause in human females represents the final emptying of the ovaries. Stocking of the ovaries begins soon after conception, and in almost all primates it stops soon after birth (Anand Kumar 1974). As oocytes are produced during embryonic development, they are quickly surrounded by a single layer of cells, known as the *follicle;* often several oocytes may be grouped together within one follicle. In time, the cells of the follicle proliferate to form many layers. Oocytes are used up in reproduction, and a female's reserves are also continually reduced by the decay of follicles

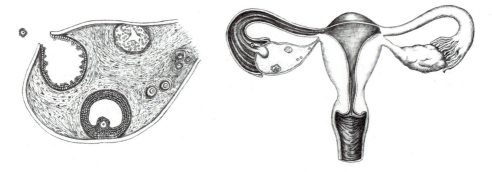

Figure 6.2   Front view of internal reproductive organs of female primates (here, *Homo sapiens*) (right) and sequential stages of follicle development (left).

and the oocytes they contain: traces of decayed follicles can be found in the ovaries of even immature females. In fact, the majority of follicles ultimately decay (Koering 1974).

During the preovulatory, or *follicular,* phase of the physiological cycle associated with ovulation (Figure 6.2), several follicles begin to enlarge. Follicles in both ovaries show this activity up to *ovulation,* the moment at which a follicle ruptures and releases a mature oocyte. In most primates, only a single follicle in one ovary ruptures in the end. The luteal phase follows ovulation. *Luteinization* of the ruptured follicle increases the size of its cells and transforms it into an organ called the *corpus luteum.* The corpus luteum produces hormones, notably progesterone, that prepare the uterus to receive the fertilized oocyte and are of critical importance to the early growth of the embryo (Ryan and Hopper 1974). It is unclear how long the presence of the corpus luteum is needed to maintain pregnancy. In humans, for example, we know that it is not essential to any specific function after 6 weeks (Koering 1974). If the oocyte is not fertilized, the corpus luteum degenerates, shrinking away until only a scar remains to mark its site. This process of degeneration apparently stimulates growth in several follicles, causing the whole cycle to start up again.

Ovulation is spontaneous in the primates, as in most medium to large mammalian herbivores and carnivores: it does not have to be triggered, or induced, by copulation. *Induced ovulation* occurs in a few rodents, lagomorphs, and small carnivores, species with short life spans that produce large litters and are subject to heavy predation (Conaway 1971; Martin 1975; Weir and Rowlands 1973). It permits a female to take advantage of favorable breeding conditions very rapidly provided a mate is available. *Spontaneous ovulation* does not allow such a quick response, and it occurs whether or not mating has taken place. However, species in which females ovulate spontaneously usually exhibit a well-developed social system, males and females interact regularly, and opportunities for fertilization are rarely missed.

Menstruation

The lining of the uterus undergoes a period of growth in the preovulatory phase of the sexual cycle. It subsequently degenerates and is sloughed off if fertilization does not take place. In some primates, this breakdown is accompanied by bleeding from the vulva, or *menstruation.* The strepsirhines do not show external traces of blood, and little or no blood is visible in New World monkeys (Hershkovitz 1977), gorillas, or orangutans (Graham 1981a). In many Old World monkeys and in the chimpanzee, however, the flow is readily apparent (Graham 1981a; Rowell 1972a). The degeneration of the uterine lining usually begins a fixed number of days after ovulation if the oocyte is not fertilized. This sequence of uterine growth and breakdown may take place in the absence of the cycle associated with ovulation, especially in seasonally breeding species.

Estrus

Estrus, often called "heat," refers to a short period during the sexual cycle when a female primate is willing to copulate and, indeed, may solicit copulation (Rowell 1972a). At that time, she is sexually receptive. Estrus in primates is usually, but not always, accompanied by ovulation. In other mammals the two events may be widely separated. For example, females of many temperate-zone bat species go into estrus and mate in the autumn, store sperm through the winter, and ovulate and conceive in the spring (Wimsatt 1960).

In practice, primate patterns of copulation are usually more complex than suggested by a simple dichotomy between the states of estrus and a long period when the female is not receptive. For example, captive and wild chimpanzees and gorillas have been observed copulating during the early months of pregnancy, and captive and wild chimpanzees are receptive throughout the period of sexual swelling, although the frequency of copulation peaks around the time of ovulation (Graham 1981b). It has been suggested that in some New World monkeys there is not even a weak peak in the frequency of copulation during the sexual cycle (see Subsection 6.2B).

6.1B   Variability in the Sexual Cycle

Several features of the female sexual cycle vary within and between species. These variations have important consequences for the rate at which females reproduce. In particular, the time it takes a female to resume her sexual cycle after giving birth is a major determinant of the interbirth interval. Additional influences are the number of times a female cycles before conceiving, gestation length, and, to a minor degree, the length of the sexual cycle itself. Many female primates experience a *postpartum amenorrhea,* a period after giving birth when they do not cycle. The reason for this interval is that the burden on the mother of meeting all an infant's nutritional needs is usually too great for her to carry a fetus at the same time. Immediate determinants of the interval are not well understood. Postpartum amenorrhea is associated with high levels of prolactin, a hormone produced by lactating females (Short 1976; Nadler et al. 1981), and when an infant dies and its mother stops lactating, she quickly begins to cycle again. The association is not a simple causal relationship, however: even when the infant survives, the mother resumes her sexual cycle before it is fully weaned, and at least in captivity, females of some species cycle and conceive just a week or so after giving birth, long before the onset of weaning.

In many animals, including some primates, the sexual cycle occurs only in a particular season (Delany and Happold 1979; Reiter and Follett 1980). In practice, seasonal acyclicity may often be difficult to distinguish from postpartum amenorrhea, but the two have quite different implications for the interbirth interval. For example, while females exhibiting postpartum amenorrhea begin to cycle again shortly after the death of an infant, females that are seasonally acyclic usually do not resume their cycle before the next mating season. Different environmental cues trigger the onset of sexual cycles. In some strepsirhines, the immediate cue is a change in *day length* (Petter-Rousseaux 1974; Van Horn 1975). It has been argued that macaque breeding seasons are also regulated by photoperiod, although there is no experimental evidence yet to support this contention (Van Horn 1980). Seasonal variation in the food supply may be important in some species through its effect on the composition of body tissue. There is evidence suggesting that human females stop cycling when their body-fat-to-lean ratio falls too low (Frisch 1982), and work is currently in progress to examine this relationship in nonhuman primates (Burns 1984).

Food directly affects reproduction in certain mammals, in which ovulation is triggered by the consumption of hormonelike substances called *phytoestrogens,* which are seasonally present in certain plants (Labov 1977). Phytoestrogens have not yet been looked for in primate diets. In the long run, it is the timing of gestation and lactation, not the sexual cycle as such, that is important for reproductive success. Natural selection will ensure that the sexual cycle is timed to provide the young with the greatest chance of surviving the successive phases of prenatal and postnatal development.

Many primates are *seasonally polyestrous.* Although conception occurs only in particular seasons, females cycle again if they do not conceive at once. All lemurs are seasonally polyestrous, even in the rain forest where climatic variations are not strong (Martin 1972, 1975). Among other primates, seasonality in the sexual cycle is *facultative.* In other words, it varies within species depending on local conditions. For example, in the Himalayan foothills rhesus monkeys mate in September and October before the onset of winter and give birth in April or May. Further south in India, where the climate is more equable, the season of conceptions extends from September to February and births from March to July (Lindburg 1971). At another site in India, Southwick, Beg, and Siddiqi (1965) found one birth peak between March and May and another, smaller one in September and October.

Variations in the length of the sexual cycle probably contribute little to interbirth interval. Except for the New World primates, differences within and between species in cycle length are not great, and the few data we have suggest that females usually cycle only once or a few times before conceiving. This means that the cumulative effect of minor variations in cycle length is likely to be small. The cycle length is between 30 and 55 days in strepsirhines. Among tarsiers the mean length is 24 days. In Old World monkeys it ranges from 21 to 40 days, and in apes from 29 to 37 days (Butler 1974). In New World monkeys the cycle tends to be much shorter, with a range of 11 to 27 days (Hershkovitz 1977; Eisenberg 1978; Glander 1980). Within species, females may have consistently different cycle lengths, and a particular female may also experience cycles of variable length. Most of the variation takes place in the follicular or preovulatory phase (Rowell 1972a).

Rowell (1970) investigated factors influencing cycle length in female baboons (*Papio cynocephalus anubis*), and her findings provide a tantalizing glimpse of the complexity and multiplicity of factors affecting reproduction patterns. Her specific goal was to find out whether changes in social context affected cycle length. Results showed that, under laboratory conditions, adult females removed from a social situation had a significantly longer follicular phase immediately after isolation than before it, although the average length of cycles during isolation was not significantly different from that before. After being returned to the social group, six out of seven females again had a lengthened cycle immediately thereafter; cycles then shortened gradually over the next few months to their original length. The main point to emerge from this experiment is that it was apparently social *change*, rather than social context as such, that temporarily influenced cycle length. The mechanisms involved remain unclear.

The sparse evidence concerning variations in the number of times a female cycles before conceiving suggests that, regardless of species, females generally conceive shortly after resuming their sexual cycle. For example, Tutin and McGinnis (1981) found that seven chimpanzee females required an average of 3 to 4 cycles, with a range of 1 to 11 cycles. Harcourt, Stewart, and Fossey (1981) recount that three out of four gorillas conceived during the first cycle after their postpartum amenorrhea, while the fourth probably took 4 cycles. Finally, J. Altmann and associates (1977) report that baboon females become pregnant again during the fourth cycle on average and rarely take more than 9 cycles.

In summary, some aspects of variation in the female sexual cycle contribute importantly to the interbirth interval and thus to the overall pattern of reproduction in a given individual or species. In particular, the time it takes for a female to resume her cycle after giving birth is highly variable within as well as between species. A second important source of interspecific variation in the interbirth interval derives from differences in the number of months out of the year in which a female can cycle. In highly seasonal environments, females may cycle and conceive in only one or two months. While the immediate factors triggering the onset of the sexual cycle may vary, in all cases the cycle is presumably timed to maximize the offspring's chances of surviving.

6.1C   Gestation

Differences in gestation times and fetal growth rates were long believed to result simply from differences in placentation (Kihlstrom 1972). The living eutherian mammals exhibit a range of placental types, with the variation cutting across orders and, in some instances, across infraordinal groups as well (Luckett 1974, 1975; Hershkovitz 1977). Today we know that gestation length is determined not by any single factor but rather by a complex of variables. As we saw in Section 1.4, other things being equal, gestation time increases with increasing body and relative brain size. Within a given class of body

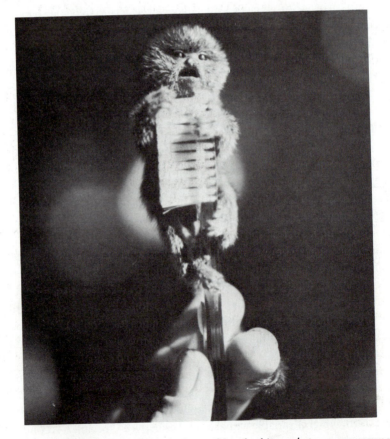

Figure 6.3   Scarcely bigger than the head of a toothbrush, this newborn pygmy marmoset (*Cebuella pygmaea*) is alert and able to cling. (Courtesy of M. de Solni.)

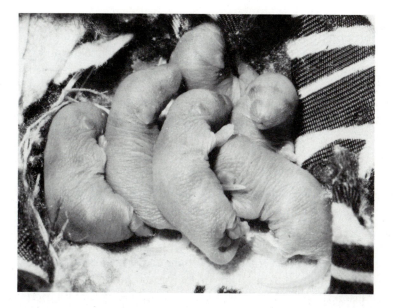

Figure 6.4   The eyes of two-day-old mice *(Mus musculus)* have not yet opened, they have no hair, and their movements are poorly coordinated. (Oxford Scientific Films/Animals Animals.)

and brain size, fetal growth rates are lower and gestation times longer in species with lower relative metabolic rates, and growth rates may be influenced by placental structure (Eisenberg 1981). Finally, some differences in gestation length are attributable to differences in the developmental state of the young at birth. A distinction can be drawn between mammals that give birth to *precocial* infants and those that give birth to *altricial* infants. Each of these types is characterized by a set of closely related features (Martin 1975; Portmann 1965). Precocial infants are usually born in small litters at an advanced stage of development: their brain is relatively large, they are mobile at or shortly after birth with eyes and ears open, and they have at least a moderate covering of hair (Figure 6.3). Their parents generally have medium to large bodies, high brain-to-body-size ratios, and females have a long gestation period and do not construct nests for their young. Altricial infants, in contrast, are usually born in large litters at an early stage of development and are completely helpless for some time afterwards (Figure 6.4). Their parents generally have small bodies and brains, and females have a short gestation period and do build nests for their newborns. The notably short gestation period and long period of postnatal development of carnivores is attributable in part to this distinction between the altricial and precocial complexes: carnivores show many features of the altricial type, while primates and artiodactyls are typically precocial.

Although primates are generally considered precocial, elements of the altricial and perhaps more primitive pattern are seen in some strepsirhines (Martin 1975). For example, while most infant primates are carried from birth, clinging to the fur of their mother or some other adult, a number of strepsirhines construct leaf nests for their infants, and others use tree hollows or bundles of epiphytes as safe shelters. Animals exhibiting this behavior include members of the families Cheirogaleidae and Galaginae, and also the aye-aye, *Daubentonia madagascariensis,* and the variegated lemur, *Varecia variegata.*

6.1D   Lactation

All mammals nurse their young. Energetically costly to the mother, nursing is critical for the survival and growth of her offspring, providing not only nutrients but also maternal antibodies as well. These antibodies confer on the infant passive immunity to diseases during its early development (Buss 1971).

The length of time a mother nurses her young and the constituents of maternal milk vary considerably among species. Large mammals tend to nurse their offspring longer than small mammals do, but they wean them at a much smaller body size relative to adult size than do small mammals, some of which wean their young at a weight close to that of subadults (Eisenberg 1981). The fat, protein, and carbohydrate content of milk varies among mammals (Figure 6.5). The protein content is generally highest in those with rapid postnatal growth rates. The constituents of milk presumably represent a compromise between what is in the best interests of the infant and what the mother is able to provide.

Fine distinctions among or within species in the duration of lactation or in the importance of milk at different developmental stages are hard to make under natural conditions. Time on the nipple is only a rough measure of milk intake, because young primates often hold the nipple in their mouths, apparently for comfort, without actually sucking. Furthermore, the composition of milk changes during a female's lactational cycle (Jenness 1974) so that at different times an infant may derive different nutritional benefits from nursing. Finally, although young primates begin sampling plant food at an early age, it is often unclear how much of it is swallowed and dietarily important, and how much is chewed and then spat out.

6.2   Case Studies

In this section we compare patterns of reproduction among primates differing in size and phylogenetic history. Since physiological and behavioral aspects of reproduction are well documented for few primates, our choice is limited; even for the five species discussed here, the information is patchy. Note also that these species are not necessar-

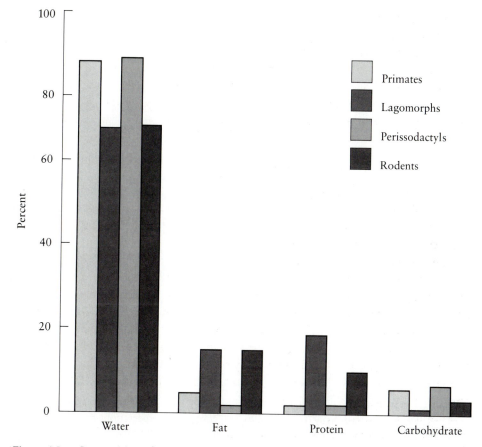

Figure 6.5   Composition of various mammalian milks. (Data from Shaul 1962.)

ily representative of the taxonomic groups to which they belong; for example, not all small New World monkeys resemble the tamarin in reproductive pattern, nor are all medium to large Old World monkeys like the baboon. Our primary purpose here is to illustrate the diversity within the order and to look at how the general reproductive trends just discussed are manifested in individual species in their natural habitats.

6.2A   Demidoff's Bush Baby and the Senegalese Bush Baby (*Galago* spp.)

Demidoff's bush baby *(Galago demidovii)* and the Senegalese bush baby *(G. senega-lensis)* (Figure 6.6) are strepsirhines in the loris family (Lorisidae; see Table 1.1). Demi-doff's bush baby lives in the African rain forest; the Senegalese bush baby is found to the

Figure 6.6   The Senegalese bush baby *(Galago senegalensis)* is more than twice the size of its congener, Demidoff's bush baby *(G. demidovii)* yet, contrary to expectations based on this size difference, the Senegalese bush baby has a much higher reproductive rate.

north, east, and south of these forests in drier bush, woodland, or gallery forest. Demidoff's bush baby has been studied in the laboratory and in the forests of Gabon by Charles-Dominique (1971, 1974, 1977; Charles-Dominique and Martin 1972) and in Congo Brazzaville by Vincent (1968, 1969). The Senegalese bush baby is known from laboratory research and fieldwork in dry, thorny forests of the northern Transvaal (Doyle, Anderson, and Bearder 1971; Doyle, Pelletier, and Bekker 1967; Bearder and Doyle 1974). Both species are small; the average weight of Demidoff's bush baby is 60 g, and of the Senegalese bush baby 165 g (Charles-Dominique 1977; Napier and Napier 1967; Martin and Bearder 1979). In neither is there marked sexual dimorphism. Although the life history characteristics of these species differ in predictable ways from those of larger primates, there are differences between them that are not predictable from their weights.

In the laboratory, Demidoff's bush babies give birth for the first time at about 1 year of age. Charles-Dominique (1977) has kept captive individuals in good condition and reproductively active to at least the age of 6 years. Butler (1974) gives a figure of $10\frac{1}{2}$ years for longevity but does not cite a source. Longevity will only be well established when known individuals have been watched in the wild or bred and maintained in captivity for many years.

The sexual cycle lasts 38 days on average (Vincent 1969) and includes a period of about 3 days when the vagina is open and mating occurs. Cyclic opening and closing of the vulva, the entrance to the vagina, is typical of a number of strepsirhines (Butler 1974). In immature, gestating, and sexually inactive animals, as well as in active animals during most of their sexual cycle, the vaginal opening is minute or may be closed off completely. In Demidoff's bush baby, opening takes 12 to 24 hours and is accompanied by swelling and redness of the vulva. Gestation lasts from 111 to 114 days.

The young show some altricial features: they are covered with a sparse coat of hair, the eyes are partially closed, and only the hands have a developed grasping capacity. (Most primates use both their hands and their feet to cling tightly to their mother's fur as soon as they are born.) For the first few days the infant is kept in a leaf nest. The mother begins to carry it away from the nest when she becomes active at dusk, parking it in the vegetation, where it stays immobile while she forages and feeds. Offspring are weaned by about their 45th day.

Females give birth throughout the year in the laboratory. Periodically, a captive female exhibits a *postpartum estrus* and conceives just a few days after she has given birth, but in the wild there is no evidence that females reproduce continuously. Although births occur throughout the year, there are seasonal breeding peaks, and females apparently give birth only once a year. Examining 85 wild females caught at different times of the year, Charles-Dominique (1977) found that 11 were lactating and 31 were pregnant. None were simultaneously lactating and pregnant. Periods of anestrus are difficult to document in a small, nocturnal primate that visually signals its reproductive state briefly and not very noticeably, and the determinants of its interbirth interval are unclear. The presence of breeding peaks shows that sexual activity is somewhat synchronized among females (as of course it is between males and females), probably corresponding to seasonal changes in the environment.

It is not known how many times on average a female cycles before she conceives. Dissections of 18 pregnant females by Charles-Dominique (1977) showed that twinning is rare: only 4 animals had two embryos, while 14 had a single embryo. Age at first reproduction has not been reported for animals in the wild. Longevity in the wild is about six years at most, with many animals dying before they reach the age of four.

In the Senegalese bush baby, females first reproduce at about the age of one year in the laboratory. A maximum longevity in captivity of nine years has been reported for a female who was finally "sacrificed" (Butler 1974). During the sexual cycle, which lasts about 30 days, the vagina is open for 5 days. Gestation lasts approximately 120 days and, as in Demidoff's bush baby, the young are altricial in some features and kept in a leaf nest for a few days after birth. They are weaned 60 to 80 days later.

Most Senegalese bush babies live in strongly seasonal environments and births are clumped, probably to coincide with periods when food is most abundant. In the Transvaal, for example, all births occur during the rainy seasons between late October and early November and between late January and early March. In contrast to the annual breeding cycle of Demidoff's bush baby, however, this pattern represents *two* successive gestation periods for each female. This pattern is consistent with reports of postpartum estrus and conception in captive animals (Butler 1960) and results in two interbirth intervals—a long one during winter in the Southern Hemisphere and a very short one between the beginning and end of the southern summer. Duration of the short interbirth interval is determined by the length of gestation and the number of days that the female takes to achieve a fertile mating after *parturition* (giving birth). The long interbirth interval comprises a postpartum amenorrhea, the winter months during which females are acyclic, and an unknown number of cycles before conception takes place. This pattern of birth spacing, coupled with the observation that most females give birth to twins or, more rarely, triplets, suggests a reproductive rate substantially greater than that of Demidoff's bush baby.

In summary, these two strepsirhines are similar in several respects: The young are born at a relatively early stage of development from 111 to 120 days after conception; females give birth for the first time at about the age of one year; and in captivity animals can live for 10 years or more. There is also a striking difference between them: the Senegalese bush baby's reproductive rate is much higher than that of its *congener* due to its larger litter size and shorter average interbirth interval. (Congeners are species belonging to the same genus.) The Senegalese bush baby is more than twice as big as Demidoff's bush baby, and its higher reproductive rate is the opposite of what we would predict on the basis of this size difference. This contradiction will be discussed in Section 7.5.

6.2B   Cotton-Top Tamarin *(Saguinus oedipus)*

Tamarins (*Saguinus* spp.) comprise a group of nine species (Hershkovitz 1977) in the Callitrichidae, one of the three families of New World monkeys. They are among the smallest of the higher primates, adults of both sexes weighing scarcely more than 250 g (Figure 6.7). In the past few years, biomedical researchers have become increasingly interested in the diminutive callitrichids, and this has led to a spate of laboratory studies of their reproductive biology and also to the initiation of research on their behavior in the wild (see Kleiman 1978a). Those callitrichids that have been studied in the laboratory (primarily *S. oedipus, S. fuscicollis, Callithrix jacchus,* and *Leontopithecus rosalia*)

Figure 6.7 Despite superficial similarities between tamarins (*Saguinus* sp., left) and red squirrels (*Sciureus* sp., right), both the way in which they move about the forest and the way in which they exploit it are very different.

share many behavioral and biological traits distinguishing them from other New World monkeys as well as from the primates of the Old World. The following account focuses on the cotton-top tamarin *(S. oedipus),* but laboratory data from other species are included when none are available for the cotton-top tamarin.

In the laboratory, female tamarins are sexually active for at least six years, and their reproductive pattern is characterized by a short sexual cycle, twinning, rapid maturation of the young, and a postpartum estrus (Eisenberg 1978). Together these features enable females to achieve a reproductive rate as high as the Senegalese bush baby's, and under favorable conditions in captivity they fulfill this potential (M. S. Cebul, pers. com. 1979).

Laboratory-housed females reach sexual maturity at about 18 months. Like other New World primates, they show no sign of menstruation or marked behavioral changes indicative of estrus (Hampton and Hampton 1978; Hearn and Lunn 1975), although more subtle behavioral cues may be involved (Kleiman 1978b). The cycle associated with ovulation is much shorter than that of any strepsirhine or Old World primate, but not unusual in New World species (Eisenberg 1978; Glander 1980); it lasts about 15 days and ranges from 14 to 18 days (Hampton and Hampton 1978; Preslock, Hampton, and Hampton 1973). At ovulation, an oocyte is usually released from both ovaries, and after the 140-day gestation period, females habitually give birth to twins.

The young are not born at a particularly early stage of development, and the double birth presents the mother with a major physical burden. The combined weight of the neonates usually exceeds 10% of her body weight and can be as high as 28%. This is far above the values for most of the Cebidae, whose young at birth weigh less than 10% of their mother's weight and sometimes as little as 5% (Kleiman 1977). Yet tamarins do not build nests, and infants must be carried as well as nursed throughout the early months of life.

In captivity, successfully breeding tamarins are usually housed in groups containing one male, one female, and their immature offspring, and the father and older offspring carry the infants most of the time when the infants are not being nursed by the mother (Cebul 1980). In this way, the burden of infant care is shared. While the older, usually female offspring of many primates assist their mothers, the extensive paternal care exhibited by adult male tamarins is rare outside the Callitrichidae. A recent study has shown that marmoset (Callithrix jacchus) males helping raise their young exhibit distinctive hormone levels (Dixson and George 1982). This behavioral and endocrinological association has yet to be investigated in caretakers other than the father.

There are no published data on the interbirth interval in the cotton-top tamarin, but mating and conception are common in other callitrichids within the first week after giving birth, and many callitrichid females are continuous breeders in the laboratory (Hearn 1978; Rothe 1978). However, some tamarins show seasonal breeding peaks even under laboratory conditions, where changes in the environment are minimal. For example, in the Northern Hemisphere the lion tamarin (Figure 6.8) (Leontopithecus rosalia) usually gives birth between March and August (Kleiman 1978b), and there is similar evidence of seasonality in S. fuscicollis (Gengozian, Batson, and Smith 1978).

The caretaking role of fathers may be important for the survival of offspring. In this respect the social system, which provides the father with access to the infants, enhances the reproductive rate of females. In some contexts, however, social factors may be indirectly responsible for reducing the reproductive rate. In captivity, only the female with the highest social rank gives birth when more than one adult female is housed with a male, even if these other females copulate with the male. Social interactions seem to regulate endocrinological function, rendering subordinate females anovulatory

Figure 6.8   A lion tamarin *(Leontopithecus rosalia)* threatens the photographer. (Courtesy of A. Young.)

as long as they are in social proximity to the female of highest rank (Epple 1970). Subordinate *Callithrix jacchus* females likewise stop cycling in the presence of a dominant female, but if each subordinate is removed from the group and placed alone with an adult male, they start cycling again and may conceive shortly thereafter (Hearn 1978).

Field data suggest that tamarins in the wild do not always and perhaps do not usually achieve the high reproductive rate reported for captive animals. During a field study in Panama, Dawson (1978) found that cotton-top tamarin females bred seasonally, with births peaking between November and February. Females usually gave birth to twins but rarely did both survive. Neyman (1978) reported similar findings for cotton-top tamarins in Colombia. He noted that at a distance juveniles were indistinguishable from adults after about 10 months, suggesting that the surviving young mature rapidly. Males were often seen carrying infants at both study sites, and the pattern of births indicated that, in the wild as in captivity, the highest ranking female inhibited sexual activity among other mature females in the group. Only a single infant or pair of infants was found in any social group at one time, even when other adult females were present.

At both study sites, each group contained one reproductively active pair. In contrast, data from an ongoing study of tamarins (*S. fuscicollis* and *S. imperator*) in the Manu National Park, Peru, show that in these species social groups sometimes contain more than one reproductively active female (Terborgh 1983; Terborgh and Wilson in press). Emerging disparities between the pictures of tamarin social organization painted by field and laboratory studies will be discussed in Subsection 8.2B.

In summary, female tamarins can achieve a high reproductive rate in the laboratory through rapid maturation, a short interbirth interval, and habitual twinning. They are assisted in the care of infants by males and older offspring. There are not enough data to determine whether and how often they reach these high rates in the wild. The evidence so far suggests that wild animals mature rapidly and produce twins but that only one twin usually survives; the interbirth interval is probably longer, on average, than in captivity, because breeding is seasonal. In some circumstances, subordinate females apparently experience anovulatory cycles in the wild as in captivity.

6.2C   Savanna Baboons (*Papio cynocephalus*)

The common or savanna baboon (*Papio cynocephalus*) is found over most of sub-Saharan Africa. Some classifications divide savanna baboons into separate species (e.g., Napier and Napier 1967), but we follow Thorington and Groves (1970) and consider them as geographic races or subspecies of a single species (Figure 6.9). Long an important subject in biomedical research, the baboon has been well studied in the laboratory (e.g., Kriewaldt and Hendrickx 1968; Hendrickx and Kraemer 1971), and studies at two sites in Kenya, namely Gilgil and the Amboseli Game Park, provide field data on the life history patterns of olive (subspecies *anubis*) and yellow (subspecies *cynocephalus*) baboons, respectively (for Gilgil, see Strum and Western 1982; for Amboseli, see Altmann 1980; Altmann, Altmann, and Hausfater 1978; Altmann et al. 1977).

Baboons are strongly sexually dimorphic with respect to body size (Napier and Napier 1967), several other morphological features (Crook 1972), and also dietary habits (Post, Hausfater, and McCuskey 1980). At about 13 kg, females weigh approximately half as much as adult males, and the maturation time of the two sexes reflects this difference: males take twice as long to reach sexual maturity (full development of their secondary sexual characteristics). At Amboseli, females reach sexual maturity and begin to cycle at the age of about five years and become pregnant for the first time a year later. (Many primates experience a period of "adolescent sterility" upon reaching sexual maturity.) Unlike the preceding case studies, where field and laboratory data fit together quite well, captive female baboons reach sexual maturity almost two years before wild baboons (Gilbert and Gillman 1960). The maturation rate is influenced by resource abundance (Sadleir 1969a), and the discrepancy between the data from Amboseli and the laboratory probably reflects the richer food supply in the latter environment. Findings at Gilgil support this proposition. There, age at first reproduction increased from 5.7 years to 7.7 years as the estimated supply of food declined. This decline was probably due to increased competition with other mammals (Strum and Western 1982).

Unlike either bush babies or the tamarin, adult female baboons exhibit clearly visible signs of estrus and menstruation (Hausfater 1975; Rowell 1969). Both events appear to be closely synchronized with ovulation, and both are marked by changes in the *sexual skin*. Zuckerman, Wagenan, and Gardiner (1938) coined this term to describe naked

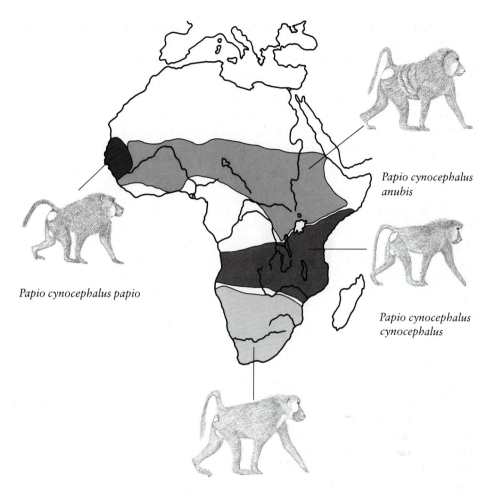

Papio cynocephalus anubis

Papio cynocephalus papio

Papio cynocephalus cynocephalus

Papio cynocephalus ursinus

Figure 6.9   The geographic races or subspecies of the savanna baboon, *Papio cynocephalus*.

areas of skin on the rumps of adult females of some species of Old World primates that respond by swelling or heightened color to injections of the hormone estrogen. No strepsirhines or New World monkeys have a sexual skin. Among the Old World species that do, the skin differs in anatomical origin and in the relation of its swelling to menstruation and ovulation. It is unlikely, therefore, that the sexual skin is *homologous* in all these species, that is, derived from a common ancestor with this characteristic (Rowell 1972a). Although the sexual skin presumably conveys information about the female's reproductive state to other animals, it is not the only or perhaps even the major signal. In the laboratory, male baboons usually react sexually to females before they have seen the

Figure 6.10   Close-up view of an adult female baboon's *(Papio c. cynocephalus)* sexual swelling during the peak of estrus. (Courtesy of C. Saunders.)

swelling, and they do not respond with sexual behavior to females who are fully swollen but do not show typical estrous behavior.

The different races, or subspecies, of baboons show minor variations in the size and color changes of the sexual skin. The following description is of yellow baboon females, although it broadly applies to the other subspecies as well. The sexual skin is confined to the *perineum,* the region between the anus and vulva, and it forms a heart-shaped swelling shortly after menstruation that gradually increases in size until ovulation occurs (Figure 6.10). Thereafter the swelling subsides until the end of the sexual cycle. Naked skin also covers the hips, but it does not undergo swelling during the sexual cycle. However, it does change color from black to red during pregnancy. After the female has given birth, it fades to pink and then slowly to black again during the months that she is nursing her infant. Rowell (1972a) notes that the changes in the appearance of these lateral patches of skin are much more striking than those of the sexual skin except at close range; in other words, the female baboon seems to signal her pregnancy more clearly than her receptivity.

The sexual cycle of the baboon is completed in about 33 days (Hendrickx and Kraemer 1971). Gestation takes about six months, ranging from 154 to 183 days (Napier and Napier 1967). Only single births have been seen. The interbirth interval at Amboseli between 1971 and 1974 averaged 22 months but varied widely. Much of this variation could be accounted for by differences in infant survival. After the birth of an infant, the mother underwent a period of postpartum amenorrhea that ended rapidly if the infant died or was stillborn. On average, females began to cycle again within three and a half weeks, with a range of one and a half to five weeks, after this happened. There did not appear to be a relationship between the age of the infant at death and the time it took the mother to resume cycling, but the sample size may have been too small to detect one. If the infant survived, postpartum amenorrhea lasted much longer, about 12 months, although the range was wide, from 6 to 16 months.

The interbirth interval at Gilgil varied from 19 to 28 months. Because most infants at Gilgil survived at least the first 6 months of life, little of this variation could be attributed to differences in infant survivorship. Two other factors played an important role in generating the observed variation. First, ecological conditions at Gilgil changed over the 10 years (1971 to 1981) when data for Strum and Western's analysis were collected, during which competition for food increased and its abundance decreased. The apparent effect of this change on age at first reproduction has already been noted. There was also a strong correlation between measures of resource abundance and the interbirth interval: as resources became scarcer, the interbirth interval became longer. The second factor influencing the interbirth interval at Gilgil was a female's age. The interval was greatest in young females at the onset of reproduction, decreased sharply from the age of about 8, and rose again as females approached the age of 16. Those estimated to be older than 20 years still had a slightly shorter interbirth interval than those in the youngest adult age class.

Note that baboons live for many years, and even these long-term studies have yet to span enough time to provide reliable estimates of longevity in the wild. In captivity, baboons live for 28 years or so (Napier and Napier 1967). The pattern of variation with age in the interbirth interval and reproductive rate that Strum and Western report from Gilgil is similar to that found in many mammals (Pianka 1974a; Caughley 1977), probably including most primates. Unfortunately, though, the data needed to establish age-specific patterns are rarely available for primates.

J. Altmann and her associates (1977) monitored the effects on interbirth interval of the time spent cycling by a female before she became pregnant. In a sample of 32 intervals, they found that the interval ranged from 1 to more than 18 months. The five longest intervals each included a skipped cycle, suggesting that conception had taken place but had been followed by an abortion. None of the other intervals lasted more than $8\frac{1}{2}$ months, and on average, females became pregnant only 4 months after cycling was resumed.

In Subsection 6.1B we noted that seasonality of sexual cycling can also influence interbirth interval. Although there is a slight birth peak among baboons in Kenya (Altmann and Altmann 1970; Hausfater 1975; DeVore and Hall 1965), the effect of cyclical changes in the environment on particular phases in the cycle of reproduction has yet to be analyzed in detail.

In summary, just as we would predict, baboons take longer to mature, live longer, and have a longer gestation period and interbirth interval than bush babies or tamarins. The extent and causes of variability in certain aspects of baboon life histories are probably known as well as or better than they are for most other primates. For example, maturation is demonstrably sensitive to long-term changes in food supply. Among the factors influencing the interbirth interval are survival of the previous offspring, overall food availability, and the age of the mother. A low interbirth interval carries different biological implications in different contexts, accordingly. It can mean that a female's infants are not surviving, so that she is giving birth frequently but contributing little or nothing to subsequent generations. Alternatively, her infants may be surviving and an abundance of food permits a shortening of the period of postpartum amenorrhea. Finally, it may be a consequence of the female being between 8 and 16 years of age, when, other things being equal, the interbirth interval is reduced compared with that of younger and older females. In short, there are at least three possible causes of variability in the interbirth interval, and although each has different long-term implications, they may all result in an immediate increase—or decrease—in the interval. The moral of the story, thus, is that the biological significance of changes in life history parameters must be interpreted with great caution and with as much contextual information as possible.

6.2D   Chimpanzee *(Pan troglodytes)*

Common chimpanzees are classed in the superfamily Hominoidea, which comprises the Great Apes of Africa and Asia, the Lesser Apes of Southeast Asia, and ourselves (Andrews and Cronin 1982). Like other apes, chimpanzees live at low population densities and have a restricted distribution in African rain forests north and east of the Zaire River and in more open country to the south and southeast. They have been studied under relatively undisturbed conditions at a number of locations (e.g., Itani and Suzuki 1967; Izawa 1970; Kano 1971; Nishida 1968, 1979; Reynolds and Reynolds 1965; Sugiyama 1968, 1969; Suzuki 1969) and are best known from research done at Gombe Stream National Park in Tanzania. In this mosaic environment of grassland, woodland, and rain forest, animals have been provisioned to varying degrees and studied more or less continuously since 1960 (e.g., Goodall 1965, 1968; Goodall et al. 1979; Bygott 1979; Pusey 1979; Teleki 1973; Teleki, Hunt, and Pfifferling 1976; Tutin 1979; Tutin and McGinnis 1981). From Gombe Stream comes the most complete information about the life history pattern of wild chimpanzees, covering a period of almost 20 years. The following overview is derived from syntheses of these data (Teleki, Hunt, and Pfifferling 1976; Tutin and McGinnis 1981). The reproductive biology of chimpanzees in captivity has been the subject of many detailed studies (e.g., Yerkes and Elder 1936; Young and Yerkes 1943; Graham 1970, 1981a, 1981b).

Figure 6.11   A female chimpanzee *(Pan troglodytes)* with a six-month-old infant eats while an adult male touches her in a friendly gesture. (N. Nicolson/Anthro-Photo.)

Chimpanzees are among the largest primates, by far the largest considered in this chapter. At Gombe Stream adult males weigh about 39 kg, and females 30 kg (Wrangham and Smuts 1980). Both sexes are characterized by a long period of maturation and a long life span. In the wild, females begin to cycle sexually between the ages of 11 and 13 years. The interval between the onset of the sexual cycle and first known conception was 13, 28, 30, and 34 months for four females at Gombe. The physiological causes of this interval, known as *adolescent sterility* in humans, are not known. Combining the range of variation for the onset of the sexual cycle with the range for the period of adolescent sterility, and including a 224-day (eight-month) gestation period, age at first reproduction among wild females is potentially as low as $12\frac{1}{2}$ years and as high as $16\frac{1}{2}$. Tutin and McGinnis (1981) suggest that a typical female gives birth for the first time between the ages of 13 and 15 (Figure 6.11). In captivity, females begin to cycle at a younger age ($8\frac{1}{2}$ years on average) and give birth for the first time between 10 and 11 years (Young and Yerkes 1943).

Sexual maturity is more difficult to define in males. Goodall (1965) reports that males reach sexual maturity at 9 or 10 years, but do not show full development of secondary sexual characteristics and probably do not contribute significantly to reproductive activities for another 3 to 4 years. Tutin and McGinnis (1981) note that the only male chimpanzee of known age at Gombe to reach puberty was first observed ejaculating when he was just over 9 years old. In captivity, males reach sexual maturity 1 to 3 years earlier (Smith, Butler, and Pace 1975).

Females typically continue to reproduce until they die, probably at the age of 40 to 45. One of the oldest females in captivity was still cycling at the age of 48 (Graham 1981a), but the frequency of sexual cycles declines in captive females over the age of 35, as do conception frequency and the frequency with which fetuses are carried to term. These findings recall the decline in reproductive rate with advanced age reported from Gilgil (Subsection 6.2C). However, a low reproductive rate has not been reported among young chimpanzees in captivity, and there is little evidence of age-specific differences in reproductive rate in the wild. (This may be because the sample size is still too small and the age estimates of animals still too uncertain.)

The female sexual cycle lasts about 35 days, terminating with a 2- to 4-day menstrual flow. Alone among the Hominoidea, the chimpanzee has a sexual skin. While other apes show only the usual mammalian vulval enlargement when receptive, the chimpanzee female's perineal skin changes color and swells dramatically during the sexual cycle, reaching a volume as great as 1400 ml at its peak. This takes place at about the midpoint in the cycle and lasts about 10 days before collapsing in a period of 48 hours. Maximum perineal swelling is associated with estrous behavior (Yerkes and Elder 1936), but since the period of maximum swelling is long and copulation occurs at other times, estrus is not clearly defined, as it is in some primates (Graham 1981a). Sexual swelling can usually be observed in young females one to three years before the onset of the sexual cycle.

Only single births have been observed in the wild among chimpanzees, although twins are occasionally born in captivity. For about three years after parturition, wild females with surviving offspring do not show any perineal swelling. Thereafter, the sexual cycle is resumed, but females rarely conceive until the third or fourth cycle (with a range between 1 and 11). Although a mother usually begins to restrict the infant's access to the nipple during its second year, weaning is not completed until the infant is five years old, by which time the mother is usually pregnant again. Average interbirth interval between two consecutive offspring when the older one survives is five and a half to six years (with a range of four years and five months to seven years and seven months). If the first offspring dies within this period, the mother conceives again within 4 to 8 months and gives birth to another infant 12 to 16 months after the death of the first.

In combination, these features of the female chimpanzee's life history pattern add up to an extremely low reproductive rate. Even her long reproductive life span, estimated to be about 25 to 32 years, does not make up for this. The total number of offspring to which a female usually gives birth is probably no more than 5 or 6. In captivity, the young are usually taken away from the mother some time during the first year of life, and she stops lactating and quickly resumes her sexual cycle. As a result, a captive female may give birth to as many as 12 offspring in her lifetime (Graham 1981a), a reproductive rate that could not be approached by females under natural conditions.

The length of the various phases in the chimpanzee's life varies considerably, as it does among baboons. This variation is found not only between animals in captivity and in the wild but also among wild animals and among captive animals. The causes of this variation are generally unknown, although the survival of offspring clearly plays an important role in determining the length of the interbirth interval; there is also some evidence that the interbirth interval may increase in old females. The importance of fluctuations in the food supply can only be inferred from studies of other primates.

6.3   Commentary

In Section 1.4 we looked at major life history trends among primates and other mammals, and in this chapter we have examined more closely the reproductive patterns of primates. What emerges from this examination is a picture of variation within as well as among species. Much of the variation among species is in the direction predicted by differences in body and relative brain size: at one end of the scale, we have chimpanzees giving birth to five or six offspring in a lifetime; at the other, Senegalese bush babies giving birth to four young in a single year. The causes of variation within species are not well understood, but several factors, including food supply, survival of the previous offspring, the age of the mother, and even changes in her social environment are involved.

Whatever the causes of variation, its consequences for a female's overall pattern of reproduction and her contribution to the next generation can be great. For instance, taking (1) the lowest age at which chimpanzee females give birth for the first time in the wild, (2) the shortest interbirth interval recorded in the wild, and (3) the longest estimated life span, the number of offspring to which a female could in principle give birth is as high as seven. Taking (1) the highest age at which females are known to give birth for the first time, (2) the longest interbirth interval, and (3) the shortest estimated life span, the number of offspring drops to three. If we used values for chimpanzees in captivity, the range of variation would be even greater. The point is not that such wide variations are likely to occur often but rather that the possibility exists. Under what circumstances will natural selection favor animals that produce many offspring over animals that produce few?

A growing body of theory attempts to predict these circumstances and thus to explain the adaptive significance of differences in life history patterns among species. The scope of the theory is great, covering not only individual life history patterns but also the aggregate of those patterns, population dynamics. In this chapter we have focused on patterns of reproduction, just one aspect of individual life histories. Before we can evaluate life history theory in relation to the primates, we need data on patterns of mortality and, more generally, on patterns of population growth and decline. These data are presented in Chapter 7, and at the end of that chapter we consider life history theory and its application to the primates.

PART III

PRIMATES IN POPULATIONS

Introduction

 IN Part II we considered the attributes of individuals. We turn now to the attributes of populations made up of those individuals. Do primate populations fluctuate in size? If so, how great are the fluctuations and how often do they happen? How do changes in the birth rate, the death rate, and the migration rate contribute to fluctuations? How does a population's composition by age and sex affect birth and death rates? How does the environment influence those rates?

These are the kinds of questions posed by demographers (Caughley 1977) and the kinds of questions we shall try to answer in Chapter 7. Demography is the study of population statistics and the changes in those statistics over time.* We shall refer to such changes as the dynamics of populations. In Chapter 6 we considered individual life history patterns. A population's dynamics are the sum of its members' life histories (Cole 1954). Life history theory has been developed to explain the evolution of individual life histories and of emergent patterns at the level of the population. Chapter 7 evaluates some of the ideas central to this theory in light of the data on primates.

Demography is important. Insight into population dynamics is critical to the study of the relationship of animals to their environment and to other species in that environment (Chapter 10). Moreover, demography plays or should play a key role in the study of social behavior (see Section 9.4).

---

* Strictly speaking, demography is the study of human populations but it is commonly applied to animal populations too, and is used here in this broader sense.

CHAPTER SEVEN

Demography

7.1 Introduction

WHAT is a population? How long does it take to characterize its dynamics? There are no simple answers to either of these questions. Concerning the first, for the time being let us simply describe a population as a spatially distinct subset of a species. Concerning the second, let us note Slobodkin's (1961) suggestion that a population should be studied for a period 10 times the length of one generation. Even if this is an overestimate, it means that describing a single population of large mammals could take up a researcher's entire working life and perhaps involve some posthumous labor as well.

Take the chimpanzee *(Pan troglodytes),* for example, which has an estimated mean generation length of 19.6 years (Teleki, Hunt, and Pfifferling 1976). *(Generation length* is calculated as the elapsed time between the birth of a female and the birth of her median offspring.) A study of just five generations would take 98 years. It is scarcely surprising, therefore, that the demography of most long-lived mammals is poorly documented even by standards less rigorous than Slobodkin's. Primates, among the longest-lived of all, are no exception.

Section 7.2 introduces concepts and methods used in demography, together with some of the associated problems. For a full account of these topics, the reader is referred to Caughley's (1977) excellent discussion of demographic theory and practice and also to the handbook on field methods published by the National Research Council (1981). Section 7.3 summarizes data for 10 primate populations, including 4 that were and, in 3 cases, still are artificially provisioned and 6 that are provisioned minimally or not at all (Tables 7.1 and 7.2). For none of these populations are there comprehensive data, and for some the data are minimal. The latter are included because they contribute in some particular way to our understanding of primate population dynamics. In Section 7.4 we review what is known about some of the environmental factors precipitating the changes in population parameters indicated in Section 7.3. Finally, Section 7.5 relates the material presented in the preceding sections to life history theory.

7.2   Concepts, Methods, and Problems

7.2A   Population

Defining the term *population* is difficult because it can refer to several different entities. Some authors use it synonymously with *species*. Others use it to refer to subsets of a species that are partially or completely isolated from one another by geographic barriers. Ecologists often equate the term with the animals they are studying. In the following discussion, *biological population* will be used to denote the second of these three usages, and *study population* to denote the third.

Figure 7.1 shows four areas of different size containing four study populations drawn from a single biological population. When an animal disappears from study pop-

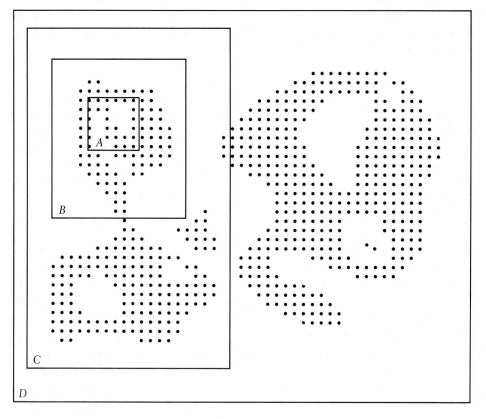

Figure 7.1   Boundaries of study populations *(A–C)* and the biological population *(D)*. (Adapted from Caughley 1977)

ulation *A, B,* or *C,* it may not be possible to tell whether it has left the study area or died, since cadavers are quickly dismembered by scavengers and rarely found by field researchers. The problem of distinguishing between deaths and departures decreases as the size of the study area increases. The larger it is, the greater its area relative to its perimeter, and the lower the percentage of animals in the study population likely to cross this perimeter (Caughley 1977). Most primate study populations are very small, and despite human efforts with axes and rifles, many of them are drawn from widely distributed biological populations. Distinguishing deaths from departures is correspondingly difficult. Movement *into* the study area is easily detected so long as all members of the study population are marked or individually recognized.

A small study population presents a second problem when the goal is to determine broad demographic patterns. The smaller it is, the greater the impact of random sampling events. While a change in the sex ratio of a social group may have important social consequences (Section 9.4), it is unlikely to have much effect on the sex ratio of the larger biological population to which the group belongs. But if the study population consists of just that social group and perhaps one or two others, the change will appear to be an important demographic event.

7.2B   Population Parameters

Having delineated our population as best we can, we may now describe it. There are many attributes, or parameters, that could form part of the description, but only a few of them are used commonly (Caughley 1977). Fertility by age, or *age-specific fertility,* is the mean number of female live births per female over a given interval of age. Some researchers use the term *fecundity* instead of *fertility.* Fecundity is the number of female offspring to which a female is *biologically capable* of giving birth over a given interval of age. Field workers record fertility, not fecundity, and fecundity may be higher than fertility (Section 9.4). Thus, the terms should not be used interchangeably.

Survival by age, or *age-specific survival,* is the proportion of animals alive at a particular age that survive to the next age class. The *frequency distribution of animals by age* refers to the proportion of animals in different age classes in the population. *Sex ratio* is usually measured as the proportion of females to males in a population. A sex ratio of 0.5 means that for every 100 males in the population there are 50 females. The sex ratio may be measured for the whole population or for animals at different ages. *Numbers* or *density* are both measures of population size. The latter is an estimate of population size in relation to the area in which the population lives.

Once the values of these parameters have been estimated, a secondary series of statistics can be derived. These include fertility rate, birth rate, mortality rate, and rate of increase for the population as a whole. The *fertility rate* is the average number of female offspring to which an adult female gives birth in a year; the *birth rate* is the average number of offspring, regardless of sex, to which she gives birth in a year. The *mortality rate* is the proportion of the population that dies in one year. The rate of population increase is discussed in Subsection 7.2F.

Several factors acting singly or together may change the fertility or mortality rate. For example, the fertility rate may go up because (1) more food is available and females are producing more offspring, (2) females produce a disproportionately high number of female offspring in a given year (fluctuations due to such random sampling events are most likely to affect small populations), or (3) the proportion of females in each age class in the population has changed; if young females reproduce more rapidly than old ones, for instance, an increase in the number of young compared with old females will result in an increase in average fertility for the population as a whole. While difficult to interpret, average birth and fertility rates are commonly used measures for primate populations because female ages are rarely known.

A population's size and structure are changed by *immigration* and *emigration* as well as by births and deaths. Here these terms refer to movement into and out of the study population and not, unless otherwise stated, to the frequency with which animals enter or leave the biological population. In all mammals, members of one or both sexes leave the vicinity of their mother as they reach maturity. In social species, animals often join a neighboring group after leaving their *natal group*, the group into which they were born. When the study population comprises more than one social group, individuals may *circulate* among these groups. An animal that moves to a group outside the study population is said to have emigrated. In the opposite direction, it has immigrated.

If a group splits and one of the resulting new groups enters or leaves the area occupied by the study population, this event is also scored as emigration or immigration by however many individuals constitute the group. Immigration and emigration in this limited sense have different demographic, genetic, and ecological implications from migration into and out of the biological population as a whole. It is important that the distinction be kept clear. The biological boundaries of primate populations are rarely established, and movement across them has not been well studied.

7.2C   Aging Animals

In the best of all possible worlds, all members of a population would be marked at or shortly after birth so their ages would be unequivocally known throughout their lives. In practice, however, only in provisioned populations has this been done. In other studies of wild populations, animals have been aged according to a combination of criteria based on physical appearance. When this method is used, the possibility of mistaking ill health for old age is obvious, but some animals do not even have to be sick to look old. Moreover, the bigger looking of two juveniles does not always turn out to be bigger when weighed and measured, and even when it is, it is not necessarily the older.

The eruption sequence of teeth is generally considered the most reliable index of age in young animals, but even this criterion is not infallible. We realized this in our research on rhesus macaques in Pakistan. First, animals were assigned approximate ages according to their size and general appearance. They were then trapped and examined. Certain animals classified as old turned out to have virtually unworn teeth, indicating that they were in fact young adults; none of these old-looking individuals were sick as far as we could tell. Again, even though all offspring were born during a six-week period in the year, by the time individuals of known birth dates reached the age of two years, the largest of them were as big as or bigger than the smallest of the three-year-olds. Moreover, the timing of the eruption of teeth varied enough among individuals for some two-year-olds to be dentally indistinguishable from three-year-olds (Melnick 1981; J. Phillips-Conroy and D. Melnick, pers. com., 1981).

The real issue is not whether errors in aging occur, since they undoubtedly do, but rather their importance. Caughley (1977) argues that errors do not balance each other out and can lead to unwarranted conclusions. Assume, for example, that a population has only two age classes containing 100 and 10 individuals, respectively. If 10% of each class is wrongly aged and placed in the other, the apparent frequencies become 91 and 19. The ratios 100:10 and 91:19 differ sufficiently to distort population statistics that require an estimate of the population's composition by age. The bias is not reduced if the number of age classes is increased.

7.2D   Age Distributions

The age distribution of a population refers to the proportion of animals in each age class established by the observer. If age-specific fertility and survival rates remain constant, the proportions of individuals belonging to different age classes will remain the same, generation after generation (Wilson and Bossert 1971). For different sets of rates, the proportions themselves will be different. In nature there are good years and bad years, and fertility and survival rates vary so that most populations probably never achieve a *stable age distribution*. Note that a population with a stable age distribution may be increasing, decreasing, or unchanging in size. The important point is that the pattern of survival and fertility does not change and so the proportion, although not necessarily the absolute number, of individuals in each age class remains the same.

Calculating the stable age distribution of a population answers the following question: if the fertility and survival rates remain at their current values, what proportion of individuals will eventually be in each age class in the population? A stationary age distribution, which is a special case of the stable age distribution (Caughley 1966, 1977), occurs when fertility and survival rates are balanced so that the population is neither increasing nor decreasing and the actual number as well as proportion of individuals in each age class stays the same.

One of the reasons for making these distinctions will become clear when we consider the construction of life tables and their applications by some researchers. The other is to emphasize that the shape of an age distribution does not automatically indicate whether a population is growing, declining, or stable. Caughley (1977:123) makes the point bluntly: "The diagnostic features of age distributions of declining, stationary, and increasing populations must be determined separately for each species or group of species. ... Without prior knowledge of this kind, interpretation of an age distribution is an exercise in clairvoyance."

7.2E   Life Tables

The probabilities of dying by (or surviving to) different ages are usually organized in the form of a life table; a life table for females may also show the probabilities of giving birth at different ages. Conventionally, a life table represents mortality rates as if they were progressively depleting a cohort of animals born into a population at the same time. Since mortality rates often differ for males and females, life tables are usually calculated for each sex.

Data on death and survival may be presented in four different forms in a life table:

1. the probability at birth of surviving to age x, conventionally denoted as l_x;
2. the probability of dying during the age interval x to $x + 1$, conventionally denoted as d_x;
3. the mortality rate, or the proportion of animals alive at age x that die before age $x + 1$, conventionally denoted as q_x; and
4. the survival rate, or the proportion of animals alive at age x that survive to age $x + 1$, conventionally denoted as p_x.

In practice, cohort life tables are rare. Researchers lack the resources to monitor the lives of a large number of individuals all born at the same time, and they rely instead on techniques requiring less massive data sets (Caughley 1977). Two such techniques concern us. One involves calculating the proportion of animals of a given age that die before reaching the next age class. The procedure is repeated for all age classes and provides estimates of q_x. The other method involves collecting data at a single point in time on the number of animals alive in different age classes. Age-specific fertility and survival rates are then calculated directly from this distribution. Life tables drawn up in this second way are valid only when the rate of increase in the population is already known and the age distribution is stable. In practice, the method is usually applied only to populations whose rate of increase has been close to zero for some time—in other words, to populations that are presumed to have a stationary age distribution. Since populations with stable age distributions, let alone stationary ones, are probably infrequent in nature, the method is rarely likely to be appropriate. Moreover, a population with a zero growth rate does not necessarily exhibit a stationary age distribution. In short, in most instances the preferred method of calculating a life table relies on the determination of q_x for all cohorts.

7.2F Population Growth Rates

A population's growth rate can be estimated in a variety of ways, of which we shall briefly consider four. The *intrinsic rate of increase,* symbolized r_{max}, is the rate at which a population grows when it has a stable age distribution and unlimited resources (Wilson and Bossert 1971; Caughley 1977). As the maximum rate at which a population could increase under optimal conditions, r_{max} is more often a measure of the potential than the actual growth rate. It is calculated from age-specific schedules of mortality and fertility. The *survival-fertility rate of increase,* r_s, is the rate at which a population would grow if it had a stable age distribution appropriate to its life table at the time of measurement. Unlike r_{max}, r_s can be measured when resources are limited.

The *observed rate of increase,* symbolized $\bar{r}$, is the rate at which a population changes over a period of time. Its measurement does not require a stable age distribution, a constant rate of increase over this period, or unlimited resources. When the value of $\bar{r}$ is positive, the population is growing; when negative, the population is decreasing. Only general interpretations of this sort are possible, however, because populations may have the same $\bar{r}$ for different reasons. For example, a positive $\bar{r}$ in two populations each containing 100 individuals can be achieved in several ways. In one, a large number of females may produce a small number of offspring, whereas in the other, a small number of females may produce a large number of offspring. A population with high fertility and high mortality may have the same $\bar{r}$ as another with low fertility associated with low mortality, and so on.

A fourth measure of population growth is the *net reproductive rate (R_o)*. R_o is the increase per individual per generation, and it is usually expressed as the average number of female offspring each female will have contributed to the population by the time she dies. Like r_{max} and r_s, it is calculated from age-specific fertility and mortality schedules, and like these two measures of growth, it assumes a stable age distribution. When a population is growing, R_o will be greater than one; when it is declining, R_o will be less than one.

7.2G How Long Is Long Enough?

Slobodkin (1961) suggested that the study of 10 generations should be enough to characterize a population's dynamics. In the same discussion, he distinguished between evolutionary time, measured for species in millions or fractions of millions of years, and physiological time, measured for individuals in seconds, minutes, hours, or days. Be-

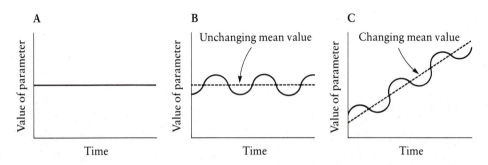

Figure 7.2   Schematic illustration of the difference between a population parameter that is static (**A**), at a stable equilibrium (**B**), and unstable (i.e., undergoing directional change) (**C**).

tween these extremes is ecological time, during which it can be assumed that the ecological world and its constituent populations are in a steady state. In other words, during ecological time the number of animals in a species or region remains about the same. At one end of the spectrum ecological steady states are disrupted by the evolution of species. At the other end minor perturbations may produce marked but brief changes in the state of individuals. According to Slobodkin, study of a population over a period of about 10 times its generation length should be short enough to avoid having to take evolutionary change into account and long enough to smooth out the effects of short-term perturbations.

While Slobodkin's argument makes a certain amount of intuitive sense, I cannot agree with him. He seems to assume not only that ecological systems are at equilibrium but also that they fluctuate little. (The parameters of a population at equilibrium may fluctuate, but they do so about a mean that does not change. Only in a static population do parameters not fluctuate [Figure 7.2].) Neither of these assumptions is well supported, and where data are available, they suggest that the second in particular is dubious (Wiens 1977; Western and Van Praet 1973; Bormann and Likens 1979). Generation length is thus not the only parameter to be considered in designing demographic research. Records of natural cycles in the environment are also needed. At Amboseli in Kenya, for example, the height of the water table changes in an 80-year cycle (Western and Van Praet 1973). When it rises, it has the effect of increasing the salinity of the environment. High salinity, compounded by elephant damage, significantly changes the distribution and composition of vegetation. There is less grass and thicket cover and the number of fever trees *(Acacia xanthophloea)* declines dramatically. Perhaps 90% of the fever trees in the Amboseli basin living before 1950 have died (Western and Sindiyo 1972) (Figure 7.3). Changes like these affect the resources available to herbivores. Resource availability, in turn, influences the fertility and mortality rates of herbivores and, indirectly, of the carnivores that prey upon them. A 10-generation study of the Ambo-

Figure 7.3   Dead and dying trees litter the landscape as baboons *(Papio c. cynocephalus)* set off on their morning progression at Amboseli National Park, Kenya. (Courtesy of C. Saunders.)

seli baboons would characterize their population dynamics only if those 10 generations spanned at least 80 years.

In sum, in the best of all possible worlds—and in an equilibrial world at that—a population should be studied for a period equal to or greater than the periodicity of cycles of critical resources. Lacking evidence concerning such cycles, the pragmatic approach is to study the population as long as possible (see also Anderson, Turner, and Taylor 1979 for an extended discussion of this issue).

7.3   Case Studies

7.3A   Provisioned Populations

Cayo Santiago Rhesus Macaques *(Macaca mulatta)*
A small colony of rhesus macaques was established in 1938 on the island of Cayo Santiago (Figure 7.4) off the southeastern coast of Puerto Rico, but effective monitoring of the population began only in 1956, when the 17-ha island contained about 200 monkeys (Altmann 1962; Carpenter 1972). Although the island is still covered by vegetation, animals depend for food on prepared monkey chow supplied at three feeding stations.

Figure 7.4   Cayo Santiago Island. Puerto Rico is in the background. (Photography by A. Richard.)

Table 7.1 Demographic studies of selected primate populations

Study Site	Species	Years of observation[a]	References
	Provisioned Populations		
Cayo Santiago, Puerto Rico	*Macaca mulatta*	26	Hausfater 1972; Sade et al. 1976; Altmann 1962; Koford 1965, 1966
Chhatari, Aligarh District, India	*Macaca mulatta*	20	Southwick and Siddiqi 1976, 1977; Southwick 1980
Takasakiyama, Japan	*Macaca fuscata*	29	Masui et al. 1975; Koyama, Norikoshi, and Mano 1975; Itani 1975
Gombe Stream, Tanzania	*Pan troglodytes*	19	Teleki, Hunt, and Pfifferling 1976; Goodall 1983
	Unprovisioned Populations		
Dunga Gali, Pakistan	*Macaca mulatta*	4	Melnick 1981
Polonnaruwa, Sri Lanka	*Macaca sinica*	14	Dittus 1975, 1977a, 1977b, 1979
Amboseli National Park, Kenya	*Cercopithecus aethiops*	19	Struhsaker 1967a, 1967b, 1973, 1976
Sankaber, Ethiopia	*Theropithecus gelada*	4	Dunbar 1980; Dunbar and Dunbar 1975
Barro Colorado Island, Panama	*Alouatta palliata*	50	Carpenter 1934, 1962; Collias and Southwick 1952; Altmann 1959; Chivers 1969; Mittermeier 1973; Smith 1977; Froelich, Thorington, and Otis 1981; Otis, Froelich, and Thorington 1981
Berenty, Madagascar	*Propithecus verreauxi*	20	Jolly 1966a, 1972b; Jolly, Gustafson, Oliver, and O'Connor 1982b; Richard 1978a

[a] Number is approximate, and observation may have been continuous or intermittent. Descriptions in the text are sometimes based on data covering shorter periods.

Table 7.2   Demographic parameters of populations listed in Table 7.1

Study site	Species	Study population	Biological population	Approximate size of study population	Average birth rate	Observed growth rate ($\bar{r}$)	Mortality during first year of life[a]	Adult sex ratio
				Provisioned populations				
Cayo Santiago	*Macaca mulatta*	Total island population	Total island population	1975: 347	1973–74: 0.542	1973–74: 0.0705	1973–74: 0.176[b]	1965: 1.88
Chhatari	*Macaca mulatta*	2 social groups	Larger than study population, undefined	1962: 49 / 1975: 112 / 1978: 149	$\bar{x}$ = 0.907 Range = 0.727–1.00 N = 15 years	1962–75: 0.064	$\bar{x}$ = 0.154 Range = 0.0–0.444 N = 15 years	$\bar{x}$ = 2.7 Range = 1.8–3.6
Takasakiyama	*Macaca fuscata*	3 social groups	Larger than study population, undefined	1953: 180 / 1953: 220 / 1973: 1535	$\bar{x}$ = 0.58 Range = 0.40–0.88 N = 16 years	1950–73: 0.093	$\bar{x}$ = 0.131[b] N = 15 years	$\bar{x}$ = 2.5 Range = 2.0–2.7
Gombe	*Pan troglodytes*	1 community	Larger than study population, bounded by park boarder	1960: 60 / 1972: 38 / 1980: 57	$\bar{x}$ = 0.221 Range = 0.06–0.44 N = 15 years	1965–80: –0.003 / 1972–80: 0.055	$\bar{x}$ = 0.273 N = 15 years	$\bar{x}$ = 1.55 Range = 0.6–2.8 N = 16 years
				Unprovisioned populations				
Dunga Gali	*Macaca mulatta*	7 social groups	Larger than study population, undefined	1979: 291	1978–79: 0.38	1978–79: –0.059[c]	1978: 0.460	1979: 2.0
Polonnaruwa	*Macaca sinica*	18 social groups	Study population not completely isolated, but migration restricted	1971: 446 / 1975: 380	$\bar{x}$ = 0.688	[d]	$\bar{x}$ = 0.461	1971: 2.3

Study site	Species	Study population	Biological population	Approximate size of study population	Average birth rate	Observed growth rate ($\bar{r}$)	Mortality during first year of life[a]	Adult sex ratio
					Unprovisioned populations (continued)			
Amboseli	Cercopithecus aethiops	5 social groups	Larger than study population, undefined	1963: 113 1971: 85 1975: 65	1963–64: 0.72 1975: 0.82	1963–75: −0.047	Unavailable	1963–64: 2.1 1971: 1.2 1975: 1.7
Sankaber	Theropithecus gelada	1 band	6 bands; contact possible with population on adjacent ridge but movement not documented	1972: 289	$\bar{x} = 0.373$[c] $N = 7$ years	1972–75: −0.013, 0.137[f]	$\bar{x} = 0.045$	1972: 2.72 1975: 2.40
Barro Colorado Island	Alouatta palliata	Total island population	Total island population documented	239–1116[g]	$\bar{x} = 0.364$ Range = 0.26–0.51 $N = 7$ years	1932–51: −0.027 1932–67: 0.029 1951–67: 0.096 1951–59: 0.153	d	$\bar{x} = 2.54$ Range = 1.8–3.8 $N = 7$ years
Berenty	Propithecus verreauxi	9 or 10 social groups	Larger than study population, almost completely bounded by sisal fields	42–58[h]	$\bar{x} = 0.43$ Range = 0.38–0.5 $N = 4$ years		1963–64: 0.555 1970–71: 0.428	$\bar{x} = 0.87$ Range = 0.49–1.23 $N = 5$ years

Note: [a] Some of these figures are for females only, some for both sexes combined.
[b] Females only.
[c] Figure for only three social groups.
[d] See text.
[e] See Figure 7.14.
[f] First figure for Main band, second for Abyss.
[g] See Table 7.4.
[h] See Table 7.5.

Table 7.3 Life tables for selected primate populations

Macaca mulatta (Cayo Santiago)			Macaca fuscata (Takasakiyama)		Pan troglodytes (Gombe Stream)			Theropithecus gelada (Sankaber)		
Age	Survival (Females)	Annual birth rate	Age	Survival (Females)	Age	Survival Males	Females	Age	Survival	Annual birth rate
0– 1	1.000		0– 1	1.000	0– 1	1.00	1.00	0– 1	1.000	
1– 2	.824		1– 2	.840	1– 2			1– 2	.955	
2– 3	.804		2– 3	.806	2– 3	.75	.71	2– 3	.924	
3– 4	.788	.128	3– 4	.790	3– 4			3– 4	.905	
4– 5	.671	.516	4– 5	.537	4– 5			4– 5	.878	.474, .327[a]
5– 6	.628	.750	5– 6	.398	5– 6			5– 6	.878	
6– 7	.573	.800	6– 7	.398	6– 7	.62	.61	6– 7	.878	.263
7– 8	.499	.800	7– 8	.346	7– 8			7– 8.5	.878	
8– 9	.499	.667	8– 9	.294	8– 9			8.5–10		
9–10	.499	.624	9–10		9–10			10–11	.878	.368
10–11	.405	.924	10–11	.211	10–11			11–12		
11–12	.343	.500	11–12		11–12	.52	.57	12–13	.580, .635[b]	.111
12–13	.343	.667	12–13		12–13			13–14+		
13–14	.286	.508	13–14	.205	13–14					
14–15	.286	0	14–15		14–15					
15–16	.286	0	15–16		15–16	.48	.53			
16–17	.190	1.00[c]	16–17	.205	16–17					

Table 7.3   Life tables for selected primate populations (continued)

Macaca mulatta (Cayo Santiago)			Macaca fuscata (Takasakiyama)		Pan troglodytes (Gombe Stream)			Theropithecus gelada (Sankaber)		
Age	Survival (Females)	Annual birth rate	Age	Survival (Females)	Age	Survival Males	Survival Females	Age	Survival	Annual birth rate
17–18	.190	0	17–18	.205	17–18					
18–19	.143	0	18–19		18–19	.48	.53			
19–20	.071	0	19–20	.096	19–20					
20–21	.071	1.00ᶜ	20–21		20–21					
21–22	.071	0	21–22		21–22					
22–23	.000	0	22–23	.035	22–23	.36	.47			
			23–24		23–24					
			24–25		24–25					
			25–26	.009	25–26					
			26–27		26–27	.21	.47			
			27+	.000	27–34					
					34+	.12	.09			

Note: Survival is to the beginning of the specified interval, that is, l_x.
ᵃ Second figure includes nine postpuberty juvenile females.
ᵇ First figure is for males, the second for females.
ᶜ Most females are dead by this age. Those few surviving in the 16–17 and 20–21 age classes all happened to give birth in 1973–74.

The population has increased rapidly since monitoring began, and the average crude annual growth rate has been 13% to 16% (Koford 1966). Despite repeated culling, today it numbers close to 1000 animals. Sade and his colleages (1976) calculated age-specific survival and birth rates among females in 1973–74 (Table 7.3). From these they estimated r_{max}* and the stable age distribution (Figure 7.5). The similarity between the actual and the stable age distribution indicates that the life table of this population has remained stable for several years.

Females on Cayo Santiago produce their first offspring between the ages of 3 and 4 (Figure 7.6). The birth rate peaks between 6 and 11 years and declines rapidly thereafter.

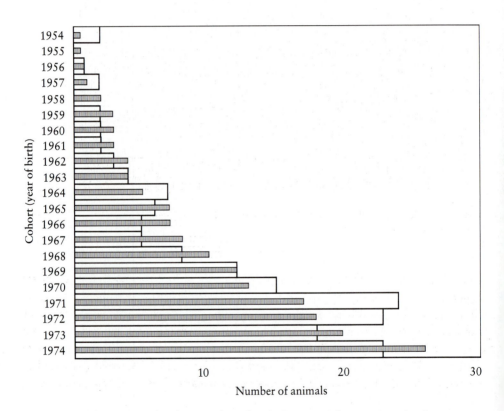

Figure 7.5   Observed age distribution of 164 female *Macaca mulatta* on Cayo Santiago on January 1, 1975 (open bars), compared with the theoretical stable age distribution of females (dotted bars) computed from the 1973–74 life table. (Adapted from Sade et al. 1976.)

---

* Their estimate is probably better referred to as r_i ; although food is available in unlimited quantities on Cayo Santiago, it is distributed in hoppers to which access may be limited.

Figure 7.6   Young rhesus mother *(Macaca mulatta)* on Cayo Santiago Island yawns while her infant nurses. (Courtesy of D. Sade.)

In 1973–74 almost no females gave birth after the age of 14. Mortality among infants is high relative to all except old adults; about 20% die before the age of $2\frac{1}{2}$ years. Sex-related differences in mortality emerge among juveniles; male mortality outstrips female between the fourth and fifth year, reaching a peak at 6 years and leveling off thereafter as animals reach adulthood (Koford 1966). From the age of about 10 years, mortality rates for both sexes increase again, and most (i.e., 93% in 1973–74) females are dead by the age of 20. Males generally die younger than females (S. R. Schulman, pers. com., 1982).

Chhatari Rhesus Macaques *(Macaca mulatta)*

A provisioned population of rhesus macaques has been studied at Chhatari-do-Raha in the Aligarh District of northern India. This population forms part of a larger biological population in Aligarh District. When the study population was censused for the first time in 1959, it numbered 49 animals living in two social groups. Since 1962, complete

Figure 7.7   Rhesus monkeys *(Macaca mulatta)* mingle with people on a road in Aligarh District, northern India. (Courtesy of C. Southwick.)

censuses made three times a year have shown a steady increase in the number of animals in the study population, while the size of the biological population has declined sharply (Southwick and Siddiqi 1976, 1977; Southwick 1980). It is unclear to what extent the growth of the study population can be attributed to immigration from areas where animals are not protected.

Members of the study population occupy a rural schoolyard and grove of trees. The schoolyard borders a crossroads where buses stop and is otherwise surrounded by farmland. The monkeys are fed nuts, wheat, rice, and cooked food by people at the bus stop and steal crops from the fields, although raids are usually cut short by irate farmers. Thanks to the school watchmen, the Chhatari population has been partially protected from trapping, unlike most populations in India until recently. In short, provisioning and protection are less systematic than on Cayo Santiago, but nonetheless these monkeys' lives mesh tightly with those of the local people (Figure 7.7).

In a period of 15 years, the Chhatari rhesus macaques have more than doubled their number to 115 animals, representing a crude average annual growth rate of about 6%. Southwick and Siddiqi (1977) speculate that if there were no trapping, the growth rate would approach that on Cayo Santiago. Between 1961 and 1973, the average birth rate was 0.907, ranging between 0.727 and 1.00. In other words, in some years every adult female gave birth, and on average 9 out of every 10 did so.

Figure 7.8   Japanese macaques *(Macaca fuscata)* crowd together at a provisioning site at Takasakiyama. (Courtesy of H. Simon.)

Mortality rates among adults and juveniles cannot be estimated with confidence, for despite the efforts of the watchmen, trapping probably has a significant effect on the population, particularly the juvenile sector. In contrast, infants were never trapped, and their disappearances could be counted as deaths. On average, about 16% died during the first year of life, but like the birth rate, the mortality rate fluctuated widely from year to year (see Table 7.2).

Although the adult sex ratio among the Chhatari rhesus varied from 1.8 to 3.8 between 1961 and 1975, there is no general trend over the years. In contrast, the adult sex ratio among unprotected animals in Aligarh District declined steadily from a high of 1.9 in 1961 to a low of 1.0 in 1975. The reasons for the more rapid decline of females than males in unprotected groups is unclear.

Takasakiyama Japanese macaques *(Macaca fuscata)*
Observation of the Japanese macaques at Takasakiyama began in 1953 with a single group containing about 180 individuals. Since then animals have been copiously provisioned, and although not confined to a small island like Cayo Santiago, these Japanese macaques come to the feeding ground almost every day (Figure 7.8).

Like the Cayo Santiago and Chhatari macaques, the Takasakiyama macaque population has grown steadily (see Table 7.2), increasing its size three and a half times over 10 years; the average crude annual growth rate was 10.2% over the 10-year period (Itani 1975). As the population has grown, the original group has split three times. In each case, between 70 and 100 animals splintered off from the parent group when it reached a membership of six or seven hundred. As of September 1973, the population consisted of about 1535 animals and was growing linearly. In contrast to geometric growth, linear growth indicates that mortality rates are increasing or fertility rates are decreasing while the population size increases. Masui et al. (1975) present age-specific mortality rates but no fertility schedule. Using data from Itani et al. (1963), Dittus (1975) calculated the average birth rate to be around 0.723.

The pattern of mortality among Japanese macaques (see Table 7.3) is similar to that among the rhesus macaques on Cayo Santiago. About 16% of those born die before the age of 2, and juvenile male mortality outstrips that of females and peaks during the late juvenile stage (Masui et al. 1975). Masui and his colleagues report that most monkeys die before they are 20 years old.

Gombe Stream Chimpanzees (Pan troglodytes)

Research on the social behavior and ecology of chimpanzees began in the Gombe National Park, Tanzania, in 1960, and demographic data for the population have been brought together by Teleki, Hunt, and Pfifferling (1976) and Goodall (1983). On a map, the park covers an area of 2.3 by 13.9 km, but the actual surface area of this hilly region is about 20% larger than this. Gombe is ecologically isolated, bounded variously by a lakeshore, the rift escarpment, and human settlement. Rich in small vertebrates and primates, the animal community contains few large mammals. This can be attributed partly to the park's relatively small size.

The chimpanzees proved difficult to habituate and observe, and so a program of provisioning with bananas was launched to entice them into the range of observers. At the peak of provisioning in the late 1960s, about 68 kg of bananas were distributed per day to members of the study population, the community within whose range the provisioning area was located. Other animals were attracted into the area by the abundance of food. Since the early 1970s, both the frequency and volume of provisioning have been reduced. The possible effect of provisioning on the Gombe chimpanzees has been discussed by Wrangham (1974) and Goodall (1983), and its effect on their demography will be considered in Section 7.4.

Since 1965, the study population has numbered between 38 and 60 animals (see Table 7.2). Births, deaths, migration, and fission of the community constituting the original study population have contributed to the fluctuations. A total of 51 infants (24 males and 27 females) were born between 1965 and 1980, yielding an infant sex ratio of 1.1. The average annual birth rate ranged widely (see Table 7.2). Goodall (1983) attributes the particularly high birth rate (0.44) in 1976 to the killing of several newborn infants within the community in 1975. Females who lost their infants mated soon afterward and so gave birth in 1976. Other reasons for the variation in annual birth rate are unclear. Teleki, Hunt, and Pfifferling (1976) looked for changes in fertility with estimated age but found no such pattern. They suggest that low birth rates in the late 1960s may have been due to the stress generated by intensive provisioning during that period.

Mortality is high among males and females up to the age of six years, declines during their juvenile and adolescent years (six to nine), and rises again in later life (see Table 7.3, calculated from data in Goodall 1983). Mortality among adult males is somewhat higher than among females, but the accuracy of the life table may be lower in the older age classes because the ages of the individuals were estimated.

Between 1965 and 1980, 12 females moved into the study population from neighboring communities, two of them accompanied by four- to five-year-old males still dependent on them. In contrast, only two females are believed to have emigrated permanently: the supply of bananas may have encouraged females born in the community to remain.

Between 1970 and 1972, the original community, which constituted the study population, split into two: seven independent males and three mothers with offspring established a separate community to the south. The social separation of these animals led to their exclusion from the study population even though they did not migrate. The demography of the splinter community is not well known. Its establishment contributed importantly to the decrease in size of the study population and to changes in the adult sex ratio between 1970 and 1972. Although the number of males and females born in the study population was about the same, the adult sex ratio varied widely (see Table 7.2). Goodall attributes much of this variation to the departure of so many males in 1970–72 and to the frequent arrival of new females in the study area.

Demographic data on the Gombe chimpanzees are particularly difficult to interpret. This is partly because the data themselves are incomplete, partly because the precise effects of provisioning are difficult to document, and partly because the study population is a segment of a wider population within which there have been extensive social and spatial realignments.

7.3B   Unprovisioned Populations

Dunga Gali Rhesus Macaques *(Macaca mulatta)*

About 290 animals belonging to seven social groups were monitored for two years (1978 and 1979) at Dunga Gali in the Himalayan foothills of northern Pakistan. Ages of a few animals were known from their birthdates, but most were estimated from the animals' general appearance and were subsequently corroborated or corrected by the evidence of dental eruption and wear patterns. The latter data were collected when animals were trapped and briefly held in captivity. Three immature age classes were delineated, zero to two years, two to four years, and four to six years (Melnick 1981). All animals over the age of six were assigned to a single class. The correspondence between ages assigned to immature animals from field observation and those assigned according to dental evidence was good for age classes spanning two years but broke down when animals were assigned to one-year age classes, even though births occurred during only a few weeks each year.

Juvenile males outnumbered females in the Dunga Gali population, with a sex ratio of 0.86, but in the adult age class the ratio shifted to 2.0. Several factors may have contributed to this. First, most males spend part of their lives outside a social group (Subsection 8.2C), living alone or with one or two other males. At such times they are difficult to see, and so they may be underrepresented in the sample. Second, mortality may be high among young males emigrating from their natal groups: males were sometimes seriously injured when they attempted to enter a new group, and traveling alone, they may have been more susceptible to predators. Third, if data on dental wear indicate age reliably, and if the teeth of males and females wear down at the same rate, there are clearly many more old females than males in the population, suggesting that females generally outlive males.

Females give birth every two years at the most (Figure 7.9), and the estimated average annual birth rate was only 0.38. In 1979, 5 out of 11 infants died; most deaths occurred during the first 6 months of life, and the rest in late winter at the age of 10 or 11 months. The sample is small and perhaps misleading, however, for in 1979 only 2 out of 10 infants died within 6 months of birth. The only other deaths recorded during the study were of a 1-year-old female and an adult female.

Little can be said about the growth rate of the Dunga Gali population. In Table 7.2, $\bar{r}$ was calculated for only three of the social groups making up the study population, and its negative value may reflect the circulation of males out of these groups rather than a population-wide trend.

Polonnaruwa Toque Macaques *(Macaca sinica)*

Polonnaruwa lies in the semideciduous forests of the north central dry zone of Sri Lanka. At the outset of Dittus's research (1975, 1977a, 1977b, 1979), the study population consisted of 446 animals distributed in 18 social groups (Figure 7.10). Prior to 1972, two of these social groups spent about half of each day foraging around a rice mill. The mill closed in 1972, after which the two groups expanded their ranges into those of neighboring groups and competed with them for conventional forest foods. Another group, dubbed Oval, fed extensively from a municipal rubbish dump just outside the study area. These exceptions are taken into account in the results presented below. The study site covered about 6 km², consisting of a peninsula of forest largely bounded by water and cultivation. A narrow forest neck connected it to more extensive forests and a larger biological population. Most of the study site lay in a religious sanctuary, and animals were not seriously harassed, although people with dogs did use the

Figure 7.9   The infant of this female rhesus monkey *(Macaca mulatta)* is about 6 weeks old. (Courtesy of W. Keene.)

Figure 7.10   Adult female toque macaque *(Macaca sinica)* grooms a juvenile while others look on. (Courtesy of R. Dewar.)

forest to graze cattle. All species of Sri Lankan primates were present, but larger mammals were absent.

Census data span the period July 1968 to January 1972, and the population was then recensused in March and April 1975. Known dates of birth and criteria such as time of first pregnancy were used to age young animals. Adults were classified into five broad age categories, based on "a host of subtle morphological changes, similar to those criteria one might use in subjectively assessing the age of humans" (Dittus 1975:130). Dittus used three methods to estimate the growth rate of the population between 1968 and 1972. Unfortunately, two of these methods can be legitimately applied only to populations with a stable age distribution. In the fluctuating, cyclone-prone environment of Polonnaruwa, it is unlikely that fertility and mortality schedules remain constant long enough for the population to achieve a stable age distribution (see also Dunbar 1979). The third method is open to question on other grounds (Dunbar 1980). The dynamics of the population during that period therefore remain an open issue. Nevertheless, by 1975 changes had definitely occurred. All groups but one showed an overall decrease of 14.5%. The one exception, Oval, showed a net increase of 60% between 1971 and 1975, an average crude annual growth rate of 12.5%.

Dittus's estimates of age-specific birth and mortality rates must be viewed with caution because the ages of animals born before the study began were estimated from their general physical appearance. Moreover, mortality rates were inferred from the age structure of the population. As noted above, this procedure is legitimate only when the age distribution of the population is stable, and it is unlikely that this was so at Polonnaruwa.

The annual birth rate (see Table 7.2) did not vary significantly between the four birth seasons from 1968 to 1972. Fertility increased with estimated age and remained high until menopause, which Dittus defined as two or more successive years without a birth. Then death followed quickly. The longest postreproductive period survived by a female was three years. Between 1968 and 1972, 90% of males and 85% of females died before reaching maturity. The probability of survival increased significantly once animals reached adulthood.

The pattern of deaths differed between the sexes. Mortality was higher in infant and juvenile females than in like-aged males but dropped below that of males as the females approached maturity. Subadult males, aged from three to five years, showed the highest mortality rates, which leveled off again as they reached maturity, but the chance of dying remained higher for them than for females throughout their adult life.

The population decline recorded in 1975 was due to a reduction in the proportion of immature animals. There was no significant decrease among adult females or adult, subadult, and older juvenile males, but the number of younger juvenile males and of older and younger juvenile females declined significantly. Overall, there was a greater

Figure 7.11   Juvenile vervet monkey *(Cercopithecus aethiops)* in the Amboseli National Park, Kenya. (P. Whitten/Anthro-Photo.)

decrease in the number of younger (32%) than older (23%) juveniles. Because the 1975 census took place some months after the birth season, it is unclear whether the drop in juvenile numbers was a result of reduced female fertility, increased postnatal mortality, or both. The exceptional growth of the Oval group during this period was associated with a high birth rate coupled with low infant and juvenile mortality rates.

Adult females outnumbered adult males in the population by more than two to one (see Table 7.2), but males predominated in the juvenile age classes and at birth the sex ratio did not differ significantly from one. Dittus (1980) suggests that these differences reflect the age-specific pattern of mortality in each sex.

Amboseli Vervets *(Cercopithecus aethiops)*
Vervet monkeys (Figure 7.11) have been studied intermittently at Amboseli since 1963. Amboseli's huge area supports the full size range of mammals endemic to the region, including at least 18 species of ungulates and carnivores; large herds of domesticated cattle also graze in the park (Altmann and Altmann 1970).

In 1963, the study population numbered about 121 animals belonging to five social groups. These animals formed part of a larger biological population, the boundaries of which are not known. Like the baboons at Amboseli (Subsection 6.2C), the vervet population has steadily decreased in size since 1963. Censuses in 1971 and 1975 revealed that although the number of groups in the main study area did not change, group size declined by about 43% between 1963–64 and 1975 (Struhsaker 1973, 1976). For several reasons, Struhsaker attributes the decline to increased infant and young juvenile mortality rather than to a lower birth rate. First, between 1963 and 1971, the decrease in numbers was most apparent among young juveniles, animals aged between six months and one-and-a-half years. Second, comparing completed birth season statistics for 1963 and 1975, he found little difference between the years: 72% of females produced infants in 1963 and 82% in 1975. Finally, between 1971 and 1975, there was a 37% decline in the juvenile age class, animals aged from a year and a half to three years, and a 30% decline in the number of adult and subadult males. Struhsaker suggests that these figures reflect the impact of high infant and young juvenile mortality rates on the next age class, old juveniles, together with their "potential" effect on the subadult and adult class. It is unclear that mortality rates are in fact higher among males than females, however, and it is possible that males emigrated rather than died: male vervets normally leave their natal group as they approach maturity.

Sankaber Gelada Baboons *(Theropithecus gelada)*

Demographic data were collected on geladas living in the Sankaber area of the Simēn Mountains National Park, Ethiopia, during two nine-month studies conducted in 1971–72 and 1974–75 (Dunbar 1980). The study area consists of a ridge bounded by precipitous slopes. Grasses and scattered pockets of treelike shrubs cover the area. From mid-May until October it is persistently cold, wet, and cloudy, with daily hail and thunderstorms. In the dry season, from November to mid-May, daytime temperatures are much higher, although it is still cold at night.

The Sankaber geladas exhibit the social structure typical of the species. Animals live in reproductive units of 10 to 15 individuals, including a single adult male (Figure 7.12). (Crook 1966; Dunbar and Dunbar 1975). Other subadult and adult males form all-male groups. A set of these reproductive units, together with one or more all-male groups, forms a band that shares a common range and often forages as a coordinated unit of 60 to 250 individuals. Bands occasionally coalesce, forming herds of six to seven hundred individuals. The Dunbars' primary study population was one band, the Main band, although some data were collected on other bands living on the ridge and surrounding slopes. In March 1972, the Main band contained 289 animals.

Dunbar (1980) used two methods to estimate the growth rates of bands. Three estimates were based on births and deaths over a 12-month period. Two further estimates were obtained by calculating the net increase in band size over longer periods on the

Figure 7.12   Two gelada baboon *(Theropithecus gelada)* harems, each consisting of females and young, cluster around their harem males. Note both the larger size of the males and their striking mantle of hair. (Courtesy of R. I. M. Dunbar.)

basis of a compound rate of increase over time, with the assumption of no migration into or out of the band. All these estimates gave a positive result of approximately the same size and a mean growth rate of about 12% per year. However, the actual increase was only 5% per year. Dunbar suggests that the discrepancy between the two values is due to the periodic emigration from the study area of splinter groups from the resident bands.

Like Dittus, Dunbar (1980) looked more closely at fertility and mortality rates within different sectors of the population. His conclusions must be viewed with similar reservations, since he too assigned adults to age classes according to their general physical appearance.

The average annual birth rate among the Sankaber geladas was .373. Note that the actual number of births fluctuated from year to year and that these fluctuations were consistent from band to band (Figure 7.13). Their causes will be discussed in Section 7.4. Very old females were the least fertile, and subadult females the most fertile (see Table 7.3). However, Dunbar notes that if the nine females experiencing adolescent sterility are included in the subadult class, then the fertility of subadults drops below that of young and old adults. The resultant pattern looks like that reported for macaques (although the reproductive status of subadult females in these studies is unclear).

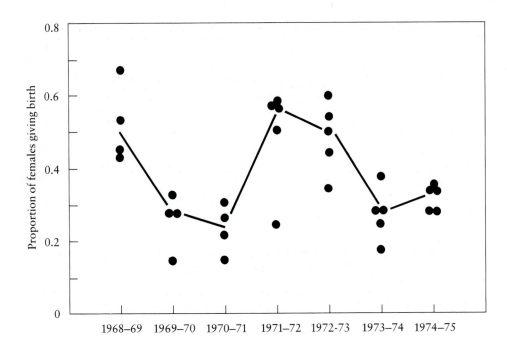

Figure 7.13   Birth rates per female in different years among five bands of gelada baboons *(Theropithecus gelada)*. (Solid line connects data points for Main band.) (From Dunbar 1980.)

Estimated age-specific mortality rates differ from those reported so far. Mortality among infants and juveniles was low: about 88% survived to the age of four years, and there were no deaths among subadults and young adults. The survival of immature males and females was not calculated separately for each sex in the immature age classes, but more female adults survived than males. This may contribute to the change in the sex ratio that takes place with increasing age. Overall, males and females were born in approximately equal numbers, although from year to year the sex ratio at birth varied from 0.5 to 3.0. Among adults, in contrast, there were between two and three females for every male.

Barro Colorado Island Howler Monkeys *(Alouatta palliata)*

Barro Colorado is a rugged island in Lake Gatun, a sector of the Panama Canal. About 15 km² in area (Carpenter 1934), it harbors the longest-studied biological population of primates in the world. In the first census in 1932, the population was estimated to be about 489 animals, concentrated in the more mature forests in the south and west of the island. Today the forests have matured to the point where they are all good howler habitat, and animals are found everywhere except for a small clearing containing the island's

Table 7.4   Census data on howler monkeys *(Alouatta palliata)* of Barro Colorado Island

Observer	Year	Percentage of island censused	Number of monkeys censused	Projected total population	Mean group size	Proportion of immatures[a]	Birth rate
Carpenter	1932	100	398	398	17.3	.41	.37
Carpenter	1933	100	489	489	17.5	.44	.51
Carpenter	1935	?	239	?	15.9	.44	.37
Collias and Southwick	1951	100	239	239	8.0	.28	.26
Carpenter	1959	100	814	814	18.5	.33	.33
Chivers	1967	16	176	1100	14.7	.36	.41
Smith	1967	33	372	1116	13.8	.28	.30
Froelich, Thorington, and Otis	1976	10	200	—	—	.36	.38–.48[b]

Source: Adapted from Smith 1977 and Froelich, Thorington, and Otis 1981.
[a] Includes infants and juveniles.
[b] Higher figure includes eight infants presumed to have been born and to have died in the six months prior to the census. Mortality rates during the first six months were assumed to be similar to those documented by Glander (1980) in Costa Rica. The lower figure excludes these infants.

research facilities (Smith 1977). The island is too small for the larger carnivores of mainland Central and South American rain forests, and careful patrols have kept out human poachers; as a result, the Barro Colorado Island howlers are without most of their potential predators.

Between 1932 and 1951 the number of social groups increased, and animals spread to all parts of the island; however, the mean size of these groups decreased drastically, as did the estimated total population (Table 7.4). Collias and Southwick (1952) found nine howler monkey skulls in the 1951 dry season and noted that out of 28 researchers polled, only three had ever found howler remains in the past. They concluded that the population had crashed during or just prior to their 1951 census. However, since the previous census had been made 18 years earlier, we do not know how long it took for the population to crash, if indeed it did. Rapid population growth between 1951 and 1959 brought the population up to an estimated 814 animals, a figure approaching current estimates. The crude growth rate during that time was 16.6% compared with 4% between 1959 and 1967. The recent decline in the crude growth rate is sufficiently marked to suggest that the population was close to saturation during the 1960s (Smith 1977).

Was the population decline recorded in 1951 a result of decreased fertility, increased mortality, or both? What has brought about the recent leveling off in the crude growth rate? We cannot answer these questions. Most of the censuses have been widely spaced, and precise demographic data are particularly hard to collect. Howlers are difficult to sex even as adults, and they have no discrete birth season, making it hard to identify cohorts of immature animals. There has been a general decline in the proportion of juveniles and infants in the population, the 1951 census marking the only anomaly, but we do not know whether this was due to decreased fertility, increased infant mortality, or both.

Studies of a segment of the Barro Colorado Island population between 1973 and 1977 show that even if the population is approximately stable in size, there are still striking fluctuations in the number of animals in each age class (Froelich, Thorington, and Otis 1981). The authors captured 75 animals from seven contiguous groups and estimated their ages from dental wear and eruption sequences. Using this sample and additional census data on the seven groups, they estimated the age distribution of the entire study population, about 200 animals in all, as it appeared in July 1976 (Figure 7.14). It is unlikely that fluctuations in the size of age cohorts are completely attributable to random sampling effects. For example, the authors suggest that the virtual absence of seven-year-old animals in 1976 may have been due to high juvenile mortality in 1970. In that year, heavy rains in the early dry season disrupted flower pollination, leading to a poor fruit crop in the wet season. If they were outcompeted by larger animals and in any case less able to survive on a low-fruit diet, young juveniles would have been most affected by the scarcity of fruit.

Froelich and his colleagues (1981; Otis, Froelich, and Thorington 1981) calculated age-specific survival for males and females from the age distribution of living animals censused and from the age distribution of 43 dead monkeys found in the study area between 1977 and 1978. The problematic assumptions associated with the first of these procedures have already been discussed. The general pattern that emerges from both methods is of high juvenile mortality in males and females, much higher mortality among subadult males than among subadult females, followed by a leveling off for both sexes in adulthood. Mortality increases in old age. Cadavers suggest that more females survive to old age than do males; the age distribution of living animals indicates a slightly higher mortality rate among old females than males, giving males an estimated maximum life span of 23 years to the females' 22 years. Females give birth for the first time at the age of about 4 years, although they do not reach full adult body weight (about 6 kg) until they are 5 or 6 years old. Between 1932 and 1976, the birth rate ranged between 26% and 51%. There are no data on age-specific fertility.

Berenty Sifakas *(Propithecus verreauxi)*
On the east bank of the Mandrare River in southern Madagascar, 200 ha of gallery forest have been protected from human disturbance by the De Heaulme family since the early 1940s (Figure 7.15). These forests, dominated by the kily tree *(Tamarindus indica)*, support populations of two diurnal primates, the sifaka and the ring-tailed lemur

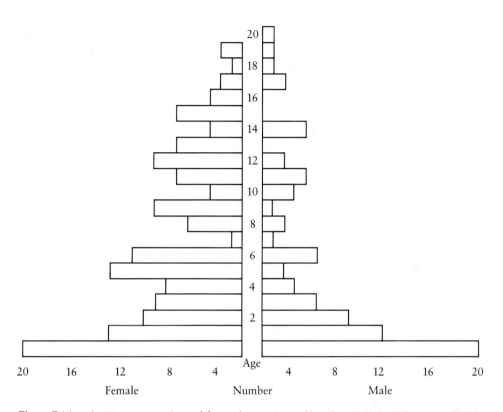

Figure 7.14   Age structure estimated from observations of howler monkeys *(Alouatta palliata)* on northeastern Barro Colorado Island in 1976. (Adapted from Froelich, Thorington, and Otis 1981.)

(Lemur catta). Neither is artificially provisioned, and both have been studied intermittently since 1963. Here we consider only the sifaka, for which the most detailed information is available. Data on the ring-tailed lemurs have been summarized by Jolly, Gustafson, Oliver, and O'Connor (1982a) and by Mertl-Milhollen et al. (1979).

The reserve is virtually surrounded by sisal fields, which the primates do not cross. Migration can occur through a narrow connecting strip of forest along the riverbank, but hunting pressure is intense and the habitat heavily degraded outside the reserve. Consequently, the density of primates in these adjacent areas is very low. One goal of a study that has just begun is to monitor movement into and out of the reserve through this corridor of forest (S. O'Connor, in progress). Until this is done, the degree of isolation of primates in the reserve will remain unclear.

In 1963, 1964, 1970, and 1971, sifakas were censused in a 10-ha section of the reserve; in 1974, 1975, and 1980, they were censused in a 100-ha area that included the

Figure 7.15   Gallery forests dominated by the kily tree, much like this forest at Berenty, line the banks of most rivers in southern and western Madagascar. (Photography by A. Richard.)

original 10-ha study site (Table 7.5). The data from these censuses reflect similar methodological problems to those from most censuses on Barro Colorado Island. None of the animals were marked, few could be identified with confidence from year to year, and several of the censuses were separated by more than one year. As a result, a life table for the population cannot be constructed despite the long period over which the population has been monitored. A final difficulty is that markedly different maximum and minimum estimates are available for 1980 (see Jolly, Gustafson, Oliver, and O'Connor

Table 7.5   Sifaka *(Propithecus verreauxi)* census data from 1963 to 1980 at Berenty

Observer	Year	No. animals censused[a]	Adult sex ratio	Birth rate
		1963–1980 in 10-ha portion of reserve		
Jolly	1963	42	0.65	0.5
Jolly	1964	47	0.71	—
Richard	1970	47	0.96	—
Richard and Struhsaker	1971	58	0.91	0.42
Gustafson	1975a[b]	47	1.17	—
Gustafson	1975b[c]	48	1.23	0.38
Jolly et al.	1980	43–50	0.43–0.61	0.42
		Census data from 1974 to 1980 in approximately half total reserve (100 ha)		
Richard and Laitman	1974	87	1.11	—
Gustafson	1975a[b]	103	1.17	—
Gustafson	1975b[c]	103	1.17	0.38
Jolly et al.	1980	99–128	0.6[d]	—

Source: Adapted from Jolly, Gustafson, Oliver, and O'Connor 1982b, Richard 1978a.
[a] Infants are excluded because some censuses were carried out before the birth season and others after.
[b] Collected between February and May.
[c] Collected between May and July following the fission of one group and several changes of membership in others.
[d] No maximum estimate given.

1982b for a discussion of the practical problems necessitating this). The number of animals in the Berenty study population has remained about the same since 1963, if the minimum estimate for 1980 is accurate. If the maximum figure is correct, then there is some indication of recent population growth.

Between 1963 and 1975, the adult sex ratio increased steadily. In 1978, I suggested (Richard 1978a) that the earlier predominance of males in the population was caused by differential migration by males into the reserve when the land around it was cleared and that a more even sex ratio was restored with successive cohorts of newborns. However, the census data for 1980 indicates a sharp drop in the adult sex ratio since 1975: males once again outnumber females by almost two to one. The adult sex ratios in three other sifaka populations censused (Richard 1978a) were 0.86 ($N = 86$), 1.14 ($N = 31$), and 1.13 ($N = 43$), respectively. Alison Jolly and her colleagues (1982b) attribute the striking imbalances at Berenty to random sampling events in a small population. We return to this issue in Subsection 7.3C.

Figure 7.16   At six months, this young sifaka still hitches the occasional ride on its mother's back. (Photography by A. Richard.)

Females probably reproduce for the first time at the age of three years, although this has yet to be well established. Thereafter, they likely give birth every two years on average (Figure 7.16) except when an infant dies during the first few months of life. A female may then conceive and give birth the following year. Sifakas mate for a few days each year, and all the infants in a given region are born within a few weeks of each other. Little is known about survivorship after the first year of life, although a few individually recognized animals are known to have lived at least 10 years and perhaps as many as 20.

7.3C   Emerging Patterns

In the introduction to Part III, we posed a number of questions about the ways in which primate populations change over time. Having reviewed the available data, let us see which, if any, we can now answer.

Do primate populations fluctuate in size? The provisioned macaque populations from Chhatari, Takasakiyama, and Cayo Santiago were all expanding rapidly. The segment of the macaque population in Sri Lanka with access to a rubbish dump grew rapidly between 1972 and 1975. The fourth provisioned population, the Gombe Stream chimpanzees, was one-third smaller in 1972 than it had been in 1960, but eight years later had almost regained its 1960 size. Of the unprovisioned populations, the number of vervets at Amboseli dropped sharply between 1963 and 1975. The status of the segment of the Sri Lankan macaque population without access to rubbish is unclear between 1968 and 1972, but between 1972 and 1975 its numbers dropped by 14.5%. The Sankaber gelada baboon population grew at an estimated rate of 12.5% per annum during the study, although the observed increase corresponded to a growth rate of less than half this. Censuses of howler monkeys on Barro Colorado Island suggest that this population experienced a major crash prior to 1951, after which it grew rapidly. The growth trend appears to have leveled off in recent years. There was no significant change in the Himalayan rhesus macaque population over two years, but a comparison of its age structure with that of rhesus macaques in India suggests that it may have been growing. Stable numbers have also been reported over a four-year period for two Himalayan rhesus populations occupying temples in Katmandu Valley, Nepal (Teas et al. 1981). If we accept Alison Jolly's minimum estimate of the number of sifakas *(Propithecus verreauxi)* at Berenty in 1980, then this is one unprovisioned population whose size has changed little over 20 years.

Plotting numbers of animals in a population over a few years provides only the crudest indication of the changes going on, but the sketchy data summarized here show that not all primate populations are static in size. Observations of animals from Amboseli, Sri Lanka, and Panama document marked declines in their numbers. Those of provisioned and protected macaque populations show that under favorable conditions these populations grow rapidly. While the validity of extrapolating from the dynamics of provisioned populations to those of unprovisioned populations may be questioned, these studies do demonstrate a capacity to achieve and sustain a high rate of population growth. It is the task of the ecologist to determine how often such favorable conditions occur in the wild.

Before concluding that primate populations are characteristically unstable, we need more data on undisturbed populations, particularly those inhabiting rain forest. Seasonal fluctuations are not pronounced in the rain forest, and it is generally assumed that the number of animals changes little from year to year. This assumption has yet to be tested in any primate population except for the howler monkeys on Barro Colorado Island, where it can be rejected.

Documenting annual changes in the size of primate populations is relatively easy. Accounting for changes, or indeed for their absence, is altogether more difficult, requiring a time depth and level of precision in demographic records that have rarely been achieved. Changes in average fertility, migration rates, or mortality may all contribute to changes in rates of population growth. The challenge is to identify the precise contribution of each.

Some of the changes in population size reported here are certainly, and others possibly, due to migration into or out of the study population. For example, Dunbar (1980) attributes the discrepancy between the estimated (about 12%) and actual (about 5%) growth rate of the Sankaber gelada baboon population to emigration from the study area. It is possible that some of the growth of the Chhatari rhesus study population was due to immigration from other parts of Aligarh District where animals were unprotected. The chimpanzees at Gombe Stream represent a special case. The community constituted the study population, so that when the original community split, the study population was reduced in size even though members of the splinter group did not in fact migrate: the change was primarily social, not spatial.

Before considering the contribution of fertility and mortality rates to population growth, two points should be made. First, we must assume that birth rates accurately reflect fertility rates because most researchers measure the former, not the latter. The birth rate will be twice the fertility rate if males and females are born in approximately equal numbers. This appears to be true on the average for most primates, although when samples are small, random fluctuations in the sex ratio from year to year may be wide (Dunbar 1980). In addition, among certain primates in Madagascar, males may consistently outnumber females at birth, an issue to which we shall return shortly.

Second, comparisons among species must be made carefully, especially if the species differ in body size. For example, the first year of life represents about one-third of the maturation time for some populations in Table 7.2 and, say, a sixth or an eleventh for others. Thus, the significance of mortality during this year may not be the same in each case. The five studies of macaques are particularly useful, because they are of closely related and similarly sized animals living under different conditions, but the demographic patterns found in macaques may not be present in all primates.

Keeping in mind these problems, we can use Table 7.2 to make a few tentative generalizations about relationships between fertility, mortality, and population growth. First, relatively low mortality rates during the first year of life characterize rapidly growing populations. These include the Cayo Santiago, Chhatari, and Takasakiyama macaques and the Sankaber geladas. This last population has the lowest mortality rate of all populations considered. Note that when the study population is part of a wider biological population, as is the case for the geladas, population growth predicted from fertility and mortality rates may be offset by emigration from the study area.

Second, high fertility rates are not always associated with population growth. Within a population, average fertility may vary widely from year to year, but the variation does not appear to correspond closely to changes in the population's growth rate. Comparing different populations, the birth rate of vervet monkeys at Amboseli is exceeded only by that of the provisioned and rapidly growing rhesus population at Chhatari, yet the number of vervets declined sharply between 1963 and 1975. Conversely, the birth rate is low among the Dunga Gali macaques, Sankaber geladas, and Berenty sifakas, yet these populations are static or growing. All three inhabit highly seasonal environments and females normally give birth every other year, the timing of the reproductive cycle tightly constrained by environmental conditions. Similarities in the pattern of births among these species cut across considerable differences in weight. Adult female geladas weigh about 13.6 kg (Napier and Napier 1967), adult female macaques at Dunga Gali about 7 kg (personal observation), and sifakas somewhere between 3.5 and 6.7 kg (Tattersall 1982).

How does a population's composition by age and sex affect fertility and mortality rates? To answer this question we need to know how the age and sex of an animal influence the probability of death and how age influences the probability of a female giving birth. In those populations for which age-specific rates are available (see Table 7.3), the general trend is similar. Mortality is high during the first year, somewhat lower in the second, and decreases further as animals reach adulthood. It is lowest among adults in their prime and then increases again with old age. Note, however, that although the general trend is the same in all cases, the actual rate for particular age classes varies widely. For example, juvenile mortality rates range from a low of 0.02–0.04 per annum in gelada baboons (Dunbar 1980) to a high of 0.28–0.31 in toque macaques (Dittus 1975). Second, in some populations there are sex differences in mortality rates. At Polonnaruwa and on Cayo Santiago, for instance, the mortality rate of adult males exceeds that of females. The higher rate recorded for males may occur because, unlike females, they leave their natal group upon approaching adulthood and at that time may be more vulnerable to predators and more likely to be killed trying to enter a new group. If this surmise is correct, we might expect that sex-specific differences in mortality would disappear in species in which both sexes disperse and be reversed in species characterized by female dispersal. Data are not yet available to test these predictions.

Reports of age-specific variation in fertility present a less consistent picture. On Cayo Santiago, the birth rate of rhesus macaque females peaks between the ages of 6 and 11 years and then declines rapidly. At Polonnaruwa, in contrast, Dittus (1975) reports an increase in fertility with age until females reach menopause, after which they live for only a couple of years. At Sankaber, the pattern is similar to that reported for the

Cayo Santiago rhesus if females experiencing adolescent sterility are included in the sub-adult cohort. Finally, Teleki and his colleagues (1976) found no correlation between fertility and age among chimpanzees at Gombe Stream.

Sex-specific differences in mortality with age lead to changes in the sex ratio over time. Among geladas, macaques, vervets, and howler monkeys, adult females consistently outnumber adult males, usually by two or three to one. Lacking life tables for each sex or data on the sex ratio at birth, we do not know for sure whether the skewed adult sex ratio is due to differences in the pattern of mortality or in the proportion of males and females in each birth cohort. However, in the few instances where we have a large enough number of births to offset the effects of random sampling events, the sex ratio at birth approaches unity or is slightly weighted in favor of males; thus, differences in age-specific mortality appear to be the more important determinant of adult sex ratios. During certain periods at Gombe Stream, chimpanzee males have outnumbered females, and the same is true for sifakas at Berenty. At Gombe Stream, the overall sex ratio at birth is one, and according to the age-specific mortality schedule for each sex, there should be more females than males in older age classes. Some of the variation in the adult sex ratio from year to year, particularly the disproportionately large number of males sometimes present, is probably attributable to differential migration by males and females. At Berenty in Madagascar, the adult sex ratio among sifakas increased steadily between 1963 and 1975, but according to the 1980 census data, males once again outnumber females. In one out of three other forests censused, males outnumbered females. A sex ratio biased toward males has been reported for another Malagasy primate, the black lemur *(Lemur macaco)* (Figure 7.17). Petter (1962) found a sex ratio of 0.71 among 10 social groups containing 96 animals, the excess of males being characteristic of all but one of the groups. Both cases may be examples of random sampling events. Alternatively, the sex ratio at birth or the sex-specific pattern of mortality or both may be uniquely different among these Malagasy primates. Without more data, they remain an enigma.

In conclusion, the primate populations described here experience periods of rapid growth and decline. The periodicity of these fluctuations is unknown, nor do we know whether most or all primate populations are characteristically unstable. We may also tentatively conclude that, in the short term, changes in the mortality rate are more influential than changes in the fertility rate in determining the growth rates of populations studied so far. An animal's age and sex influence the probability of death in ways that are broadly similar among species. There is some evidence that a female's age influences her fertility, but the pattern is less clear. Finally, because of these relationships between age, sex, fertility, and mortality, some variations in the growth rate among populations may be partly or completely attributable to differences in their composition. In Section 7.4 we go on to consider ecological factors that contribute to short-term changes in fertility and mortality in a population.

Figure 7.17   The so-called black lemur *(Lemur macaco)* is unusual both because its males outnumber females and because of the striking sexual dichromatism. The two males (right) are plain black, whereas the females (left) are predominantly reddish brown. (Photography by A. Richard.)

7.4   Environmental Influences on Fertility and Mortality

Food shortages, bad weather, predators, disease, fighting, and cannibalism have all been cited as factors that constrain or depress the growth rates of primate populations. Supporting evidence frequently comes in the form of a correlation between a food shortage and a subsequent increase in mortality, for example, or between cold weather and an absence of births. In Chapter 4 we reviewed the evidence that food may be an important determinant of fertility and mortality rates, and here we focus on other environmental determinants. Note that the demographic and contextual data are rarely of sufficient quality to reach firm conclusions, and that in any case the field-worker is often faced with a problem of confounding variables. For example, if animals die in greater numbers when food is scarce and the weather is cold, is it because they are hungry or cold or both?

In the tropics and subtropics, where most primates live, the most important effect of weather on primate populations is indirect: the availability of food is strongly in-

Figure 7.18   A young baboon *(Papio cynocephalus anubis)* plays at a water hole. In the dry season, many pools dry up and animals travel long distances in search of water. (Courtesy of R. Johnson.)

fluenced by weather conditions, particularly by the pattern and quantity of rainfall. Occasionally the effect is direct. Drought, for example, may cause death from dehydration, especially in species that regularly drink from free-standing water (Figure 7.18). (However, some primates obtain enough water from vegetable food to meet their needs.) Wrangham (1981) noted that 10 out of 26 animals in a vervet monkey group died soon after the waterhole in their home range dried up, deaths he attributed to water shortage.

For primates living in temperate areas with harsh winters like Japanese and Himalayan rhesus macaques and langurs, cold periods are also periods of food shortage, and it is difficult to distinguish the relative importance of the two (3.2). Of their combined importance there can be no doubt. Among the Himalayan rhesus monkeys, for example, births are limited to a six-week period just after the spring thaw, and most animals lose weight over the winter months (Pearl 1982).

Dunbar's (1980) study of gelada demography is particularly interesting here, because these baboons are most likely to suffer weather-induced stress during the wet season, when food is most abundant. It is thus possible to examine separately the effects of food availability and weather conditions. If food shortages were the major cause of mortality,

Table 7.6   Risk of death in different trimesters for Main band, 1974–75

	May–July	November–January	February–April	Total
Rainfall in trimester (rank order)	1	2	3	—
Observed deaths	5	2	1	8
Animal-months sampled	515	673	671	1859
Probability of death	0.0097	0.0030	0.0015	—
Expected deaths[a]	2.2	2.9	2.9	—

Source: From Dunbar 1980.
[a] Assuming constant likelihood of death in all months.

Dunbar argued, the peak in deaths should occur during the dry season months, when primary productivity was lowest. But in fact the death rate correlated positively with rainfall (Table 7.6). During the wet season, animals were frequently soaked by rain and mist in below-freezing temperatures, and many developed coughs and colds. Although nutritional stress from the preceding dry season might increase the susceptibility of some individuals to respiratory problems during the following wet season, old age was probably a more important consideration, because mortality rates among adults were highest in the older age classes. The pattern of infant mortality provided further evidence of the stressful effects of cold, wet weather, for the likelihood of death increased with the number of wet season months to which an infant was exposed (Table 7.7). Infants do not have a waterproof coat and must be particularly susceptible to the effects of repeated drenching.

In gelada females, most conceptions coincide with periods of high rainfall; thus births occur during the subsequent dry season. We have already noted the benefits to an infant of not being exposed to wet weather. If births were timed, as they apparently are in many primates, to maximize the food available to the mother when she is lactating,

Table 7.7   Risk of death for black infants (aged 0–5 months) exposed to different numbers of wet season months (May–October)

	Number of wet season months		
	0–1	2–3	4–5
Infant-months sampled	159	224	163
Infant deaths	1	2	3
Probability of death	0.0063	0.0089	0.0184

Source: From Dunbar 1980.
Note: Data for Main, Abyss, and High Hill bands in 1971–72 and 1974–75.

the birth peak should occur during the wet season, when primary production is at a peak. Cycles of food availability may influence the timing of births in another way, however, for many mammals ovulate only after they have reached a certain nutritional status. Dunbar suggests that female geladas may not be capable of ovulation until after the onset of the rains and the burgeoning of new growth after the long dry season. How soon breeding can take place is thus limited by weather acting through nutrition, while at the same time early breeding is favored because of the high, weather-related mortality among late-born young.

The ubiquity and in some instances (e.g., Seyfarth, Cheney, and Marler 1980) specificity of alarm calls among primates suggest that a significant advantage has long accrued to primates who alerted each other to the presence of predators close by. The size of social groups is greater on the average in open country than in forests. This may be a response to greater predator pressure out on the savanna (Crook and Gartlan 1966). It has even been argued that the reason animals live in groups in the first place is because of the increased protection from predators that a group offers (Alexander 1974). However, general observations of this kind tell us little about the demographic importance of predation on primates today.

The feces of open-country predators rarely contain primate remains. In a study of predation by lions, leopards, cheetahs, and wild dogs in the Serengeti, Tanzania, only the leopard was seen to prey on baboons, which comprised 4% of their kills (Kruuk and Turner 1967). In the Kafue National Park, Zambia, 2 out of 96 leopard kills were baboons, but lions, cheetahs, wild dogs, crocodiles, and martial eagles never attacked primates (Mitchell et al. 1963, cited in Altmann and Altmann 1970). Data of this type suggest that primates are not important dietary items for predators, but they say little about the importance of predation from a primate's point of view.

The contribution of predation to mortality rates among primates is difficult to assess. Many field researchers have reported episodes of actual or attempted predation upon members of their study populations, but the rate of predation can rarely be extrapolated from these reports. The observer's presence is likely to discourage predators, predation is probably infrequent in any case, and lucky is the observer who is in the right place at the right time to witness it. More often he or she knows only that an animal has disappeared suddenly and is limited to a tentative and unconfirmable verdict of "lost, presumed eaten." This does not mean that predation is necessarily insignificant for primate population dynamics. Because most primates live at low densities in relatively small social groups, three or four successful kills in a year could cause major changes in the composition of a group or even in the structure of the whole population.

The Altmanns' (1970) account of predation on the Amboseli baboons is one of the more extensive, perhaps because interactions with predators are more frequent in open country primates (Figure 7.19). During 1006 hours of observation of the focal group,

Figure 7.19   Lioness in front of a baboon sleeping grove at Amboseli, Kenya. Although lions rarely prey on baboons, the baboons were alarm-calling when this photograph was taken. (Courtesy of C. Saunders.)

which contained about 40 animals, 103 reactions by baboons to predators or supposed predators were observed. Of these, 71 were false alarms; that is, the observers were confident that no predator was present. However, in one attack a leopard succeeded in killing an adult male and a large juvenile male, and in another an infant was killed by an unidentified predator. A total of 12 animals from the focal group were known or presumed to have died during the 14-month study. Predation thus accounted for at least 25% of all deaths.

Predation on forest primates seems to be much rarer or else is much harder to detect. In 1627 hours of observation spread over a year, Oates (1977a) saw no episodes of predation on *Colobus guereza* at Kibale, Uganda. He found no evidence of predation by mammals, although he cites two other reports of *Colobus angolensis* remains discovered in leopard excrement. He did find the remains of a large monkey together with some brown-and-white-barred feathers that had probably belonged to a crowned eagle *(Stephanoaetus coronatus)*. Five times he saw an adult male charge a bird of prey that had flown low and perched near the group, and on many occasions the monkeys roared as a raptor flew overhead. These observations suggest that birds of prey are the guereza's chief predators but that successful attacks are rare.

The toque macaques studied by Dittus (1975, 1977a) in Sri Lanka lived in an area from which most large predators had been eliminated. Jackals *(Canis aureus)* elicited alarm responses from the monkeys, although the jackals themselves seemed uninterested. Of the cats, leopards were absent and only the fishing cat *(Felis viverrina)* and the jungle cat *(Felis chaus)* were big enough to be a threat, but no interactions between them and the monkeys were seen. Fifty-one episodes of predation by leopards in an area where they were sympatric with macaques did not include any attacks on the latter (Muckenhirn and Eisenberg 1973). Dittus saw five successful predatory attacks at Polonnaruwa, all by domesticated dogs. In Pakistan, rhesus macaques fled from domesticated dogs that sometimes seemed to be stalking them. Leopards occasionally traveled through the area, but were never seen near the monkeys. Apart from dogs, the only animal eliciting alarm was the diminutive yellow-throated marten *(Martes flavigula)*, but the martens themselves seemed indifferent. Potential predators on rhesus monkeys in the Asarori forest in northern India were the leopard, tiger, jackal, and several species of raptors (Lindburg 1977). Yet only five contacts with any of the cats were witnessed in 517 hours of observation spread over a year, and only one of these resulted in a kill, an adult male taken by a young tigress. An unsuccessful attempt by a raptor to snatch a small juvenile from a treetop was seen, and animals generally responded with alarm to the presence of such birds overhead.

In summary, in many areas where primates have been studied, their nonhuman predators have been eliminated. Populations most susceptible to predation live in open

Figure 7.20   Skulls of macaques and colobines hunted over a period of several years and kept for ritual purposes by people in the Mentawi Islands, Indonesia. (Courtesy of A. Mitchell.)

country conditions or near human settlement; some macaque populations in particular face an evolutionarily recent but possibly significant predator in the domesticated dog. The impact of nonhuman predators is slight, however, compared with that of human hunters. Many primate populations in the New and Old World are declining fast, and in many instances predation by people plays a major role (Figure 7.20) (Rainier and Bourne 1977).

Disease and predation work together to limit population growth, since many diseased animals are probably picked off by predators before they die of the disease itself. In at least four cases, however, disease has been invoked as the direct cause of deaths in a wild population.

Collias and Southwick (1952) suggested that the 1951 population crash (if crash it was) among the howler monkeys of Barro Colorado Island resulted from an outbreak of yellow fever in the population. A search for wild animals that had been infected by yel-

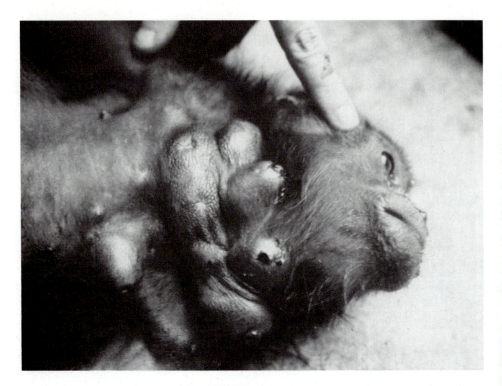

Figure 7.21   Howler monkey *(Alouatta palliata)* with a severe botfly infection. Note the holes bored by fully developed botflies emerging from the monkey's neck. (Courtesy of K. Milton.)

low fever in Panama in 1949–50 revealed that of various species of monkeys collected, howlers most frequently showed evidence of immunity to yellow fever. Presumably these animals had had the disease and recovered, although when those with immunity were exposed experimentally to yellow fever virus, many contracted the disease and died. This may have been because a different strain of the virus was used in the experiments. Three human deaths from yellow fever in Panama in 1949 provided additional, if circumstantial, evidence. One of the victims contracted the disease on a farm about 12 miles northeast of Barro Colorado Island, even though yellow fever had not been endemic in the human population there since 1905. By implication, the disease was raging through the howler population in 1949, and these people caught it from contact with infected monkeys.

Smith (1977) suggested that botfly infestations represent a more continuous constraint on population growth among the Barro Colorado Island howlers. After a botfly

Table 7.8   Number of botfly larvae *(Dermatobia hominis)* infecting howler monkeys

Season	Males	Females[a]		Infants	Juveniles	Total
		Without infants	With infants			
October–April	2.0 ± 0.29	2.8 ± 0.39	1.3 ± 0.53	2.0 ± 0.91	1.6 ± 0.45	2.2 ± 0.21
May–July	3.9 ± 0.61	6.9 ± 0.67	4.4 ± 0.87	3.1 ± 1.47	5.3 ± 1.86	4.9 ± 0.43
t-value	3.19[b]	5.03[b]	3.20[b]	0.52	3.00[c]	6.32[b]

Note: Values are means ± one s.e.

[a] Difference between the means for females with and without infants in October through April is nearly significant ($t = 1.920$, $df = 75$) and in May through July is significant ($t = 2.078$, $df = 28$).

[b] Statistically significant at $p < 0.01$.

[c] Statistically significant at $p < 0.05$.

lays its eggs on a mosquito's belly, the eggs hatch in response to the warmth to which they are exposed when the mosquito alights on a howler monkey to feed. The larvae burrow through the monkey's skin, enter the bloodstream, and migrate usually to the neck tissue (Figure 7.21). Smith reported that a botfly larva infecting a captive juvenile monkey remained in it for 29 days before dropping out as a mature pupa. The extent of parasitism varied seasonally among all adult monkeys and with the presence or absence of nursing young among adult females (Table 7.8). The rise in the number of infestations toward the end of the dry season (running from February to June) may have resulted from increased mosquito activity at that time. During 11 months of study, Smith found three monkeys dead or dying, all at the end of the dry season. The extent of parasitism could not be determined in two because they were too decomposed, but the third, a juvenile, was infected with 25 botfly larvae. Many of the holes occupied by the larvae were secondarily infested with larvae of another species, which were eating through the body wall and the peritoneum and even into the liver. The heavy botfly infestation was probably the cause of the secondary infection and the monkey's eventual death.

Females without infants had more botfly infections than those with infants. Smith suggested three possible reasons: (1) the infestations could be the cause of the absence of infants either through their death or through an earlier failure of the female to ovulate and conceive; (2) the poor health of females resulting from other causes could lead to both infertility and heavy levels of parasitism; and (3) the presence of an infant could disturb the activity of mosquitoes. He argued for the first, suggesting more generally that botflies limit howler monkey population density on Barro Colorado Island through increased mortality and reduced fertility.

At Gombe Stream, the causes of death are known or presumed for 45 out of 54 chimpanzees who died or disappeared and were presumed dead between 1965 and 1980

(Goodall 1983). Diseases appear to have caused or contributed to 22 of these deaths. In 1966, an epidemic of a paralytic disease (probably poliomyelitis) broke out in the study population about a month after an outbreak of polio in the human population bordering the park. Between August and December that year, 7 animals died and at least 6 were partially crippled from the disease. Five disappeared in 1968 during an outbreak of a respiratory illness tentatively diagnosed as pneumonia. Both these epidemics occurred under a regime of heavy provisioning, when large numbers of animals often crowded together in the provisioning area. At that time, diseases could have been transmitted more rapidly and more widely than would have occurred under natural conditions. Other diseases contributing to mortality were various forms of gastroenteritis, ulcers, sores, and "wasting."

Dunbar (1980) cites parasitic infestation as the main cause of death among geladas apart from the rigors of the wet season. Infestations, manifested as swellings of various size in different places on the body, were rare among immature animals and most common among young and old adults. Animals that avoided or survived infestation as young adults were less likely to be infected later in life. Of animals sampled in 1971–72 and 1974–75, about 10% were visibly infected at both times.

Swellings are caused by tapeworm larvae. These larvae, which develop in muscle tissue just below the skin of herbivorous mammals, originate in eggs excreted by primary carnivorous hosts onto vegetation eaten by the herbivores. Sometimes the swellings are so big that an animal has difficulty moving, and ultimately they burst, becoming "foul suppurating masses" (Dunbar 1980:497) that are clearly painful and presumably a severe physiological strain. Of eight deaths among adult animals in Dunbar's sample of infected animals, two died a few days after their swellings had burst. Three others whose swellings burst were unable to keep up with the group and were left behind. Two of these subsequently recovered. Ohsawa (1979) likewise concluded that parasitic infections and old age were major causes of adult deaths in a nearby population of geladas.

In conclusion, the environmental determinants of fertility and mortality rates in primate populations are poorly understood. Sample sizes are small, and it is often impossible to be sure an animal has died, let alone to assign a cause of death. As a result, evidence about the determinants of mortality is usually anecdotal. This is unfortunate because theoretical considerations (Section 7.5) suggest that how a population is regulated has important implications for the evolution of life history traits. In particular, factors that affect a population regardless of its density are distinguished from factors whose effects depend upon population density. When factors of the first kind predominate, population regulation is *density-independent,* and when factors of the second kind predominate, it is *density-dependent.*

The importance of density-dependent and density-independent regulation in primate populations has received little attention. Smith (1977) suggested that mortality due to botfly infestations in howler monkeys might increase with rising population density. Strum and Western (1982) have argued that among the Gilgil baboons, increasing population density has intensified competition for limiting resources and slowed maturation rates and hence the population growth rate. In contrast, the cold and wet winter on the Simēn Plateau in Ethiopia may be an important cause of death among gelada baboons regardless of population density. In the absence of more data, however, exactly how the environment regulates primate populations remains unclear.

7.5   Evolution of Life History Traits

In 1950, Dobzhansky suggested that natural selection operates in fundamentally different ways in the tropics than in temperate areas. In the latter, he argued, physical factors act independently of population density to limit numbers. A severe storm, for example, may kill large numbers of animals regardless of population density. In temperate areas such catastrophes are common, and natural selection favors those that take advantage of favorable conditions by maturing rapidly and having many offspring at once. In tropical areas the environment is stabler and catastrophes are rare. Animals live at consistently high density, and an individual's survival depends on its success in a variety of biological interactions with others. For example, it must compete for limited resources with conspecifics and with members of other species that exploit these resources. The higher the population density, the more intense this competition will be and the likelier it is that some animals will not obtain enough food for growth, maintenance, and reproduction. Population regulation is density-dependent, and natural selection favors slow maturation and the production of a few offspring upon which one or both parents lavish considerable care.

Dobzhansky's idea that different environments select for different life history traits has persisted and developed into what is today called K and r theory (MacArthur and Wilson 1967; MacArthur 1968). This theory posits a basic distinction between nonseasonal, stable environments and seasonal, fluctuating ones. Nonseasonal, stable environments are predictable, whereas seasonal, unstable ones are unpredictable. (We shall see shortly that these may not always be reasonable assumptions.) The parameters of *predictable* environments rarely exceed the tolerance limits of species occupying them, and there is a good chance that animals will survive from one breeding season to the next. The parameters of *unpredictable* environments frequently fluctuate beyond the tolerance limits of species, making survival to the next breeding season uncertain.

Predictable environments favor large individuals that devote much of their energy to rearing one or a few offspring at a time. The young require parental assistance to avoid predators and to compete successfully with other immature and adult animals. Species with these life history characteristics are said to be *K-selected*. Unpredictable environments favor individuals that produce large numbers of offspring early in life. Corollaries of this reproductive pattern are small body size among adults and a short life span. Individuals mature rapidly and give birth to large numbers of offspring, but expend little energy rearing them. Species with these life history traits are said to be *r-selected*.

Unpredictable environments are often patchily distributed, and many *r*-selected species have efficient ways of scattering their young over a wide area, whereas *K*-selected species tend to be more sedentary (MacArthur and Wilson 1967; Gadgil 1971; Horn 1978; Diamond 1975). When conditions fluctuate locally and catastrophes are correspondingly localized, selection favors the dispersal of offspring, because some will find habitable patches and survive, even though many will not. The *r*-selected species with efficient dispersal mechanisms are colonizers, the "supertramps" of the plant and animal world. They are the first to invade newly available habitats, from which they are eliminated only when the competitively superior but more slowly reproducing and dispersing *K*-selected species arrive.

In its original formulation, *K* and *r* theory paid little explicit attention to the age of animals most likely to die under particular conditions. The importance of age-specific patterns of mortality emerged when a different set of predictions was made about characteristics likely to evolve in fluctuating environments. *Bet-hedging theory*, from which the predictions are derived, postulates that such environments will favor individuals that live a long time and have a few offspring at a time throughout their adult lives. They are likely to leave more descendents by spreading out their reproductive effort than by producing all their offspring at one time, because there is a high probability that all the offspring born in a particular year may die. Data can be produced to support both the bet-hedging and *K* and *r* theories (Stearns 1976). The apparent contradiction arises because of different assumptions about the pattern of mortality. *K* and *r* theory assumes that mortality rates are highest or most uncertain among juveniles in *K*-selected species and among adults in *r*-selected species. Bet-hedging theory leads to opposite predictions by considering populations in which mortality rates are greatest or most uncertain among juveniles (Horn 1978).

Table 7.9 summarizes the main tenets of *K* and *r* theory and bet-hedging theory. The importance of age-specific patterns of mortality is obvious: they predict life history characteristics better than the environment does. In species where juvenile mortality is high, adults produce few young at a time and spread their reproductive effort over many seasons, regardless of environmental stability. In species with high or uncertain adult mortality, we find the opposite reproductive pattern. Primates produce few young at a time and spread their reproductive effort over many years. Like other large mammals (Caughley 1966, 1977), they also show the age-specific pattern of mortality expected

Table 7.9   Summary of tenets of life history theory

Environment	Probability of dying	Individual life history traits	Adaptive explanation	Population corollaries
Stable	High among juveniles	Slow maturation, long life span, few young, high level of parental care (K-selected)	Competition	Stable size; regulation density-dependent; slow dispersal
Unstable	High among adults as well as juveniles	Rapid maturation, short life span, many young, low level of parental care (r-selected)	Taking maximum advantage of favorable conditions that are unlikely to persist	Unstable size; regulation density-independent; efficient dispersal mechanisms
Unstable	High among juveniles	Few young at a time, reproductive effort spread over many seasons (many traits characteristic of K-selected species)	Bet-hedging	

with this reproductive pattern. Mortality is high among juveniles, and low among adults until they reach old age (Section 7.3).

Age-specific patterns of mortality determine life history traits. In order to identify the environmental parameters favoring particular life history traits, then, we need to study the effect of the environment on age-specific patterns of mortality. This effect is importantly regulated by body size. Differences in size are associated with differences in maturation rate and ability to withstand environmental fluctuations. Separately and in combination, these factors influence the pattern of mortality under specified conditions. In stable environments births occur year-round, and immature animals always make up a high proportion of the population. Smaller than adults, these animals are at a persistent competitive disadvantage, and the mortality rate in juvenile age classes is high.

In unstable environments, the situation is more complex. Consider the seasonal cycle in temperate latitudes. The young of small mammals mature fast, and only those born late in the birth season do not reach adult size before the onset of the next winter. Even as adults, however, small mammals are vulnerable to particularly long or particularly intense periods of cold, and such events may have catastrophic effects, killing a large proportion of the adult population as well as any subadults present. Large mammals are better able to withstand the adverse effects of year-to-year climatic fluctuations, but they cannot raise their young to maturity in a single spring and summer. Because of their small size, the young face their first winter at a competitive disadvantage compared with adults and are less likely to survive. As a result, even in seasonal environments mortality is likely to be high in immature age classes and low in all but the oldest adult age classes among large mammals.

The significance of seasonality is thus different for small and large mammals. In seasonal environments, parameters critical for the survival of small mammals frequently fluctuate widely enough to exceed the physiological tolerance limits of many individuals. Seasonal fluctuations are less likely to exceed the tolerance limits of large mammals. In short, the dichotomy between tropical and temperate environments proposed by Dobzhansky (1950) and subsequently elaborated in K and r theory has greater predictive value for small than for large mammals.

Among the primates, African bush babies (*Galago* spp.) constitute one of the few genera with small-bodied members distributed in tropical and temperate zones (Figure 7.22). Senegalese bush babies weigh about 165 g, compared with the Demidoff bush baby's 60 g, yet Senegalese bush baby females have four offspring each year and Demidoff females only one (Subsection 6.2A). This is counter to expectations based on body size, and may be attributable to the effects of K- and r-selection. In the northern Transvaal, where the Senegalese bush baby has been studied, there are huge seasonal changes in temperature. During the summer it reaches 38°C in the shade, while during winter months the temperature commonly drops below freezing at night. Rainfall and temperature vary markedly from year to year as well as seasonally. For example, precipitation ranges from a maximum of 900 mm in a year to a minimum of just 280 mm. Thus, the northern Transvaal may be an unpredictable environment for Senegalese bush babies, whereas tropical rain forest is a highly predictable one for Demidoff's bush babies. According to K and r theory, mortality should be high among Senegalese bush baby adults in bad winters regardless of population density and persistently high among Demidoff's bush baby juveniles. In the absence of data on mortality for these two species, the contrast between their patterns of reproduction cannot be taken as support for K and r theory; however, the example does show how the theory might be usefully applied to small primates (see also Martin and Bearder 1979).

The dichotomy between seasonal and nonseasonal environments is not useful for the study of adaptive differences in the life histories of large mammals because mortality is high among juveniles and low in most adult age classes in both environments. Increased environmental stress reinforces this pattern. For large mammals, including most

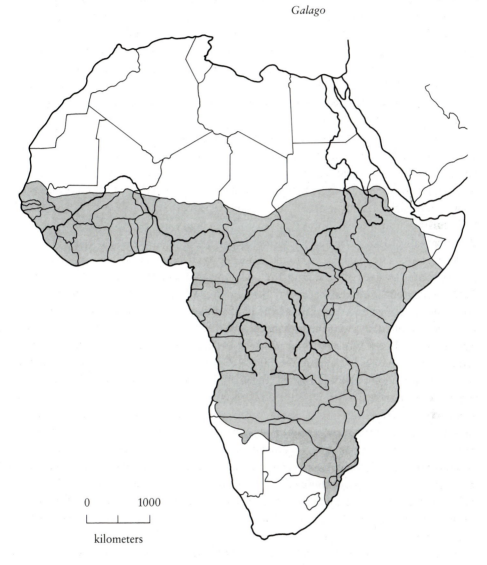

Galago

0   1000

kilometers

Figure 7.22   The distribution of species of bush baby *(Galago)* in Africa. (Adapted from Wolfheim 1983.)

primates, a seasonal environment is not unpredictable in the sense in which we have defined this term, and the causal links suggested by K and r theory cannot be usefully applied. Bet-hedging theory, with its emphasis on the age-specific pattern of mortality, appears more promising.

Let us conclude by briefly considering the population corollaries listed in Table 7.9 in relation to primate populations. According to K and r theory, corollaries of primate-like life history traits are stable population size, density-dependent population regulation, and slow dispersal. Sections 7.3 and 7.4 reviewed the evidence bearing upon the first two of these corollaries. It can be summarized as follows. Major fluctuations in the size of some primate populations have been documented, although we do not know whether density-dependent or density-independent factors are more important in population regulation. Concerning dispersal, most or all members of one or both sexes leave their natal group as they approach maturity, but reports of what happens thereafter are rare. How far do primates disperse? How often do they colonize new areas? How often do individuals or groups migrate between biological populations?

In Pakistan, most male rhesus monkeys dispersing from their natal group join neighboring groups (Melnick, Pearl, and Richard in press). Short-range dispersal of this kind is common among primates (Hrdy 1977; Lindburg 1969; Packer 1979a; Harcourt 1978), although much longer movements have been documented (e.g., Galdikas 1979; Dawson 1978). We know little about the frequency with which primates enter previously unoccupied habitat (but see Chivers and Raemaekers 1980; Jones 1980), although it is generally assumed to be low. The limited evidence thus suggests that the nature of dispersal differs among the primates and that a simple dichotomy between sedentariness and dispersal does not adequately describe it. This conclusion is in keeping with recent objections raised to the distinction between K-selected species as sedentary and r-selected species as efficient dispersers (Hamilton and May 1977; Horn 1978). Hamilton and May argue that selection will favor parents who enforce dispersal of some of their offspring even from a stable environment. Offspring that remain close to their parents as they mature must compete with one another, with their parents, and also with any immigrants present. Parents that allow their entire family to remain with them thus risk leaving no descendents at all if other members of the population disperse competitively superior progeny. In short, dispersal mechanisms will be favored in K- as well as r-selected species. These arguments, it should be noted, refer not only to a pattern of behavior but also to one of its outcomes, the colonization of new areas. More attention must be paid not just to the distance over which primates disperse but also to the frequency with which dispersal involves colonization rather than circulation among existing social groups.

In conclusion, current theory does not explain the evolution of the life history traits of large, long-lived mammals such as most primates. To be fair, however, we know so little about the demography of primate populations that evaluating predictions derived from this theory is difficult. Moreover, even if the present generation of theory does prove inadequate, its importance is still great, for it helps us to identify the kinds of information needed to develop and test more appropriate models.

PART IV

PRIMATES IN SOCIAL GROUPS

Introduction

 PART II emphasized the individual and Part III the population, but primates do not spend their lives as isolated individuals in a population. They maintain networks of social relationships throughout their lives, usually within spatially cohesive social groups. An important aspect of ecological research is the study of how social life influences and is influenced by such basic activities of the individual as the acquisition of food or the avoidance of predators. For example, animals may be able to harvest food more efficiently in social groups because they can help one another locate food; on the other hand, an animal may be unable to eat certain foods because of competition with other members of the group.

The order Primates exhibits a wide range of social organization, and two broad questions concern us in Part IV. First, what factors give rise to different kinds of social organization, and second, what are the ecological consequences for the individual of living in a social group? The answers to these questions are not always the same. For instance, while added protection from predators may be a cause and a consequence of living in a social group, the higher risk or parasitic infestation posed by life in social groups is unlikely to have favored their evolution. These questions provide the starting point for Chapter 9. By way of introduction, Chapter 8 considers how some of the general features of primate social organization vary and illustrates their diversity through case studies.

CHAPTER EIGHT

Diversity of Primate Social Organization

8.1 A Plethora of Terms

DISCUSSIONS of primate social behavior abound with references to social interactions, relationships, networks, groups, structures, organizations, and systems. Sometimes many of these terms are used interchangeably, sometimes each means something different. Let us begin by stating which of them will be used here and how they will be used.

A *social group* is made up of animals that interact regularly and know one another individually. Its members spend most of their time nearer to one another than to nonmembers and are often hostile toward nonmembers (Eisenberg 1966; Struhsaker 1969). A social group differs from an *aggregation* in several features. The makeup of an aggregation may change almost continuously and anyone can join. Members do not interact in a patterned way and probably do not know one another individually. Aggregations are temporary, dissolving when their members scatter. A *social network* is made up of animals that interact regularly and know one another, but do not necessarily spend most of their time nearer to one another than to nonmembers. Social groups or social networks have been found in all primates studied so far.

A *social interaction* is a single act of communication between two or more individuals. Usually these individuals are familiar with one another. A description of an interaction specifies what the participants are doing together (the interaction's content) and how they are doing it (its quality). A *social relationship* is the sum of the interactions between two individuals over time; its description includes the content and quality of the component interactions and how they change with time (Hinde 1976). *Social structure* refers to the content and quality of relationships among all the members of a social group, and *social dynamics* to the way in which that structure changes with time.

Social organization is the most inclusive term used in this chapter. Our understanding of social organization has changed in the last 10 years as our conception of the social group has altered:

This static unit is coming to be seen as but a steady-state phase of longer-term processes of troop formation, dissolution, or membership replacement. [Using a dynamic approach] a typology of primate troops yields to a typology of the social processes that result in the formation of different types of troops, with recognition of the possibility that similar forms of social organization may in fact be convergences arrived at through different pathways of formation. (Reynolds 1976:74)

The specification of these pathways of formation is a recent and exciting departure in primatological research, but it is not yet developed sufficiently for primate social organizations to be described in such terms. In this chapter, then, social organization refers to a heterogeneous set of spacing and social behaviors that characterize a population of animals of known age and sex (Table 8.1). In Chapter 9 we consider some of the determinants of social organization. Here our goal is descriptive. We shall try to answer two questions about primate social organization. First, how does it vary among species? Second, how does it vary over time and space within a species?

8.2   Case Studies of Primate Social Organization

Interspecific variations in primate social organization have long preoccupied researchers (e.g., DeVore 1963; Crook and Gartlan 1966; Jolly 1972a; Eisenberg, Muckenhirn, and Rudran 1972; Crook 1970; Crook, Ellis, and Goss-Custard 1976; Hladik and Hladik 1972; Hladik 1975; Clutton-Brock and Harvey 1977a). Variations among populations of a single species have also been remarked upon and studied (e.g., Oppenheimer 1977; Richard 1978a), as have changes over time in the social organization of single populations (e.g., Jolly 1972b; Oates 1977b; Hausfater, Altmann, and Altmann 1982). These studies consider some or all of the attributes listed in Table 8.1, which provides a framework for describing the primate species discussed in this section.

The classification of primates by social categories presents as many or more problems as their classification by dietary categories (Chapter 5). Such typologies have been used extensively, however, and will be discussed at length in Chapter 9. Here species are ordered by taxonomic position, and the aim is not to produce a typology but rather to emphasize the flexible and changing nature of primate social organization.

8.2A   Strepsirhines

Nocturnal species
The nocturnal strepsirhine leads a generally solitary life (Charles-Dominique 1977; Charles-Dominique et al. 1980; Martin 1972; Bearder and Doyle 1974). However, the

Table 8.1   **Attributes of social organization**

Class of attributes	Attributes
Spatial distribution of population	• Degree of clumping of individuals • Degree of overlap in areas used by individuals or sets of individuals
Composition of spatially and/or socially delineated groups within population	• Number of animals • Age of animals • Sex of animals • Biological relatedness of animals
Relations among groups	• Degree of tolerance • Defense of geographic boundaries
Social structure of groups	• Social relations among females • Social relations among males • Social relations among males and females • Social relations among adults and young

opposite of solitary is gregarious, not social, and the isolation of an animal seen moving alone through the forest night after night is spatial rather than social. The nocturnal strepsirhine maintains a network of social relations through occasional meetings with other animals, long-distance communication by vocalizations, and olfactory signals that persist for a long time. Each adult female in these solitary species occupies a distinct home range. *Home range* refers to the area in which an animal or group of animals habitually moves, feeds, and rests (Burt 1943). Its occupants do not defend its borders against intruders, and it is not necessarily used exclusively by one animal or group. Among the nocturnal strepsirhines, the home ranges of neighboring females may overlap extensively or not at all. Lorisines are the most solitary of primates. They sleep singly by day except for mothers, who curl up with their dependent offspring. Galagines and probably the nocturnal Madagascan strepsirhines forage alone at night, but by day two or three females with overlapping home ranges and their offspring may sleep together. Male galagines sleep alone.

Despite the daily spatial associations among females in these species, the primary social unit is not an all-female group. The home range of one or more females is encompassed by that of a male (Figure 8.1), who patrols the borders of their home ranges and visits them every few days, mating with any who are in estrus. In most species there are few extra males. Since the sex ratio is about equal at birth, it is likely that many young males die in fights for the possession of areas containing females. In Demidoff's bush

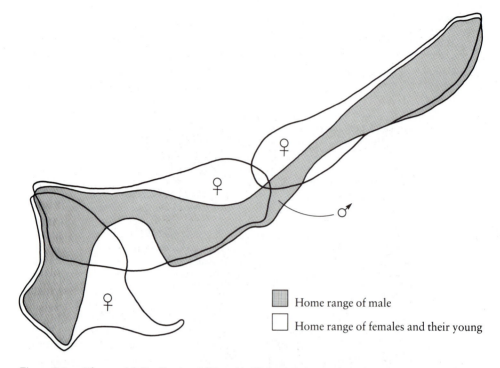

Figure 8.1   The spatial distribution of Demidoff's bush baby *(Galago demidovii)* in a sector of the Makokou forest, Gabon. Like many nocturnal strepsirhines, the home ranges (unstippled) of several females are encompassed by the home range (stippled) of one male. (From Charles-Dominique 1977.)

baby, however, four categories of adult male can be distinguished by body weight and behavior. These are (1) large males whose home ranges overlap those of females; they are usually called central males; (2) smaller, lighter, males living between the home ranges of females and large males; they are usually called peripheral males; (3) young adults just past puberty who wander vagabond-style through the forest; and (4) small, possibly immature males whose presence is tolerated by a central male. In some cases, these are the sons of the females with whom the central male associates and may be his offspring. When a central male dies or disappears, a male from one of the three reproductively inactive groups is quick to fill the vacancy. The mobile life and unpredictable fortunes of young males are in marked contrast to those of young females, who spend their lives in or near the area they were born; in fact, potto *(Perodicticus potto)* mothers sometimes leave their home ranges to their daughters and establish new home ranges for themselves nearby (Charles-Dominique 1977).

Indris *(Indri indri)*
In contrast with the nocturnal strepsirhines, the diurnal forms (found only in Madagascar) are gregarious. The first major study of the indri (Figure 8.2), done by Pollock (1975), indicated that these animals are monogamous. Social groups consisted of a reproductively active male and female and their immature offspring. Each group had al-

Figure 8.2   The indri *(Indri indri)* is the largest of the living primates on Madagascar, but 2000 years ago there were several primate species on the island much bigger than the indri. Today, the indri is itself threatened as more and more of its habitat is destroyed. (Courtesy of J. Pollock.)

most exclusive use of a patch of forest, and the adult male defended its borders against neighboring groups. An area that an individual or social group uses exclusively and defends is called a *territory* (Noble 1939). Subsequent changes in Pollock's study population suggest that the composition of social groups and their territorial boundaries may be less stable than initially thought. It is also possible that indri groups are not always nuclear families but groupings of siblings or of parents, offspring, and the future mates of those offspring (J. Pollock, pers. com. 1983).

Neighboring indri groups rarely come face to face at their borders, but when they do, singing battles and displays between the males ensue. The indri's song consists of a spellbinding chorus of eerie howls that echo 1 to 2 km through the forest. Indris sing not only during intergroup encounters but also as part of their daily routine. Calling by one group often induces another to reply in what may turn into a 30-minute singing session. Calls probably declare the occupation of a territory, reunite temporarily separated group members, and perhaps transmit information about the reproductive status of calling individuals, for adolescents participate in the adults' song and their voices can be singled out by the human ear.

Social dominance is defined by the outcome of agonistic encounters which involve aggression or submission or both. In many social mammals, members of a group can be ranked in a linear dominance hierarchy by the outcome of these encounters. Indris can be ranked in this manner even though individuals within a social group interact infrequently. Dominance among indris is asserted primarily in feeding contexts, when adult females and young animals regularly displace males from food-bearing parts of trees. In fact, males are habitually quite isolated, tagging along at some distance from their mates and young, devoting more time to territorial defense and less to feeding than do females and immature animals (Pollock 1979).

Ring-tailed lemurs *(Lemur catta)*
The social behavior of the ring-tailed lemur contrasts vividly with that of the lethargic indri. They have been studied in gallery forest at Berenty in southern Madagascar (Jolly 1966a, 1972b; Sussman 1974; Budnitz and Dainis 1975; Jones 1983). Skittish and short-tempered, ring-tailed lemurs live in groups of 5 to 22 animals and give the impression of being in perpetual social motion. Groups contain several males and females, but the core is the females, their young, and one or two males. While females apparently remain in their natal group throughout their lives, males tend to be spatially peripheral to the group as adults and transfer, perhaps several times, from one group to another. In the process of making such transfers, they sometimes join with two or three other males also in social limbo to form a temporary all-male group.

Jolly (1972b) documented a change in the distribution of social groups between 1964 and 1970. In 1964, groups had exclusive use of most of their home range, although there was little evidence of boundary defense; groups usually avoided each other at a distance.

Figure 8.3   The striking ringed tails of these lemurs *(Lemur catta)* almost certainly play a role in visual communication and they are known to be used to direct scent signals at an antagonist: the animal rubs scent on its tail from glands on its wrists and then waves its tail over its back and head toward the opponent. (Photography by A. Richard.)

In 1970, in contrast, groups shared over half their home range with one or more neighboring groups. They avoided one another in overlapping areas by using resources at different times of day, a pattern Jolly called time-plan sharing, but even so there were many more encounters and battles between groups in 1970 than there had been 6 years earlier. The reason for this change is not known, nor is it clear whether it was temporary or lasting.

Within the social group, there is a social dominance hierarchy among members of each sex, and females are dominant over males. The female hierarchy is stable. A female's dominance position does not correlate closely with other aspects of her social behavior. For instance, females approach and groom one another freely, and if preferences based on social rank exist, they have yet to be detected. Budnitz and Dainis (1975:229) characterized the female hierarchy as follows: "Female spats (fights) are usually instantaneous squabbles over a specific action or object such as a right of way or a tamarind pod. Within an hour of a spat, it is not uncommon to find the winner and the loser curled up together during a siesta."

A male's position in the male dominance hierarchy has a more pervasive effect on his behavior, although the hierarchy is less stable. Low-ranking animals often lag behind the main core of the group and move with their heads and tails down. High-ranking individuals, in contrast, carry their tails curled up in a question mark over their backs and walk with a swagger (Figure 8.3). Agonistic interactions among males are more common and more varied in form than among females although, as among females, they do not result in physical injuries except in the mating season. At that time, social order collapses in an uproar of howling, chasing, and mating. Access to females is decided in free-for-alls, and the dominance status of males is apparently irrelevant. The only ones who consistently fare worse are subadults, but even they manage to mate when the older males are otherwise occupied; Budnitz and Dainis (1975) watched one such young male mate with a female while his elders fell out of a tree fighting over her!

Sifakas *(Propithecus verreauxi)*

The sifaka lives in forests running along the western half of Madagascar. Its habitats thus include the spiny, xerophytic forests of the south and the lush, mixed evergreen and deciduous vegetation of the tropical northwest. A day spent with a group of sifakas recalls a day with indris; animals interact little, move at a leisurely pace, and spend most of their time eating or sleeping. Sifaka groups, which vary more in size and composition than indri groups, contain from 2 to 12 members. The adult sex ratio of sifaka groups ranges from 0.25 (one female to four males) to 5.0 (Richard 1978a).

The distribution of sifaka social groups varies according to habitat. In the spiny southern forests, sifakas live in territories that overlap little, and ritualized battles regularly occur along the boundaries. These battles involve no physical combat but much leaping back and forth by the two groups across their common boundary. In the northwest, in contrast, the home ranges of neighboring groups overlap extensively, and no attempt is made to defend a fixed periphery; when groups come into contact with one another, they may both move off quickly in opposite directions, ignore each other, or advance to conduct a formal battle. As in ring-tailed lemurs, most sifaka females spend their lives in one group, whereas sifaka males are more mobile, moving in and out of groups especially during the mating season.

Sifakas interact little, and when they do, they are more likely to groom than threaten one another. A dominance hierarchy is apparent only during confrontations over food. Adult males usually give way to adult females, and immature animals give way to adults. During the brief annual mating season, social order dissolves among sifakas as it does among ring-tailed lemurs. Males leave their groups and loiter on the periphery of other groups containing estrous females. Sometimes these forays result in severe or even fatal injuries (Figure 8.4) as resident males and intruders battle with one another; sometimes the intruder is allowed into the group without a fight to mate if the female will have him. The subsequent actions of successful intruders vary. Some (possibly those who were low ranking in their old group) permanently attach themselves to the new group, while others (possibly those already with high rank) return to their old group after mating (Figure 8.5). Losers who are driven off by the resident males usually return to their old group. More observations are needed on this aspect of social dynamics, but it seems

Figure 8.4 (left)   This male sifaka *(Propithecus verreauxi)* lost part of his nose and several patches of fur in a fight during the mating season. (Photography by A. Richard.)
Figure 8.5 (right)   A sifaka *(Propithecus verreauxi)* bears its teeth as another one slides on top of it. (Photography by A. Richard.)

that male access to new groups and estrous females depends on a combination of on-the-spot fighting ability or prior knowledge of that ability in other males, and the preference of the females (Richard 1974).

8.2B   New World Monkeys

Howler Monkeys *(Alouatta* spp.)
Descriptions of the social behavior of the mantled howler *(A. palliata)* come from Panama (Baldwin and Baldwin 1976a, 1976b; Bernstein 1964; Carpenter 1934; Chivers 1969; Collias and Southwick 1952) and Costa Rica (Jones 1980); descriptions of the red howler *(A. seniculus)* come from Venezuela (Neville 1972; Rudran 1979).

The size of howler monkey social groups varies from 4 to 20 animals within a single population. On Barro Colorado Island, the dramatic drop in the average size of groups from 18 to 8 between 1935 and 1941 was associated with a major population decline (Subsection 7.3B). However, the average group size decreased between 1959 and 1967

Figure 8.6   A male howler monkey *(Alouatta palliata)* steadies himself by wrapping his grasping tail around the tree trunk while grooming a female. She displays the rhythmic tongue movement commonly given by both sexes before mating. (Courtesy of K. Glander.)

even though the population grew during that period. Chivers (1969) attributed the change to a general restructuring of the population's distribution. He found not only that the monkeys were living in smaller groups than before but also that the groups were occupying smaller home ranges. Where before the home ranges of neighboring groups had overlapped considerably, now each had almost exclusive use of its home range and defended it by means of frequent, loud calls. Earlier studies had indicated that these calls served to space groups out, but the rate of calling in those years was low. Chivers suggested that the distribution of the Barro Colorado howlers changed in response to increasing population density.

Howler social groups usually contain one or two adult males and several adult females (Figure 8.6), although occasionally as many as six males coexist in a single group. The size and composition of a group rarely stay the same for long, not just because of births and deaths but also because of frequent movement in and out of the group. In

Costa Rica, most juvenile males and females leave their natal group and enter a new one as young adults. Some adults also leave, but none have been seen joining another group. Migrating adult females, who are often pregnant or accompanied by offspring, may join up with extragroup males to establish new groups in previously unoccupied areas (Jones 1980). In Venezuela, red howlers of both sexes migrate (Rudran 1979); however, juvenile males migrate more frequently than females, and fully adult males may move into and out of several groups, alternating periods of gregarious living with solitary bouts of varying length.

Howler monkeys interact little with one another. In Costa Rica, status differences among group members are seen during encounters over access to food, but these monkeys hardly count as the firecrackers of the primate world. Usually one animal simply supplants the other without agonistic behavior. Males rank higher than females, and the highest-ranking adults of each sex are normally the youngest, and the lowest-ranking the oldest. Juveniles enter the adult hierarchy at the bottom, and within the space of a year are either "up or out" (Jones 1980). Those who fail to achieve high rank in their natal group leave to try their luck elsewhere.

In Venezuela, the picture of group structure and dynamics among red howlers is less complete, but two striking differences have already emerged. First, while adult males peacefully coexist within social groups, there are frequent, fierce, and sometimes fatal fights between resident males and intruders. The number of combatants on each side influences the outcome of these fights. Cooperation between two or more invading males is usually necessary to evict resident males, and conversely, cooperation among resident males usually prevents incursions by extragroup males. The second distinguishing feature is *infanticide,* the killing of infants. Over a two-year period, Rudran (1979) recorded 40 births and 10 infant deaths or disappearances. Of these 10, at least 4 were a result of infanticide by males who had recently invaded the group. The social relations of howler monkeys in Panama have yet to be studied in detail, but the absence of even anecdotal accounts of fighting suggests that they are more like their pacific conspecifics to the north than their congenerics to the south.

Spider Monkeys *(Ateles* spp.)

The social organization of spider monkeys markedly contrasts with that of howlers. The most complete information comes from a study of long-haired spider monkeys *(A. belzebuth)* in Colombia (Klein 1971, 1974; Klein and Klein 1973, 1975, 1977). Three distinct and mutually exclusive social networks were identified in the study area, each containing about 20 adults. Most members of each network were split up into small subgroups of 3 or 4 animals, although subgroups occasionally contained as many as 22 individuals. Subgroups and lone adults belonging to the same social network were sometimes separated by 1.5 km for two days or more. The composition of subgroups varied. The majority contained adult males and females in varying proportions. About one-third were made up of females with or without dependent young. While adult males commonly associated with one another in bisexual subgroups, they rarely did so when no females were present. The composition of subgroups changed from day to day. Members of each social network shared a home range, and there was a 10 to 30% overlap between the home ranges of neighboring groups. Interactions between members of different networks were usually hostile, particularly between adult males: proximity often led to growling, whooping, and charging but not, it seems, to pitched battles. In contrast, social relations among the members of a social network were peaceful. Only 13 supplantations and 60 intense agonistic interactions were seen in over 628 hours of observation.

There have been a number of brief studies of red spider monkeys *(A. geoffroyi)* (Figure 8.7) in Panama (Carpenter 1935), Costa Rica (Freese 1976), and Mexico (Eisenberg and Kuehn 1966). Carpenter's account matches the Kleins' in all respects but one: subgroups reassembled each evening at one of several habitual sleeping trees. Eisenberg and Kuehn found monkeys living in small, cohesive groups in a mangrove swamp. The distribution of spider monkeys in Santa Rosa National Park, Costa Rica, resembled that of the Panamanian population on some occasions and the Mexican population on others. The variation may have been due to seasonal fluctuations and local differences in habitat (Freese 1976).

Tamarins *(Saguinus* spp.*)*

Like most callitrichids, cotton-top tamarins reproduce in captivity most successfully when housed in monogamous pairs with their immature offspring. If more than one female is housed with a male, at best only the most dominant female reproduces and at worst the females fight. If more than one male is housed with a female, the males fight. These findings have given rise to the widespread assumption that callitrichids typically live in stable, monogamous pairs (e.g., Kleiman 1977). Yet in a three-year study of saddle-backed tamarins *(S. fuscicollis)* polyandry as well as monogamy has been observed, and the composition of some groups has changed frequently (Terborgh and Wilson in press). Preliminary findings point to similar patterns in the moustached tamarin *(S. mystax)* (Garber et al. 1984). A mating system is said to be *polyandrous* if females typically copulate with two or more males (Stacey 1982). While polyandry has not been seen in wild cotton-top tamarins, their social groups are larger and more variable in composition than laboratory findings might lead us to expect.

Dawson (1978) studied tamarins *(S. oedipus)* living at different elevations on the Pacific slope of Panama in the Canal Zone. Vegetation ranged from rich evergreen forest at lower elevations to scrubby deciduous woodland higher up. Social group size varied from 2 to 12 individuals, with an average of 6. In evergreen forest, groups defended

Figure 8.7   A female spider monkey *(Ateles geoffroyi)* and infant. Note the distinctive facial markings of the infant, which lessen with age. (Courtesy of R. Mittermeier.)

small, almost exclusive territories. In the uplands they had larger home ranges with un-defended boundaries. The home ranges of adjacent groups overlapped considerably, and they used areas in common by time-plan sharing.

Dawson's (1978) groups contained about the same number of males as females. Their composition by age and sex changed little during the 15-month study, but the identity of group members changed frequently. Dawson suggested that most of these changes were due to migration by the male and female transients associated with most groups. A social group consisted of a single, reproductively active male and female pair, their off-spring of that year, and then a varying complement of young, transient animals. These animals frequently transferred to neighboring groups and occasionally moved 7 km to a new group. Sometimes an animal left the group only to return at a later date. Extragroup animals were rarely seen, suggesting that the time spent in transit between groups was short. Lowland groups had fewer transients than highland groups, and their turnover of members was lower.

This distribution pattern implies that there may be two socially distinct segments in cotton-top populations. On the one hand, there are male and female pairs with long-term stable bonds who mate and raise young together; on the other hand, there is a pool of peripatetic youngsters seeking mates and space to establish their own social unit as they shuttle between existing ones. The single, reproductively active female dominates other females associated with the group, and the male dominates transient males. Most dominance interactions occur during competition for food. Grooming is a common ac-tivity during rest periods, but no one has yet studied the choice of grooming partners by different group members.

Neyman's (1978) data on cotton-top tamarins in mixed deciduous and evergreen forest in Colombia conflict with the preceding account in two aspects. Although most transient females were young and *nulliparous* (i.e., they had yet to breed) a few were clearly *parous*. (The distinction between parity and nulliparity can be made in many primates, including tamarins, by looking at nipple length; nipples of parous females are longer than those of nulliparous ones.) Likewise, some transient males were apparently old, as evidenced by heavily worn teeth. The existence of these older transients could mean that the male–female bond is not a lifetime commitment or that bonded individ-uals become transients when their mates die. Until more data are available, we cannot tell whether the two field reports describe real differences or, perhaps more likely, whether they describe different aspects of a more complex and variable form of social organization than simple monogamy.

8.2C   Old World Monkeys

Guerezas *(Colobus guereza)*

The black-and-white colobus, or guereza, monkey has been the subject of several field studies, the longest of them in the Kibale Forest, Uganda (Oates 1977b). For the most part, similar findings are reported for populations in other forests. Guerezas are distrib-uted in social groups that typically contain about 10 animals, including 2 to 4 adult fe-males, their immature offspring, and 1 adult male. All-male groups and groups with several adult males as well as females and young also occur, but they are much rarer and at Kibale at least, last no more than a month or so (T. Struhsaker, pers. com. 1983). The typical one-male social group is very cohesive, its members usually within a few meters of one another. Female membership in groups is stable, but there is a turnover of males. At Kibale, for instance, the adult males changed in at least three out of seven social groups over a four-year period. The home ranges of the guerezas of the Kibale Forest overlap extensively, unlike those of guerezas studied in other habitats (Marler 1969; Dunbar and Dunbar 1974b) where groups defend exclusive territories. Still, Kibale groups did react aggressively toward one another, with subadult and adult males playing primary roles in the hostilities. In encounters between particular groups, one group usually supplanted the other. The reasons for one group's dominance over another are unclear. It may reflect the relationships between the groups or just between the adult males. Simple differences in the size of two groups or in the number of males they contained did not determine the outcome of these interactions.

Most social relationships of adult females in a guereza group take the form of frequent grooming bouts. Aggression is rare, and no dominance hierarchy has been detected. Mothers allow other females to handle their infants within the first week after birth. The relationships of males to each other and to females in the group are more variable. In the best-studied group at Kibale, the single adult male remained spatially close to females but initiated little grooming with them. The few intragroup agonistic interactions observed mostly involved him displacing another individual at a feeding or resting position. The one subadult male in the group was spatially peripheral, rarely groomed other animals, and was rarely groomed by them. In groups with more than one adult male, spatial positions differed relative to the females in the group. One male was usually spatially central, the others remaining on the group's periphery and rarely interacting with the females. At Kibale, aggression was rare among adult males. At Bole in Ethiopia, in contrast, the central male frequently attacked peripheral males.

So far we have considered guereza social organization at a single point in time. By adding time depth, the preceding account takes on new meaning. The Dunbars (1974b) watched a multimale bisexual group in the Bole Valley split into two one-male groups. They suggested (1976) that the presence in one forest of small and large one-male groups and multimale groups was best explained by positing different stages in a developmental sequence of guereza social groups. As births add new members, a small group becomes a large one containing several adult males. Multimale groups eventually either divide into small one-male groups or else one male forces out one or more of the other resident males.

Multimale groups are also created when adult males join a group that previously contained only one adult male (Oates 1977b). Oates never saw the actual entry, but each time more than one animal seemed to be involved. The resident adult male usually chased away approaching males, but if several came at once and hung around for a while, the resident male had difficulty fending them off. The intruders were generally peripheral to the group, but eventually one of them would take it over and drive the others off if they did not leave of their own accord. Sometimes, the original resident male reasserted himself and drove off all the intruders.

Red Colobus (Colobus badius)

The red colobus monkey has been studied intensively in the Kibale Forest (Struhsaker 1975; Struhsaker and Oates 1975; Struhsaker and Leland 1979; T. Struhsaker and L. Leland pers. com. 1983) and in the Tana River forests in Kenya (Marsh 1979a, 1979b). The Kibale population is distributed in social groups with from 19 to 80 members, about 50 on average (Figure 8.8). The distance between group members is often large, and a big group may be spread out over as much as 100 m.

Red colobus groups at Kibale typically contain at least two adult males, several adult females, and their young. The adult composition of groups is generally stable, the juvenile component less so. While maturing males usually remain in their natal group, females often transfer as juveniles and less frequently as subadults and adults. A few males have disappeared from the Kibale study groups or been sighted as solitaries, but their fate is unknown. Just three male transfers into two groups have been recorded, and none have transferred into the three main study groups in 13 years. (In contrast, juvenile females have been readily accepted into the study groups on several occasions, on both a temporary and a long-term basis.) The social organization of the Tana River population differs from that of the Kibale animals in that most social groups contain only one adult male and both males and females transfer between groups.

At Kibale, the spatial organization of groups is complex. Three social groups use one section of the forest, their home ranges overlap extensively or completely, and they aggressively exclude other groups. The largest of the three always manages to displace the other two, but when these two meet, both of them about the same size, the outcome is not consistent.

The social relationships of forest monkeys that live in large groups are particularly difficult to study. Animals up in the trees are hard to observe in any case, and the greater the number of adults of the same sex, the harder it is to tell them apart. One solution is to capture and mark them, but in practice this is not always feasible. At Kibale, where groups contain at least twice as many females as males and subjects are not marked, the social relations of males are better understood than those of the females. Adult males are able to displace other group members from resting, feeding, and grooming positions and can be ranked in a stable, linear hierarchy of dominance, although aggression is quite rare. During intergroup encounters, the males of each group exhibit strong cohesion, providing a united front against males of the opposing group. Males approaching subadulthood are actively and frequently harassed by the members of this adult male coterie. The process whereby a young male wins acceptance from adult males and makes the transition from harassee to harasser has not yet been studied.

Figure 8.8   Two adult male red colobus monkeys *(Colobus badius tephrosceles)* and a juvenile female in Kibale, Uganda. (Courtesy of L. Leland.)

Rhesus Macaques *(Macaca mulatta)*

The social behavior of provisioned, free-ranging rhesus macaques on Cayo Santiago Island has been studied longer and more intensively than that of most primates (e.g., Berman 1980; Chapais and Schulman 1980; Drickamer and Vessey 1973; Koford 1963; Missakian 1972; Sade 1965, 1967, 1972a, 1972b). However, this review focuses on the rhesus macaques studied by a team of researchers including the author, at Dunga Gali in northern Pakistan between 1978 and 1981. Their social behavior was studied in detail by Pearl (1982), and this account draws heavily on her work. The similarities in the social organization of the Cayo Santiago and Dunga Gali populations are more striking than the differences (Pearl and Schulman 1983). The latter appear due in large part to the marked seasonality of the environment in Pakistan and its effect on the life histories of the Dunga Gali animals.

Seven social groups were censused in the Dunga Gali forests, of which one (Kong's group) was observed regularly and another (Shadow group) intermittently. Groups ranged in size from 18 to 70 animals and included several adult males and females as well as young of both sexes. As on Cayo Santiago, females formed the stable core of the group, and males left their natal group. However, unlike Cayo Santiago, where males habitually leave at three to four years, in Dunga Gali the age of departing males ranged from three to eight years. Lone males or pairs of males were occasionally sighted in the study area, but we do not know how long they remained outside a bisexual group. We do know that all but 1 of the 11 males that left their natal group during the study were subsequently spotted in neighboring social groups (Melnick, Pearl, and Richard in press).

Kong's group had a home range of about 8 km^2, but it overlapped with parts of the home ranges of six other groups, and Kong's group did not have exclusive use of any of it. During most of the year, neighboring groups avoided one another or one group retreated as another approached. By contrast, when standing pools of water and streams had dried up just before the onset of the monsoon season, several groups would congregate for hours or days at a time. Initially these congregations looked to us like huge, undifferentiated herds of 150 to 200 animals, but as individuals became individually recognizable, we realized that they were composed of two or three spatially discrete groups that moved, fed, and rested alongside one another. Juveniles from adjacent groups sometimes played with one another, although adults remained apart, and many male transfers occurred shortly after such periods of congregation. The immediate factors precipitating these gatherings are not certain, but the valley where they took place contained one of the few streams still running in the area at the time, and it is likely that animals were attracted by the availability of drinking water.

Within the social group, adult males and females could be consistently ranked according to the outcome of dominance interactions, and no rank reversals were seen. Agonistic interactions were much less common than on Cayo Santiago, however, and so the rank relations of every pair of animals could not be determined. For instance, in Kong's group all females showed submissive behavior to the highest-ranking adult male, Kong; however, because interactions were not witnessed between every female and the second-ranking male, it is not possible to say whether all adult males were dominant to all adult females.

The ranking of females relative to one another generally followed the same rules as on Cayo Santiago. Namely, daughters ranked below all females to whom their mother

Figure 8.9   A mother is groomed by two of her daughters while she grooms Kong, the highest ranking adult male in a group of *Macaca mulatta* living in the Himalayan foothills. Notice the wounds inflicted on Kong's face during his fights with another male. (Courtesy of W. Keene.)

was subordinate and above all females to whom their mother was dominant, they did not usually outrank their mother, and among sisters social rank was inversely correlated with birth order: maturing females rose in rank above their elder sisters at about the age of three years. On Cayo Santiago this is also the age at which females give birth for the first time, whereas in Pakistan they do not give birth until they are five or six years old.

Pearl (1982) reports one exception to these rules in Pakistan. The daughter of the sixth-ranking female in Kong's group was orphaned when she was between four and five years old. By the time she reached the age of six years, she ranked just below the third-ranking female in the group. This rise in rank was not witnessed, but observations during the early part of the study suggest that a close bond had long existed between the young subordinate and the older, more dominant female. This is unusual. As on Cayo Santiago, close bonds (signified by frequent proximity and grooming) were most common between members of the same *matrilineage,* or maternal-descent group (Figure 8.9). Matrilineages were much smaller in Pakistan than on Cayo Santiago, probably because females were older when they gave birth for the first time, gave birth less frequently, and died younger. The largest matrilineage in Kong's group consisted of the highest-ranking female, her 5 known or presumed offspring, and 2 grandchildren. The other four matrilineages in the group contained only 3 or 4 individuals. On Cayo Santiago, in contrast, a living female may have as many as 30 descendents.

Little can be said about adult male relations within Kong's group, since there were only two adult males, one high ranking and central and the other lower ranking and peripheral, and they rarely interacted. The latter, Wayne, joined the periphery of the group in July 1978. He neither challenged nor attacked Kong but rather submitted to him consistently. He was slowly accepted by the group as a whole, and during the mating season that autumn he consorted with two of the three estrous females. Kong consorted with the third, the highest-ranking female in the group.

An episode in May 1979 illustrates not only a different style of entry, or attempted entry, into a new group by a male but also the degree to which a high-ranking male's status may depend on the support of females. In early May, Kong's group was approached by another group containing a large, aggressive male, Dave. Over the next several days, this group frequently moved alongside Kong's group, and Dave repeatedly approached and challenged Kong, inflicting severe wounds on him in the ensuing fights. After these fights, Dave was often seen in the center of his own group being groomed by an adult female, tolerated by the two other adult males present. The tide apparently turned when several females in Kong's group converged on Dave and repeatedly threatened him and chased him away. Dave was last seen in the vicinity of Kong's group on July 1. Ten days later he was seen once more, this time alone.

Patas monkeys (Erythrocebus patas)

The behavior of wild patas was first described by Hall (1965), who studied animals living in woodland and thorn scrub in Uganda (Figure 8.10). A patas population has also been the subject of a recent two-year investigation in Kenya (Chism, Olson, and Rowell 1983; Chism, Rowell, and Olson 1984; J. Chism and T. E. Rowell, pers. com. 1984). Briefer observations of patas have been made in the open savanna habitat around Lake Chad in Cameroon (Struhsaker and Gartlan 1970; Gartlan and Gartlan 1973).

In Uganda, bisexual social groups of patas contained from 9 to 31 animals (Hall 1965), but in Kenya groups were larger, with as many as 75 members (Chism et al. 1984). Bisexual groups at both sites normally contained only one adult male, although during the breeding season in Kenya, other males sometimes entered the group and mated. At both sites, many males were seen alone or in all-male groups.

Observations of individually identified members of bisexual groups in Kenya suggest that female patas spend their lives in the group in which they are born, whereas males leave their natal group at or before adolescence, typically at about three years old. After leaving, they live alone or in shifting all-male groups until they reach adulthood at about five years old (Chism et al. 1984). (Females, in contrast, reach sexual maturity at about 2.5 years, the shortest maturation time for any cercopithecine for which adequate data are available.) As an adult, a male spends part of his time as the resident adult male in a

Figure 8.10   A family of patas *(Erythrocebus patas)*. (Z. Leszczynski/Animals Animals.)

bisexual group and part of it alone or in an all-male group. The Kenyan study found male tenure in a bisexual group to last from a few days to nine months.

The ranging habits of all-male groups are not known. The home range of bisexual groups varies from 5 km² (Hall 1965) to 38.5 km² (D. Olson and J. Chism, pers. com. 1984). Hall (1965) reported that in Uganda, neighboring bisexual groups were either widely spaced or actively avoided one another, or perhaps both, because he saw only two intergroup encounters in 640 hours. In Kenya, intergroup encounters were observed only among adult females and juveniles who, typically, chased one another and exchanged threats, but rarely engaged in physical contact with one another. Adult males actively and aggressively participated in encounters only during the 2- to 3-month breeding season, at which times each male primarily tried to engage the other group's male in agonistic interactions such as threatening, chasing, and slapping.

Hall (1965) contrasted the widely spaced bisexual groups that he found with the aggregations of 100 to 200 animals reported by others. To explain the latter, he suggested that when water was in short supply, bisexual groups might congregate around waterholes. However, during the dry season in northern Cameroon, only all-male groups gathered at waterholes, even though bisexual groups came to drink there too (Struhsaker and Gartlan 1970; Gartlan and Gartlan 1973). The latter did not tolerate one another at close range, and their relations with all-male groups were also tense. Adult males from bisexual groups would chase all-male group members and sometimes even the entire all-male group. They themselves were never chased, perhaps because members of the all-male group never cooperated with one another to harass a bisexual group male. All-male group members did chase adult females, juveniles, and infants, and rarely did the bisexual group's male come to their assistance. Another explanation for huge bisexual aggregations, suggested by the Kenyan data, is that they simply represent bisexual groups engaged in an intergroup encounter (J. Chism, pers. com. 1984).

We do not know how new bisexual groups are formed. Hall (1965) saw temporary splits in his focal group on several occasions and suggested that when a group becomes very large, some of the females may drift away and join up with an isolated male or a member of an all-male group, but he did not witness this process.

The resident adult male is often spatially and socially peripheral to the bisexual group, and the group itself may forage dispersed over distances of up to 0.5 km. Adult females frequently threaten and harass the male when he is nearby, and his only response to these approaches is to turn and stand at bay. He is most likely to interact with other members of the group when he is sexually interested in one of the females. Sexual activity itself is usually initiated by females, who have a distinctive pattern of solicitation behavior. When mating, the male is likely to be harassed by juvenile and infant males, while his partner is harassed by other females. In both Kenya and Uganda, the adult males in bisexual groups spent much of their time scanning the environment from a vantage point such as the crown of a tall tree. This vigilance behavior has been differently interpreted by Hall (1965) and Chism et al. (1984). Although Hall did not witness any encounters between patas and predators, he inferred that the male was acting as a look-out for predators and might purposefully distract their attention from the rest of the group. In contrast, Chism et al. observed no sex differences in the behavior of adult patas during encounters with predators, and they suggest that vigilance behavior serves primarily for detecting the presence of other patas males and bisexual groups.

The extreme intolerance to their peers shown by bisexual group males is in strong contrast with the amicable relations exhibited by members of all-male groups, and the Gartlans (1973) report friendly and agonistic interactions in all-male groups indicative of consistent relationships between particular pairs of males.

Unlike the lonely life of the adult male in a bisexual group, that of a female is highly social. Females frequently groom one another and their offspring, and there is extensive caretaking of infants by other adult and immature females (Chism 1978; Chism et al. 1984). Aggression is rare. Some laboratory studies suggest that there is a dominance hierarchy among adult females (Hall 1965), but Rowell and Hartwell (1978) found no evidence of a hierarchy in a captive group living in a 100-m² outdoor enclosure.

Desert Baboons (Papio hamadryas)

Desert baboons have been the subjects of one prolonged field study (Kummer 1968) as well as a number of shorter studies (e.g., Crook and Aldrich-Blake 1968; Dunbar and Dunbar 1974a). They are found predominantly in Ethiopia, where much of their range is empty of tall trees, and they congregate for safety on cliff faces. At night, a rocky out-

Figure 8.11   A family of desert baboons *(Papio hamadryas).* (J. Popp/Anthro-Photo.)

crop may harbor a troop of as many as 150 individuals. When morning comes, the troop breaks up into bands that fan out over the landscape. Each band contains between 30 and 90 individuals. Sometimes the bands themselves split up into smaller, so-called one-male groups. The troop, whose size changes from night to night, is not a socially cohesive or highly organized unit. Is it a social unit at all? The answer is yes, for it is not a random gathering of bands. The same bands join up regularly, and bands from different troops do not tolerate one another. It is not completely clear how troops partition the terrain, but home ranges overlap, and troops are occasionally seen at the sleeping rocks of their neighbors.

The smallest and most tightly integrated social unit among desert baboons is generally considered to be a one-male group composed of several adult females and immature animals in addition to a single adult male (Figure 8.11). Kummer's (1968) observations show that this is an oversimplification, for like the guereza's one-male group, that of the desert baboon has a life cycle. Kummer used his data to reconstruct the stages whereby new one-male groups arise and existing ones are maintained and passed from one male to another or else disintegrate. He also showed that at certain stages in the cycle of change, the one-male group is more appropriately referred to as a two-male team.

The foundation of a new one-male group is laid when a subadult male starts trying to kidnap a mother and her infant. Eventually he adopts a juvenile female and conditions her to follow him by threatening her whenever she strays. At this stage, the relationship between the young male and his female is like a mother's with her infant, and only as they mature does the relationship become typically adult. Instead of picking up his female when she strays, the male threatens, chases, and herds her, punishing her with a ritualized bite on the nape or back. The one-male group reaches its greatest size when the leader is in his prime and controls females taken from aging males as well as those adopted as juveniles. He rigorously imposes spatial and social boundaries upon his females, punishing wanderers with bites. For the most part, the females follow their leader assiduously, compete with one another to groom him, and seek his protection in moments of danger.

Just as the leader's relationships with his females change as they mature, so do his relationships with other males. As a juvenile, he spends much of his time playing with and grooming his peers. As a leader, he neither approaches nor interacts with other leaders except to coordinate the band's movements and, more rarely, to fight over the possession of a female. Adult males who do not acquire females apparently maintain the bonds of their youth and frequently engage in grooming bouts with one another.

When the leader begins to age, he restricts his females less. He allows them to stray further, and the size of his unit may decrease as other males take over wanderers. The aging leader's activities gradually change from leading and mating within the unit to coordinating with his peers the movement of the whole troop.

Disintegration is not the inevitable or perhaps even the usual fate of a one-male group. Rather, another male who has previously been what Kummer called a follower gradually takes over the role of breeder and protector of females. From time to time, juvenile males abandon their play groups for a while and tag along behind, beside, or even in the midst of the females of a particular unit, engaging in furtive sexual interactions whenever possible. In this initial phase, unit leaders, especially young ones, are likely to chase off followers. As a follower matures to subadulthood, he tries to copulate with the females and forms a more persistent grooming relationship with them; later, these interactions cease and he grooms and interacts with only the leader.

The adult follower always moves with the group, bringing up the rear, and the females travel strung out between him and the leader. In a sense, he is a second, subordinate unit leader, but with the important difference that he has no access to the females, who belong exclusively to the leader. Only as the leader ages and relinquishes control over the females does the follower begin to take charge of them. At the end of the transition, all the females clearly belong to the younger male, and it is the old leader who brings up the rear. But as we noted already, his authority does not end here: old leaders still direct the movement of the unit and coordinate that of the troop. We do not know how long these two-male teams last or whether a male goes through all the follower stages with the same unit, but the tolerant and cooperative relationship between young and old stands in marked contrast to the uneasy jockeying for possession characteristic of male transitions among other primates living in one-male groups.

Savanna Baboons (*Papio cynocephalus* ssp.)

Savanna baboons have been more widely and intensively studied in the wild than any other primate. The early impression that their social organization was rigid and inflexible has given way to the recognition that it varies considerably in time and space.

At Amboseli, all female yellow baboons and most males are found in spatially cohesive groups of about 20 to 90 animals, although groups as big as 200 and as small as 13 have been seen (Altmann and Altmann 1970). Adult males spend varying lengths of time outside these bisexual groups, either alone or in all-male groups of 2 or 3. Most groups contain several adult males and females with their young. Usually there are about 4 females to every 3 males. However, like group size, the sex ratio varies widely and very occasionally adult males actually outnumber adult females. When they reach adulthood, males commonly migrate from their natal group. Once a male enters a new group, he is likely to remain in it for a period of months or years.

Group changes by females at Amboseli are rare and temporary. Hausfater (1975) noted four such episodes in the course of a year. Each time, two groups had spent the previous night in adjacent sleeping trees, and the migrant female moved off with the other group, but each time she returned to her original group within three days.

The size of the home range occupied by baboon social groups also varies widely, but generally they are huge compared with those of most primates. At Amboseli, the home range of one social group was estimated to be about 78 km² (Post 1978). Ten years ago, much or all of this area was used by seven other groups as well (Altmann and Altmann 1970). Intergroup encounters were rare during the day, but at nightfall as many as six groups would congregate to sleep. As they converged, some groups ignored one another and others responded with overt aggression. However, the most common interaction was "pushing," as one group repeatedly approached, stared at, and eventually supplanted another. Today, the density of baboon social groups at Amboseli is lower, intergroup interactions are rare at any time of day, and home ranges probably overlap less.

Large, cohesive groups containing many adult males and females and migration between groups by males have characterized the Amboseli baboons' social organization since research began on this population 20 years ago. However, a number of shorter studies suggest that these features are not universal among savanna baboons. At Debra Libanos, a dry forest north of Addis Ababa, Ethiopia, olive baboons were observed spread out in multimale units of varying size and composition. These units may have been parts of a single group, split up by poor visibility in the thick vegetation and rocky terrain (Crook and Aldrich-Blake 1968; Aldrich-Blake et al. 1971). In Sierra Leone, so-cial groups of guinea baboons sometimes fragmented into parties outwardly resembling one-male units, but party membership was not stable and males did not defend or herd females. Aside from this feature, the social organization of these primates resembled that of the Amboseli baboons (Dunbar and Nathan 1972).

Rowell (1966, 1969) studied olive baboons in Uganda that spent part of the day in gallery forest and the rest in grassland abutting the forest. Her data on social behavior came from two years of intermittent observation, and her census data covered five years. These baboons lived in cohesive social groups of about the same size as the Am-boseli groups, and neighboring groups seemed to avoid one another where their home ranges overlapped. One distinctive feature of their social organization was the small size of each group's home range, a mere 4 to 5 km². Another was the instability of group membership. Large juvenile males changed groups almost daily, young females also moved as they reached puberty, and adult males shuttled back and forth between groups, to the point where Rowell (1969) questioned whether it would be more appro-priate to consider the three social groups she watched as subgroups of a larger social unit. Only adult females remained attached to a single group throughout the study.

Dominance hierarchies are an obvious feature of the social structure of many ba-boon groups. At Amboseli, all adult females can be ranked in a linear hierarchy, and their rank relations remain consistent within and between generations (Hausfater, Alt-mann, and Altmann 1982). When a female reaches maturity, she generally achieves a rank immediately beneath her mother and above her elder sisters. The strong influence of a mother's rank on the eventual rank of her daughters has been demonstrated in other baboons (Cheney 1977) and also, as we saw, in provisioned and wild macaques (Pearl 1982; Sade 1972a; Kawamura 1958). At Amboseli, maternal age and the interbirth interval between sisters also affect rank order. Three females in the focal study group eventually ranked above their mothers. In each case, the mother was at least 15 years old and showing signs of old age. More generally, old age increases the likelihood that a fe-male will decline in rank. When the interbirth interval between sisters exceeded two years, the younger sister remained subordinate to the elder at maturity; when it was less than two years, the younger rose above the elder.

In an 18-month study of social relations among adult female chacma baboons, Sey-farth (1976) found a stable, linear hierarchy. Over half the agonistic interactions be-tween the chacma females occurred during competition for access to adult males, infants, or lactating females. During these interactions, females adjacent in rank some-times formed coalitions, supporting each other against a third animal. It is misleading, however, to suggest that female baboons spend most of their time antagonizing one an-other, for friendly interactions, especially grooming, far outnumber agonistic ones. Seyfarth found that females differed in attractiveness as grooming partners according to

Figure 8.12   Two adult male baboons *(Papio c. cynocephalus)* fighting at Amboseli, Kenya. (Courtesy of C. Saunders.)

their reproductive state and social status; lactating females received less aggression, more friendly gestures, and more grooming than those in other reproductive states, and high-ranking females received more grooming than low-ranking ones.

At Amboseli, males can also be ranked in a linear dominance hierarchy, but the ranks of particular animals change frequently and aggression is much more frequent among males than among females. The frequency and ferocity of male aggressive encounters (Figure 8.12) vary according to the reproductive state of females in the group. When one or more females were in estrus, males tended to keep away from each other, and the rate of agonistic interaction dropped. However, those interactions that did occur were often violent, and one or both participants were more likely to be wounded.

The relationship between dominance and reproductive success in savanna baboons is unclear, and until we can confidently assign paternity by using genetic markers, conclusions must be tentative. First-ranking males in the Amboseli group copulated less frequently with estrous females than did low-ranking males and failed to form *consortships* at all with certain females, even when no other females in the group were in estrus at the time. (A consortship is a close relationship that develops in some primate species between a male and an estrous female. Although it is the male who appears to maintain the consortship by persistently following the female, it is likely that both play a role in establishing the relationship [Figure 8.13].) While first-ranking males had lower rates of copulation and consortships, they timed their mating activities to coincide with the day when the estrous female was probably ovulating. By being so selective, they avoided

Figure 8.13   Adult male baboon *(Papio c. cynocephalus)* grooms his consort partner. Note the sexual swelling of the female. (Courtesy of C. Saunders.)

most of the negative consequences of prolonged consortship, such as reduced food intake because of the constant need to attend to the female and increased exposure to aggression and harassment from other males. It is not known for sure whether this selectivity actually results in a higher rate of fathering offspring; even if it does, it is unclear how much this contributes to differences in the lifetime reproduction of males, since no male occupies the first rank for very long (Hausfater 1975; see also Packer 1979b). Recent simulation models suggest that the age at which a male first achieves high rank may be important in this respect. Males reaching high rank early in life may have a higher lifetime reproduction than those reaching high rank later (G. Hausfater, pers. com. 1983).

At Gilgil, newcomers to the social group won significantly more agonistic encounters and ranked higher in the dominance hierarchy than long-term resident males in the group, but they had fewer consorts and consorted with proportionately fewer available females than males of lower dominance rank. Strum (1982) suggests that the residency status of males strongly influences mating success. Because of their familiarity and affiliations with females in the group, she argues, resident males are better able to assess complex social situations and to manipulate them to their own advantage, and they are also more likely to be viewed favorably as consorts by the females.

Social relations between males and females outside the context of consortships have been little studied until recently. Adult males are dominant to all females, and immature males dominate females within their own and younger age cohorts. At Amboseli, males rarely groom anestrous females or any other class of individuals for that matter. When a female is in estrus, one male herds her to the periphery of the group, away from all other males, follows her closely, and grooms her frequently between mountings. This animal, called the consort male, is persistently harassed by other males acting alone or cooperatively. In the process of driving off harassers, the consort male often loses his monopoly over a female. As a result, she may form a series of consort relations during a single estrus and mate with subadult males, who sneak up in a hit-and-run operation when the consort's attention is temporarily diverted (Hausfater 1975).

At Gilgil, males pay considerable attention to nonestrous as well as estrous females. Newcomers to the group follow a number of females in succession, staying close to them, grooming them, and generally trying to establish a close social relationship with them. The sequence of behavior is like a sexual consort, but there is no copulation. Eventually the female reciprocates the male's attentions, and an affiliation is formed. Such bonds are subsequently manifested in grooming preferences and in aid-seeking and aid-giving behavior during aggression and consortships (Strum 1982). Affiliations among adult males are rare, although a newcomer will approach a resident male and greet and harass him repeatedly. Such behavior is directed at a number of resident males in succession.

Males rarely if ever form coalitions with one another at Gilgil or at Amboseli. However, in 1100 hours of observation of a social group of olive baboons at Gombe Stream, Packer (1977) saw 140 solicitations, occasions on which one male attempted to enlist the support of a second male against a third. Ninety-seven of these solicitations resulted in coalitions. On 20 occasions the third male was consorting with an estrous female; in fact, males were more willing to join a coalition in that context. Six of these 20 coalitions resulted in the opponent losing his consort to the enlisting male while the solicited male continued to fight the opponent. The frequency with which adult males joined coalitions was strongly correlated with the frequency with which they successfully enlisted coalition partners, and preferences for particular partners may have been partly based on reciprocation. The genealogical relationships of the males were not known, but circumstantial evidence suggested that they were unlikely to be close relatives.

Rowell's (1966) brief account of social relations among olive baboons in Uganda provides yet another perspective. She characterizes their social relations as "relaxed and friendly." She found no evidence of a dominance hierarchy in either sex, although agonistic behaviors were so rarely seen that even if animals did recognize status differences, these differences would be hard for an observer to detect. Concerning the relationships of adult males in particular, she comments, "The general impression was, above all, one of mutual cooperation between them as they policed the environment" (1966:362). Although Rowell's account is not supported by quantitative data, it suggests that the diversity of baboon social organization may be even greater than already documented by research at Amboseli, Gilgil, and Gombe Stream.

Hanuman Langurs *(Presbytis entellus)*

Hanuman langurs provide our last case study of Old World monkey social organization. They live in a variety of habitats in India, Nepal, Sikkim, and Sri Lanka, and the limits of their range are in Pakistan, southern Tibet, and Bangladesh (Roonwal and Mohnot 1977). In the past 15 years, studies in India, Nepal, and Sri Lanka have uncovered four patterns of distribution among hanuman langurs, associated with different styles of interaction among males (Oppenheimer 1977). The third and fourth patterns described below may actually be different phases of the same pattern, and more data are needed to establish whether indeed there are populations typically characterized by each.

The first pattern consists of social groups containing several adult males and females with their young (Ripley 1967a, 1970; Jay 1965). The average size of these multimale groups varies among study sites from 13 to 37 individuals, although groups as small as 10 and as large as 98 have been reported. Males are generally tolerant of one another: the consistent outcome of agonistic interactions between them shows that there is a dominance hierarchy, but such interactions are infrequent.

The second pattern can be distinguished from the first only by studies lasting at least a year. Hanuman langurs in the Himalayas spend most of the year in groups of several adult males and females with their young. Although adult males move into and out of these bisexual groups year-round, it is during the annual mating season between late spring and fall that male membership of groups becomes particularly unstable. This is because the dominant male tries to exclude other males from the group at this time; in fact, multimale groups may briefly become one-male groups until they are reentered by the temporarily excluded males or joined by new males (Bishop 1979; Boggess 1980). Males reenter or join bisexual groups singly or in pairs, apparently living alone or perhaps in all-male groups of two or three when not attached to a bisexual group.

The third distribution pattern is characterized by the presence of one-male and all-male groups. One-male groups tend to be smaller than multimale groups, not because they contain fewer females but because they have only one adult male and no subadult males. All-male groups include subadults as well as adults and vary in size from 2 to 32 (Figure 8.14). In this pattern, the behavior of males toward one another in the presence of females is strikingly different from that found in either the first or second pattern. Males maturing in the bisexual group are subjected to increasing harassment from the resident adult male until, as juveniles or subadults, they are finally driven out. These exiles from their natal group then attach themselves to an all-male group, in which they coexist quite peacefully (J. Moore, pers. com. 1983). From time to time an all-male group will attack a one-male group and try to expel its resident male. When the attempt is successful, a new resident male emerges from the ranks of the all-male group (in a process of selection we do not yet understand), who in turn drives out the rest of the all-male group.

Figure 8.14   Male group of langurs *(Presbytis entellus)* at Mt. Abu, India. (J. Moore/Anthro-Photo.)

The fourth and probably most widely reported pattern occurs in populations composed of one-male groups, all-male groups, and multimale groups (Sugiyama 1964, 1965a, 1965b; Yoshiba 1968). In these populations, males tolerate the presence of their sons in a bisexual group. At intervals, they are attacked by all-male groups and driven out. At this point, the new resident male not only expels other members of the all-male group but also the previous resident's young sons and thus restores the group to the one-male type.

Boggess (1980) has considered the variation in langur social organization not according to male–male relations as such but according to the ways in which nongroup males gain access to group females. She suggests there are three ways. First, there may be takeovers as described in association with patterns three and four above. Second, nongroup males may temporarily associate with females without becoming group members; third, males may gradually become integrated into a group after a long, slow process of introduction. While Boggess is concerned with variation within rather than between populations, like Oppenheimer (1977) she stresses the broad variations in male relationships.

Shortly after some group takeovers in some langur populations, the new resident male may try to kill any infants present (Hrdy 1974, 1977; Sugiyama 1965a, 1965b). First taken serious notice of among hanuman langurs, infanticide by males has now been reported in other colobines (Rudran 1973; Wolf and Fleagle 1977; T. Struhsaker, pers. com. 1984), a cercopithecine and a New World monkey (Struhsaker 1977; Rudran 1979), and two great apes (Fossey, cited in Hrdy 1979; Bygott 1972; J. Goodall 1977). Both the frequency of infanticide and its biological significance have been much debated (Dolhinow 1977; Curtin and Dolhinow 1978; Angst and Thommen 1977; Chapman and Hausfater 1979; Hrdy 1979; Boggess 1980). In Chapter 6 we saw that many primate females come into estrus shortly after the loss of an infant. When a new resident male kills the offspring of the previous resident, it may have the effect of hastening estrus in his newly acquired females and hence the birth of his own offspring (but see Curtin and Dolhinow 1978). Hrdy (1974) presented a model of reproductive success for langur males in which takeovers were considered a common form of social change and infanticide was seen as an evolutionary adaptation to a situation in which males were frequently expelled from groups. Boggess (1980), in contrast, suggests that infanticide is rare and that intermale competition more commonly takes the form of gradual introductions to and expulsions from bisexual groups. The debate continues.

In contrast to the turbulent lives and relationships of male langurs, the life of a female is relatively calm. Females usually spend their lives in the group into which they are born, growing up into a spatially and socially cohesive coterie of adult females. A dominance hierarchy is discernible among these females, but most interactions are friendly, consisting of long grooming bouts. As with howler monkeys, dominance varies inversely with age (Hrdy and Hrdy 1976). Like most colobines, langur females allow other females to handle their infants almost from birth. This contrasts strikingly with the possessiveness of most cercopithecine mothers toward their newborns, and McKenna (1979) has explored the evolutionary context in which this behavioral difference between the two subfamilies of Old World monkeys may have evolved. Natal coat color differs from that of adults in many primates, and colobine newborns are particularly distinctive (Figure 8.15). Although this feature presumably helps mothers locate infants being passed from hand to hand, it would also seem to increase the vulnerability of infants to predators.

8.2D   Apes

There are only four ape genera, but the variety of their social organization is as great as that found among monkeys or strepsirhines, and this brief review includes species in all four genera.

Chimpanzees *(Pan troglodytes)*
Chimpanzees are distributed in loose communities containing between 15 and 80 individuals (Nishida 1968, 1979; Izawa 1970; Sugiyama 1973; Goodall et al. 1979; Itani

Figure 8.15   A female silvered-leaf monkey *(Presbytis cristata)* and her red-orange newborn come down to the ground. (Courtesy of K. Wolf.)

Figure 8.16   Adult male chimpanzee (*Pan troglodytes*, center) reaches out in a gesture of reassurance to the male on his left. (Courtesy of R. Wrangham.)

1979; Wrangham 1979). Communities are identified by the association patterns of adult males. Males who have friendly relations with one another and make use of the same home range are said to be members of the same community, and their home range can be considered the community range. Males often travel alone or in small groups, but reunions are frequent and generally amicable (Figure 8.16). They usually avoid males from neighboring communities; when encounters do occur, they are hostile and sometimes violent. No male, adult or immature, has ever been known to migrate from one community to another.

The large home ranges of males encompass several smaller home ranges each of which is used by an anestrous female and her young. These female home ranges overlap, but each contains a distinct core area used with particular frequency by the resident female and her young. The ranging pattern of an estrous female is more variable. She may travel widely within the community, associating briefly with several males, or she may form a consortship with a particular male and spend her time with him in a small area on the edge of the community range. Some estrous females migrate to another community. Among nulliparous females this move may be permanent, but among older ones it is temporary (Pusey 1979; Nishida 1979). Finally, estrous females are sometimes forcibly kidnapped by males from neighboring communities (Goodall et al. 1979).

It is not yet clear whether the home ranges of females are spaced independently of those of males (Figure 8.17). In other words, we do not know if the chimpanzee community is bisexual or made up exclusively of males who vie with neighboring male commu-

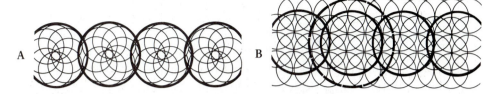

Figure 8.17   Two views of the spatial relationships between the home ranges of females and males. Circles with thick lines are male community ranges, in which a number of males cover the area equally. Circles with thin lines represent the ranging patterns of anestrous females. In **A,** females share a community range with males within which individual females occupy core areas. In **B,** females are distributed in small home ranges without reference to male community ranges. The dashed line in **B** shows expansion of a male community range: its males get to associate with more females. This would not happen if the community range expanded in **A;** rather, females would increase the size of their core areas. (Adapted from Wrangham 1979.)

nities for parts of the forest and the females living there. Wrangham (1979) noted that three anestrous females with home ranges overlapping two male community ranges associated peacefully with males of both, indicating that these females did not defend a community range as males do.

Fission of a large male community into two smaller ones has been observed once at Gombe Stream (Goodall et al. 1979; Bygott 1979). The frequency of meetings and associations between the males of the two incipient communities decreased over a period of about 18 months, and the hostility and avoidance typical of neighboring communities became more prominent.

Among male community members there is a dominance hierarchy. As an individual matures, he becomes dominant over adolescent and then adult females, and as a young adult he begins to dominate the oldest males. By middle age, he has reached his highest rank and dominates most, in some instances all, younger and older males. As his age increases and his physique deteriorates, his rank declines as younger animals begin their rise through the ranks (Bygott 1979). The precise rank achieved by a male during his life and the precise age at which he reaches it are probably influenced by many factors, including his general health, his ability to form coalitions with other males, his kinship ties, and his aggressiveness and general personality. Males groom one another frequently. In fact, grooming is more common among this class of animals than any other (Goodall 1968; Simpson 1973).

Female relationships stand in marked contrast to those of males. Anestrous females spend most of their time alone with their offspring or in nursery parties with other females and young. When together, they rarely interact aggressively and no clear dominance hierarchy is discernible. When consistent dominance/subordinance relations are apparent, usually the older female is dominant (Bygott 1979). The behavior of estrous females varies widely, as we have already noted. Some mate promiscuously with several males, the latter showing little possessiveness (Goodall 1968). Others form more protracted consortships with a single adult male (McGinnis 1979). In these cases the male usually initiates and maintains the consortship by herding the female away from other animals, thereby maintaining exclusive access to her for himself. Males rarely interact with infants, at least in part because mothers are extremely protective and shield their infants from situations where aggression is even remotely possible (Bygott 1979).

Recently infanticide has been reported (Goodall 1977; Goodall et al. 1979) during attacks by males on females and infants associated with another community and also during attacks by females on members of their own community. The explanation offered for the attacks by males is similar to that for langur male attacks. Females, it is argued, try to kill the infants of other community members to eliminate them as potential competitors, but the exact circumstances under which females kill infants have yet to be delineated.

Gorillas (Gorilla gorilla)

Gorilla populations are divided into social groups containing between 2 and 20 animals. Groups typically consist of one older male, a number of younger males, adult females, and young (Figure 8.18). Adult males are also occasionally seen travelling alone (Schaller 1963; Harcourt 1979). Young adult males are black all over, but older males develop a silver "saddle" on their backs, and they are known as blackbacks and silverbacks, respectively. This color change makes it possible to estimate the relative ages of males.

The home ranges of neighboring groups overlap (Fossey 1974), and intergroup encounters vary from "almost friendly to aggressive" (Harcourt 1979). When groups are close to each other, young females, particularly those who have yet to give birth, sometimes transfer from one group to the other. Some females transfer only once, but many transfer several times (Harcourt, Stewart, and Fossey 1976; Harcourt 1978). Males move out of groups less commonly and unlike females, they always become solitary upon leaving a group. New groups are probably formed by the transfer of one or more females from a group to a lone silverback. Unlike chimpanzees, gorilla females are not coerced into making these moves and seem to make them voluntarily.

Relationships among adult males vary. Blackbacks spend much of their time on the periphery of the group; in one group studied by Harcourt (1979), the silver-backed, dominant male and the subordinate blackback were rarely close to one another. In a second group, the silverback and blackback were often near one another, and the latter even groomed the silverback. Like the males, adult females interacted little with one another. There seemed to be a dominance hierarchy among them, but it was difficult to detect because agonistic interactions were so rare. Their closest relationships were with their offspring and with the silverback. He was a social focus for the females, and they often congregated around him (Figure 8.19).

Figure 8.18   Silverback male gorilla (*Gorilla gorilla,* left) with females and young. (A. Harcourt/Anthro-Photo.)

Gorillas rarely mate, or at least they have rarely been observed mating, and only the silverback mates with the adult females in the group. Both he and any blackbacks in the group mate with adolescent females and nulliparous young adults who have not yet given birth.

Grooming relations between the silverback and his females differed in the two groups studied by Harcourt. In one, the silverback frequently groomed other animals, while in the other he was a frequent recipient of grooming. The silverback was extremely tolerant of immature animals in the latter group, and they seemed attracted to him, following and resting near him and frequently grooming him. In another study, a silverback actually adopted an orphaned infant (Fossey 1979). In contrast to the indulgent resident silverback, lone silverbacks were known to attack groups, kill any infant present if they could, and then leave with the dead infant's mother. Fossey attributed these episodes of infanticide to intense competition for adult females brought about by an increasingly high ratio of adult males to females in the area.

Orang-utans *(Pongo pygmaeus)*
Orang-utans live in the forests of Borneo and Sumatra and have been studied in both (MacKinnon 1974; Rodman 1973, 1977, 1979; Galdikas 1979). All observers report that animals are widely scattered and interact rarely. Each adult female has a separate core area, and she usually travels accompanied by only one or two dependent young (Figure 8.20). Occasionally small groups of females with their young are seen together, but such groupings do not persist for more than a day or so. An adult male is almost always alone, his solitude interrupted only by brief periods when he accompanies or is joined by an

Figure 8.19   Silverback male gorilla *(Gorilla gorilla)* chest-beats. (A. Harcourt/Anthro-Photo.)

Figure 8.20   A young orang-utan *(Pongo pygmaeus)*. (Courtesy of A. Mitchell.)

adolescent or adult female. Individuals approaching adulthood are sometimes seen alone, sometimes with other adolescents, sometimes with their mothers, and sometimes with adult females who are not their mothers. The home ranges of adult females are stable. They overlap extensively, but each female apparently recognizes clear boundaries to her own home range and does not stray beyond them.

Male movements are harder to decipher. There seem to be two classes of males, residents and wanderers (Rodman 1973; Galdikas 1979). A resident male is regularly seen in an area about the same size as an adult female's home range. Wanderers probably travel over much larger areas, perhaps seeking an empty patch of forest. Their movements are not well known, however, because of the practical problem of following an animal far from base camp.

A simple dichotomy between residents and wanderers cannot explain certain observations of male behavior. Galdikas, for example, reported that a resident male she called Throatpouch disappeared from her study site and was presumed dead. Other males visited the vacated area, but none stayed until seven months after the disappearance of Throatpouch, when another adult male, Nick, moved in. Almost two years later there was an influx of males into Nick's range, foremost among them Throatpouch. Galdikas did not relate the story's end, perhaps because there is none yet, but her data generate tantalizing questions: why did Throatpouch leave, why was it so long before another male settled in the area, and what will be the outcome of Throatpouch's return?

The frequency of interaction among individuals varies among populations. Rodman (1979), for example, reported that animals at his study site associated with one another for less than 2% of the observation time, whereas Galdikas (1979) gave a figure of 18%. This discrepancy may be due to the way each researcher defined an association or it may reflect real differences. This account of social relations is based on observations by Galdikas. It is possible that in other areas not only the frequency but also the content of interactions and quality of relations may differ.

Female social relations vary. Sometimes a female and her young feed close to another such unit, each apparently ignoring the other. Sometimes the approach of a female unit causes another to withdraw a short distance, sometimes it causes precipitous flight, and occasionally it causes aggression. Often, however, two units travel together after making contact. What factors influence these relationships? We do not really know, but there are indications that a female prefers to travel with animals with whom she associated as an adolescent rather than with close relatives.

Encounters among male orang-utans are extremely rare, their outcome strongly influenced by the ages of the participants and by the presence or absence of females. In 41 months and 5470 hours of observation, Galdikas (1979) witnessed only two encounters between lone adult males; both were brief, involved no physical contact, and quickly ended when one of the two moved away. Loud calls may help wandering males coordinate their movements. Galdikas notes that her study area would be empty of males other than the resident for long periods and then suddenly there would be an influx of males, who called persistently. After some days or weeks, the calling stopped and the males vanished.

Encounters between adult males accompanied by females are as rare as those between lone adult males, but they lead to fierce fights between the males. Contacts between lone adult and subadult males are more frequent and less hostile, and subadults seem to be quite tolerant of one another. They have been seen traveling together without any sign of conflict, even in the presence of a sexually receptive female. Mutual antipathy seems to emerge in full strength only among adults.

Interactions between males and females vary according to the reproductive status of the female and the ages and possibly individual preferences of both participants. Most commonly, adult males and females avoid or ignore one another, but when an adult female is sexually receptive, she and an adult male may form a consort relationship. The two animals travel together for several days, during which time they mate repeatedly. Sexually receptive subadult females are also attracted to adult males but the attraction is not mutual. Even the most pointed approaches of adolescent females excite little or no interest in adult males. Subadult males are less discriminating than their elders and follow subadult and adult females zealously. Females are less than encouraging of these attentions, and when mating does occur, it is invariably without their full cooperation. Unwillingness is expressed in varying degrees, "from seemingly token resistance with some squealing to fierce battles" (Galdikas 1979).

Siamangs and Gibbons *(Hylobates* spp.*)*

Smaller than chimpanzees, gorillas, and orang-utans, siamangs and gibbons are referred to as the lesser apes. Some researchers consider the siamang a separate genus, *Symphalangus,* but here we follow Groves (1972) and drop that distinction. Features such as coat color, markings, and calling behavior have been used to distinguish nine species of gibbon and siamang (Groves 1972), but the geographic isolation of some of these populations is probably quite recent and they may not yet be reproductively isolated (Figure 8.21) (Chivers 1977). In any case, the taxonomic status of these populations does not really concern us here, and in social behavior they are similar so far as we can tell.

Siamangs and gibbons occupy many forests in Southeast Asia. Siamangs live in the mountainous regions of Sumatra and the Malay Peninsula, and their ecology and social behavior have been studied in the latter region (Chivers 1974, 1977, 1980; Chivers, Raemaekers, and Aldrich-Blake 1975; MacKinnon 1976, 1979; Raemaekers 1978). Gibbons are found in Assam, Bangladesh, Burma, and Thailand, through Laos, Vietnam, and southern China, and down into the Malay Peninsula, Sumatra, Borneo, Java, and the Mentawi Islands. They have been studied in several areas (Tenaza 1975; Carpenter 1940; Brockelman, Ross, and Pantuwatana 1973, 1974; Ellefson 1974; Gittins and Raemaekers 1980).

Siamangs (Figure 8.22) are distributed in social groups comprising one adult male, one adult female, and up to three dependent offspring. The adults are probably paired for life, although there has been no lifetime study of siamangs. One couple was monitored for eight years and had perhaps been together for eight years before that, based on the age of the oldest offspring at the start of the study. The relationship was probably terminated by the death of the male, suggesting that at least some relationships last a long time, if not a lifetime (Chivers 1977; Chivers and Raemaekers 1980).

When the young reach the age of eight or nine years, they leave their parents. This departure seems to be prompted both by the increasing intolerance of the adults, espe-

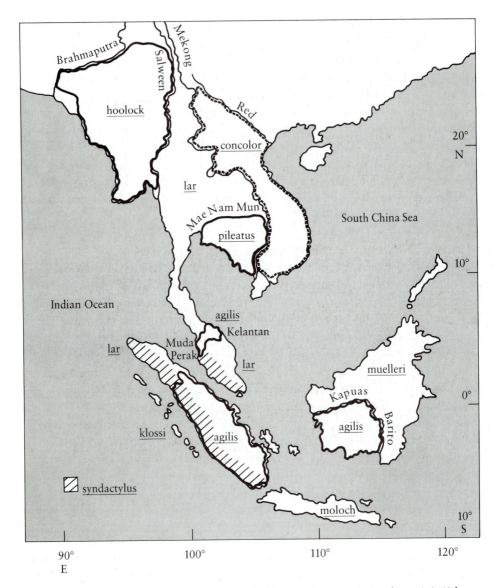

Figure 8.21   Distribution of the nine species of gibbon and siamang in Southeast Asia (After Chivers 1980.)

cially the male, and by the desire of maturing offspring to find their own mates. The young male makes known his availability by calling apart from the rest of the group, and this attracts young females from other groups within earshot. Finding a compatible mate is not an easy business, judging from the complicated saga of the successive rela-

Figure 8.22   Siamang *(Hylobates syndactylus)* hangs by its arms. (Courtesy of A. Mitchell.)

tionships formed by young adult siamangs told by Chivers and his colleagues (Chivers et al. 1975; Chivers and Raemaekers 1980). Once a young couple has finally formed a stable bond, they establish a territory in an empty patch of forest, a patch vacated by the death of its previous occupants or, with parental assistance, a corner of their parents' territory enlarged by aggressive incursions into neighboring territories.

The territories of neighboring groups overlap little. For the most part, residents maintain the exclusiveness of their territories by loud group calls, which carry for several kilometers, although residents occasionally reinforce these vocal claims in boundary disputes. Siamang groups are cohesive. Animals are rarely more than 30 m apart and usually much closer to one another than that. Altercations are rare, and pairs spend one to two hours every day lounging together, grooming one another for perhaps half this time. Infants and juveniles get to be groomed at nightfall, when they settle down with their parents.

Much of this description of siamang social organization applies equally to gibbons, although life in a gibbon group is somewhat less relaxed and less cohesive. Individuals travel farther and faster during the day, spreading out over a broad front. Social behavior both within and between groups is more frequent. Neighboring groups encounter one another regularly, and active border patrols have been reported in at least one population (Ellefson 1968). Tensions run high during these encounters, and males and sometimes females display to one another and engage in wild chases back and forth across their borders. Within the group, adult gibbons harass their maturing offspring more actively than do siamangs.

8.2E   Commentary

Most primates live in social groups. Exceptions include the nocturnal strepsirhines, tarsiers, and orang-utans. In the past, these species were thought not to have a social organization, but today we know that most solitary primates are not isolated socially. Typically, they belong to social networks. Boundaries between networks are harder to recognize than those between groups because networks are not spatially discrete. The males of many species and the females of a few live alone after leaving their natal group or a group to which they belonged as adults. Such animals may be the only primates that are truly isolated socially, but this isolation is usually temporary, for most animals subsequently join another group. Moreover, solitary animals have not been observed long enough to establish the extent of their social isolation.

Descriptions of spacing among primates usually contrast species whose members live in home ranges with those whose members occupy and defend territories. The simple dichotomy suggested by these terms is misleading (Waser and Wiley 1980). Primate spacing patterns may vary among populations, among social groups, in a population, or between males and females in a population. The behavioral mechanisms maintaining these patterns are variable, too, and particular spacing patterns and behaviors are not consistently associated with one another. In all populations of gibbons, indris, and some callitrichids, social groups have exclusive use of a patch of forest. Among sifakas, ringtailed lemurs, howler monkeys, guerezas, mangabeys, and langurs, only in certain populations do social groups occupy areas of exclusive use. Among red colobus monkeys, spacing between groups varies within a single population. At Kibale, the home ranges of three social groups overlapped completely, but they excluded other social groups from their common range. Males and females are spaced differently in orang-utans, chimpanzees, and the nocturnal strepsirhines. Chimpanzee females, for instance, have overlapping home ranges, which contain core areas of intense and perhaps exclusive use. The males of a particular social network or community share a range that does not overlap the ranges of neighboring male communities.

Relations between neighboring social units may be symmetrical or asymmetrical. Symmetrical relationships are characterized by mutual antagonism, avoidance, or tolerance. In some species such as sifakas, all three of these responses have been observed between a single pair of groups. Determinants of the quality of particular interactions are unclear. In an asymmetrical relationship, the members of one group consistently displace members of the other. The asymmetry is sometimes but not always determined by the relative size of the interacting groups.

In the past, studies of primate social behavior emphasized interactions between members of a single group and categorized interactions between groups separately. Today, several kinds of evidence suggest that intergroup relations are usefully studied as an integral part of social behavior. First, in all primates one or both sexes typically leave their natal group as they approach adulthood and join another group or establish a new group. Often the group that is joined or established is adjacent to the natal group (e.g., Melnick, Pearl, and Richard in press). Thus, some members of different groups may know one another from their infancy and adolescence spent together. Second, when new groups are created by the fission of large groups, animals may know one another from the time when their respective groups formed a single large group. When new groups are created by young animals who have left their natal group, interactions between groups may actually be between parents and their grown offspring. Third, even if animals have never belonged to the same social group, individual familiarity may be established during repeated encounters.

In summary, the dichotomy between territorial and nonterritorial species is incomplete and confusing, and relations between neighboring groups need further study. One function of these relations is to space animals according to the distribution of resources, but another is social rather than ecological. Indeed, the social component of intergroup relations in many species is almost certainly higher and more complex than we used to think.

Primate social units vary greatly in size and composition. A newly formed social group of young adult gibbons, for example, contains just 2 animals. At the other end of the spectrum are baboon groups with over 200 members. The sexual composition of social groups is also highly variable, although most contain at least one female and one male. All-male groups are found in some populations. Once viewed as unstructured and temporary aggregations, the membership of all-male groups at least among langurs may remain unchanged for months on end, and individuals form stable social relations with one another. Such groups do lack the social stability and genealogical continuity provided by the female members of bisexual langur groups. Occasionally a female is reported traveling with an otherwise exclusively male group of langurs, but this association is usually brief.

Social groups made up only of females are possible in principle but rare or nonexistent in practice among primates. Foraging groups of female spider monkeys are common, but these females are members of a larger, bisexual social network and associate intermittently with male members of the network as well. Associations of female chimpanzees are also common, but their significance remains unclear. Do females recognize social boundaries, or does each female simply tolerate and associate with others with whom she is familiar, most likely her neighbors? If females are organized into discrete social networks, do the spatial boundaries of these networks coincide with the boundaries of male community ranges? Until we have answers to these questions, the existence of independent female social units among primates remains a possibility but not a fact.

Compared with bisexual langur groups, the members of all-male langur groups are probably less closely related to one another on average. Langur daughters usually spend their lives in the same bisexual group as their mothers, whereas there is no evidence of such a persistent relationship between sons and either of their parents. This raises the more general issue of variation within and between species in the genealogical relatedness of social unit members. These differences are determined by the birth and death rate and by whether males, females, or both sexes normally leave the natal group as they approach adulthood. In monogamous species, the young of both sexes leave their natal group and it is generally assumed that brothers and sisters do not subsequently form pair bonds with one another.

In gorillas, red colobus, and howler monkeys, some males and females leave their natal group, although the proportion of each sex that leaves has yet to be established. In chimpanzees, only females move out of the area in which they were born. In most primates, however, males leave and adult females remain. The latter are thus more closely related to one another than they are to adult males in the group. The genealogical relatedness of adult males to one another is rarely known. If they were born in the same social unit, they may share one or both parents, but since the social origins of adult males are rarely known, let alone their parentage, little is known about the relatedness of nonnatal male members of a social unit.

Genealogical relationships have important implications for social behavior. This was pointed out by W. D. Hamilton (1963, 1964) who showed how the genealogical relatedness of animals might be expected to determine their social relationships. Hamilton's idea, now known as kin selection theory, has been elaborated on by E. O. Wilson (1975) and others (e.g., Alexander 1974; Boorman and Levitt 1980; Maynard Smith 1978; Trivers 1971; Vehrencamp 1979; Wade 1979) into a more sweeping set of hypotheses about the evolution of social behavior. These hypotheses are usually referred to as *sociobiology*. Sociobiology has had a major influence on interpretations of primate social behavior and on the design of behavioral studies. In particular, it has focused attention on the role of genealogical relatedness or kinship in determining social relations within primate groups. (For a review of applications of sociobiological theory to primate data, see Richard and Schulman 1982.)

Differences in the relatedness of social group members help us to understand the diversity of social relations among primate species. However, how are we to explain the initial evolution of interspecific differences in patterns of relatedness among social group members? In Subsection 9.4F we shall consider one of the few attempts that have been made to develop such an explanation.

8.3   Putting Primates in Their Social Place

In a comprehensive survey of mammalian social organization, E. O. Wilson (1975) asserts, as have others before him, that primate social organization is more complex than that of other mammals. However, the data that Wilson presents seem, if anything, to belie his claim, and Rowell (1972b) has argued that the issue cannot yet be settled. There are several reasons for this. First, though some researchers have taken a comparative approach to the study of mammalian social organization (e.g., Eisenberg 1966, 1975, 1981; Ewer 1968), methods have not yet been developed to quantify interspecific differences in social complexity (but see Pearl and Schulman 1983), and it is unclear how to define complexity in the first place. Second, even with a clear definition and appropriate methods of comparison, the undertaking would fail because data on the social organization of many mammals are still so few. Moreover, the signals that some mammals use to communicate are difficult for the human observer to detect, let alone decipher. Primates communicate by means of facial expressions, postures, and vocalizations that are relatively easy to monitor (Figure 8.23). Often the similarity of these signals to our own nonverbal signals makes them readily interpretable. In contrast, other mammals are more likely to use scents, calls, and visual signals that humans do not notice. Until we have better ways of identifying and interpreting these signals and until more detailed studies of mammalian social organization have been done, we shall have slim grounds on which to base arguments about the unequaled complexity of primate social life. In the meantime, there is enough information to suggest that the claim of uniqueness for primate social organization may be unfounded. In order to emphasize this point, we end this chapter by briefly describing the elaborate social organization of two carnivores, just as we described the social organization of primates in Section 8.2.

African wild dogs *(Lycaeon pictus)* (Figure 8.24) have been studied in Serengeti National Park, Tanzania (Kuhme 1965; Estes and Goddard 1967; van Lawick 1974; van Lawick and van Lawick-Goodall 1970). They generally live in packs of 4 to 12 adults and several young, although there are occasional reports of packs with as many as 40 adults. All researchers agree that males usually outnumber females, but that may not be

Figure 8.23   A female gelada baboon *(Theropithecus gelada)* threatens another (out of view), while her harem male comes in to support her. He gives an eyebrow threat—note white areas on eyebrows exposed. (Courtesy of R. I. M. Dunbar.)

normal. Wild dogs have been heavily hunted by humans, and females with pups may be easier targets than males. One or two females in the pack give birth each year to litters of 8 to 14 pups. After the first few months spent in and around a den, the young start to move with the rest of the pack, and many do not survive this period in their development. Adult membership of the pack is stable, although it is not yet known whether young adults leave their natal group to join another.

During much of the year the pack travels over thousands of square kilometers of grassland in search of prey. The huge home ranges of neighboring packs overlap. On the rare occasions when they meet, their interactions vary from friendly greetings and sometimes even combined hunting expeditions to hostile chases. The packs' histories probably explain this variation. New packs form when one large pack divides into smaller units, and it is likely that when these units meet, animals respond almost as though they were still members of a single pack. For about two months after the birth of a litter, the pack's activities center around the den in which the pups spend their early life. During this period, pack members eat their fill after making a kill and then return to the den, where they regurgitate part of their meal for the cubs and their mothers and sometimes for sick or crippled animals left behind. At the kill site, juveniles are given precedence by the adults, and if a litter is orphaned, the rest of the pack takes over its care. Estes and

Figure 8.24   Wild dogs (foreground) compete with hyaena for prey. (G. Rilling/Anthro-Photo.)

Goddard (1967) reported one instance in which nine orphaned cubs were reared by the eight remaining members of the pack, all of them males.

Males and females are each organized into a dominance hierarchy, but because agonistic behavior is rare, particularly among males, the hierarchy is difficult to detect. Relations between males and estrous females vary. A female may enter a consort relationship with a single male and mate only with him, or she may mate with two or three. In either case, there is no overt competition among the males. Mothers are the most aggressive members of the pack. Van Lawick described one mother who persistently drove off another female who became pregnant and subsequently killed all but one of that female's pups. She herself adopted the only survivor and would not allow its biological mother near.

Lions *(Panthera leo)*, like wild dogs, have been subjects of intensive study in Serengeti National Park (Schaller 1972; Bertram 1975, 1976, 1978; Packer and Pusey 1982, 1983). Lions form two kinds of groups, prides and nomadic groups. Prides contain 4 to 37 animals, with an average of 15 (Figure 8.25). Closely related adult females, normally sisters or cousins, form a stable core within each pride, and these females spend most or all of their lives within a fixed home range that is passed on from one generation to the next.

Figure 8.25   A pride of resting lions. (G. Rilling/Anthro-Photo.)

Attached to this core are two to four resident males, usually brothers. Their stay in the pride lasts from a few months to several years, ending with their death, voluntary departure, or forced exile. Resident males are sometimes ousted by intruding nomads, although the more of them there are, the longer they manage to keep out such intruders.

Nomadic groups are more open than prides, with animals moving in and out frequently, and they do not occupy a fixed home range. As young adults most males and many females leave their natal pride and join the nomadic sector of the population. They may subsequently form stable prides with other previously nomadic animals, and males also take over existing prides when they get the chance. Relations among members of each sex vary according to social context. Nomadic lions and lionesses tend to avoid prides, but when they do meet, the prides respond with fierce hostility. Nomads, in contrast, tolerate strangers and may establish social contacts with them. Estrous females provide a bridge between nomads and members of a pride, however: estrous females living in prides tolerate strange males, and estrous females who are nomadic are tolerated by males belonging to prides.

Relations within the pride are generally amicable except at kills. There is no dominance hierarchy, and fights over food erupt frequently: "There is a tendency for the larger of two animals to win . . . because it is stronger. . . . This fact is recognized but not necessarily accepted by the others" (Schaller 1972:135). Lionesses do most of the hunting, and only when the prey is brought down do males take advantage of their superior size and strength to barge between females and cubs to eat their fill. Lionesses cooperate closely during hunts, fanning out and converging on the prey in one concerted movement or driving it into an ambush. Males and females have little to do with one another except when females are in estrus, at which time couples sometimes form temporary consort relationships. Estrous females may also mate with several males in succession, and males display little overt competition over estrous females. In fact, female choice may be more important than male competition in determining who fathers the females' offspring.

During the first two months of life, cubs have a diet of milk gradually supplemented by meat. They are almost helpless at birth, but unlike wild dog mothers, lionesses do not stay behind to guard their offspring. Instead, they leave them hidden in thickets while they go off hunting with other females. After helping make a kill, a mother may fetch her cubs but rarely before she herself has eaten her fill. By the time the cubs arrive, there is often little or nothing left. Mothers prefer to nurse their own cubs but permit other pride members' cubs to suckle as well, and a cub may snack from as many as five females to obtain a full meal. Adult males barely tolerate cubs and certainly do not give them precedence at kills. Intruding males sometimes kill cubs present in a pride when they take it over.

These two examples provide many parallels with the social organization of certain primates. We can pick out primates that match these carnivores in many respects: group composition, the presence or absence of a clear dominance hierarchy, the presence of a stable core of adult females, the replacement of adult males in bisexual groups, the departure of young males and females from their natal group, and the variability in intergroup relations. However, the social organizations of wild dogs and lions also contain characteristics not found among primates, such as the extensive cooperation in food harvesting or the division of labor in harvesting activities and infant care. Until we have more detailed comparative data, the case for the greater social complexity of the primates must remain open.

CHAPTER NINE

Determinants of
Social Organization

9.1 Introduction

THE first modern attempt to explain the evolution of primate social groups looked not to ecology but to sexual behavior. In 1932, Solly Zuckerman proposed that the fundamental bond in permanent social groups among primates was sexual. He believed that male and female primates were sexually active throughout the year and hence mutually attracted throughout the year; he contrasted them with other mammals who were sexually active for only part of the year and who, he argued, had a less complex social life. As information accumulated, it became clear that Zuckerman's understanding of primate sociality was incomplete. Carpenter and others (e.g., Lancaster and Lee 1965) found that many primates mate and reproduce seasonally without seasonal changes in their grouping patterns. As early as 1942, with field evidence from howler, spider, and rhesus monkeys, orang-utans, and gibbons, Carpenter (1942b:199) asserted: "The long-standing erroneous belief that non-human primates are receptive throughout the menstrual cycle, and indeed during phases of pregnancy, is based first on anthropomorphic thinking and second on errors of observation. . . . To conclude that sexuality is the only enduring basis of these societies would be to oversimplify the complex facts."

Carpenter himself came up with no grand evolutionary explanations. He described the major features of primate social groups in considerable detail and proposed that every primate species had a "central grouping tendency." When a group exceeds the size typical of the species, social mechanisms are activated that cause the group to split. What determines the central grouping tendency itself? On the one hand, Carpenter (1942a:249) invoked "various kinds and degrees of social drives and social incentives"; on the other, he acknowledged in a general way that a primate's social organization must be adapted to the environment. Attempts to identify the environmental factors that shape primate social organization began in earnest in the 1960s, paralleling similar developments in research on birds (e.g., Crook 1965) and mammals, especially ungulates and carnivores (Eisenberg 1966, 1981; Crook, Ellis, and Goss-Custard 1976; Geist 1974a, 1974b; Estes 1974; Kleiman and Eisenberg 1973; Jarman 1974). Fundamental to all these efforts is the assumption that the presence of a particular form of social organization in a particular environment is the result of natural selection.

In Section 9.2 we review attempts to relate primate social organization to the environment. The perspective is historical, showing how early attempts were revised as problems were recognized and tackled (see also Wilson 1975). These problems are summarized and a few more are introduced in Section 9.3. Finally, in Section 9.4 we explore a developing area of research that both complements and extends the work reviewed in Section 9.2. The goals of these studies are broadly similar, but the scale and the methods are quite different.

9.2   Ecology and Primate Social Organization

John Crook and Steven Gartlan (1966) were the first to expound upon an adaptive relationship between primate social organization and the environment, although the adaptiveness of particular forms of social organization had already been discussed by several authors (e.g., Chance 1955, DeVore 1963, Hall 1965, Jay 1965). Based on the limited number of field reports available at that time, Crook and Gartlan proposed a qualitative classification of primate species consisting of five adaptive grades (Table 9.1), and they argued that these grades represented an evolutionary progression. Each grade was an amalgamation of broad categories of social organization, feeding behavior, activity cycle, and habitat type; the authors discussed at length the possible causes underlying each set of associations. Let us illustrate their approach by considering their explanation of Grades IV and V.

Grade IV and V species live in dry forest and savanna, and as a result they face seasonal food shortages, highly localized seasonal abundances, and a constantly high risk of predation. These circumstances favor large group size to defend against predators and to exploit locally abundant foods efficiently. The sexual dimorphism of open-country primates, particularly baboons, evolved through intense competition among males for access to females. This intensity is generated by two factors, namely the open terrain and the spatial cohesiveness of the social group, which in combination allow group members to monitor each other's activities. The large body, massive canines (Figure 9.1), and aggressive disposition of males are not the only evolutionary consequences of fierce competition; so also is the pronounced dominance hierarchy, needed to prevent frequent outbreaks of fighting. Although male physical and behavioral attributes did not evolve in response to predators, they serve as preadaptations for defense against predators. (A *preadaptation* is a feature that evolved to fulfill one function and is subsequently used for another.)

Table 9.1   A simplified version of Crook and Gartlan's (1966) scheme of the relation between primate ecology and social organization

Characteristic	Grade I	Grade II	Grade III	Grade IV	Grade V
Species[a]	*Galago* spp.	*Indri indri Propithecus* sp. *Hylobates* sp.	*Alouatta palliata Colobus* sp. *Gorilla gorilla*	*Macaca mulatta Presbytis entellus Papio c. cynocephalus Pan troglodytes*	*Erythrocebus patas Papio hamadryas*
Habitat	Forest	Forest	Forest and forest fringe	Forest fringe, savanna-mosaic	Savanna
Diet	Mostly insects	Fruit or leaves	Fruit or fruit and leaves, stems	Vegetarian-omnivore	Vegetarian-omnivore
Activity cycle	Nocturnal	Crepuscular or diurnal	Diurnal	Diurnal	Diurnal
Group size	Usually solitary	Very small groups	Small to occasionally large groups	Medium to large groups; *Pan* groups inconstant in size	Medium to large groups
Reproductive unit	Pairs (where known)	Small family group based on single male	Multimale groups	Multimale groups	One-male groups
Male mobility between groups	—	Probably slight	Yes (where known)	Yes (where known)	Not known
Sexual dimorphism and role differentiation	Slight	Slight	Slight (except in *Gorilla*)	Marked (where known)	Marked
Population dispersion	Probably territories	Territories	Territories in some species, home ranges in others	Territories in some species, home ranges in others	Home ranges

[a] Only species described in Chapter 8 are included from those listed by Crook and Gartlan.

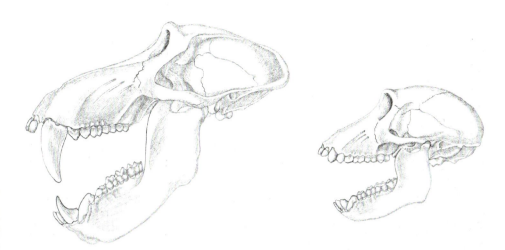

Figure 9.1   Like their close relatives the baboons, mandrill *(Papio sphinx)* males and females differ markedly not only in overall size but also in such features as the length and robustness of their canine teeth. (From Schultz 1969.)

The differences between Grade IV and V species result from differences in the seasonal abundance and distribution of their food. When food is seasonally sparse and widely scattered, it is advantageous for large groups to break up into smaller units that forage independently, thus avoiding the local overexploitation of food. Because each of these units has only one male, as much food as possible goes to females with young.

Crook and Gartlan's (1966) synthesis elicited great interest among field primatologists, but the research it inspired rapidly demonstrated its shortcomings. Since 1966 several workers have discussed these shortcomings and attempted to improve upon Crook and Gartlan's original scheme (Aldrich-Blake 1970; Crook 1970; Denham 1971; Eisenberg, Muckenhirn, and Rudran 1972; Jorde and Spuhler 1974; Altmann 1974; Clutton-Brock 1974; Hladik 1975; Milton and May 1976; Clutton-Brock and Harvey 1977a; Rowell 1979; Struhsaker and Leland 1979; Richard 1981). Let us look at some of these developments.

In 1972, John Eisenberg and his colleagues proposed a classification (Table 9.2) that differed in several ways from the classification shown in Table 9.1. First, social organization was correlated with the dietary habits of the animals, not with environmental features. In their text, the authors also discussed correlations of social organization with substrate choice (arboreal or terrestrial) and activity cycle (diurnal or nocturnal), but they generally placed much less emphasis on environmental features than Crook and Gartlan (1966) had done. Second, howler monkeys *(Alouatta palliata)* are assigned to two categories in Table 9.2 showing that social organization may vary intraspecifically

Table 9.2   A modified version of Eisenberg et al.'s (1972) scheme of the relation between primate ecology and social organization. In most cases, only species described in Chapter 8 are included from those listed by Eisenberg et al.

	Solitary	Parental family	One-male group (Minimal tolerance between adult males)	Age-graded male group (Males tolerate only males older or younger than themselves)	Multimale group
Increasing insects in diet ←	Insectivore-frugivore *Perodicticus potto*	Frugivore-insectivore *Saguinus oedipus*	Arboreal frugivore *Cebus capuchinus* *Cercopithecus mitis* *Cercocebus albigena* Semiterrestrial frugivore *Erythrocebus patas* *Papio hamadryas*	Arboreal frugivore *Ateles geoffroyi* Semiterrestrial frugivore-omnivore *Cercopithecus aethiops* *Cercocebus torquatus* *Macaca sinica*	Arboreal frugivore *Propithecus verreauxi* Semiterrestrial frugivore-omnivore *Macaca mulatta* *Papio c. cynocephalus* *Pan troglodytes*
Increasing leaves in diet →	Arboreal folivore *Lepilemur mustelinus*	Folivore-frugivore *Indri indri* *Hylobates lar* *H. syndactylus*	Arboreal folivore *Alouatta palliata* *Colobus guereza* *Presbytis entellus*	Terrestrial folivore-frugivore *Gorilla gorilla* Arboreal folivore *Presbytis entellus* *Alouatta palliata*	

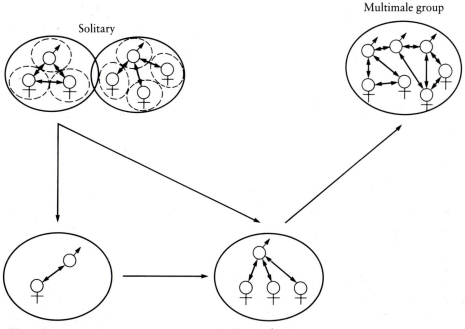

Figure 9.2   Pathways whereby new forms of social organization may arise. Dotted lines indicate subgroups with separate centers of activity that encounter one another infrequently. (Adapted from Eisenberg et al. 1972.)

(see also Crook 1970). Table 9.1 makes no such allowance; in fact, since the adaptive grades are supposed to represent an evolutionary progression, it is impossible for a species to occupy two grades simultaneously. Third, social categories are defined differently in Table 9.2. Crook and Gartlan (1966) characterized social organization by the size and distribution of social groups, the nature of the reproductive unit, the degree of male mobility between groups, and social role differentiation between males and females. They suggested an evolutionary progression but did not supply the mechanism whereby one form of social organization is transformed into another. In contrast, the social categories of Eisenberg and his colleagues are defined by the sexual composition of the group and the level of tolerance between adult males. A change in the level of tolerance between adult males, the authors argued, provides the mechanism whereby new forms of social organization arise (Figure 9.2).

The assumption that social organization is no more than an adaptation to conditions of food availability and predation was questioned by Goss-Custard and his colleagues (1972). They proposed that while the particular form of a social organization cannot be incompatible with efficient exploitation of the environment and predator avoidance, the determinants of social organization may be best found in a species' reproductive behavior. They suggested that one-male groups, for example, arise not so much because of the pattern of food distribution but because wherever possible it is in a male's reproductive interests to exclude other adult males from access to females (cf. Wittenberger 1980). Alexander (1974) developed a similar point, arguing that factors favoring the aggregation of animals may differ from those leading to the formation of social groups. Predation pressure and clumped food resources may cause animals to aggregate in groups, but the social organization of those groups is more likely to be generated by competition for mates.

In 1974, Stuart Altmann took issue with the technique of correlation and post hoc explanation used to construct and interpret classifications like Tables 9.1 and 9.2: "Correlation by itself is not adequate. For example, group size in baboons is correlated (inversely) with aridity. But many other characteristics of the environment co-vary with this environmental factor. The test of any putative ecological determinant rests on a demonstration of its mode of operation. . . . Adaptation requires a mechanism" (p. 224). Barash (1977:61) put it more bluntly: "The correlational approach poses a subtle danger in that it reflects the mental adroitness of the investigators as much as it does natural selection. Thus the presence of a consistent correlation suggests that something is going on; but, given almost any such correlation, a competent evolutionary biologist can generally point out how it is, in fact, adaptive."

Confounding variables represent one pitfall of the correlational approach. Another pitfall, Altmann (1974) pointed out, is that the approach tends to be tautologous. One starts out by assuming that animals are adapted to their environment (otherwise they would not be there) and then one goes on to prove it. To get out of this conundrum, Stuart Altmann advocated the formulation and testing of hypotheses about the relations between specific aspects of social organization and ecology in a single species. Thereafter, some hypotheses could perhaps be extended to other species, although Altmann was more interested in intraspecific variation and the possibility that some behaviors are better suited to certain environmental conditions than others. In 1970, he and Jeanne Altmann wrote: "For any set of tolerable ecological conditions, the adaptive activities of baboons tend in the long run toward some optimal distribution away from which mortality rate is higher, or reproductive rate is lower, or both" (p. 201). The hypotheses, or principles, enumerated by Stuart Altmann in 1974 were an attempt to identify those optima. In short, the Altmanns urged a move away from broad classifications based upon general correlations and toward detailed analyses of behavioral variations and their consequences within a single population.

Another criticism of the early correlational approach came from Tim Clutton-Brock (1974). He argued that the ecological categories used by Crook and Gartlan were too broad and illustrated his point with guerezas and red colobus monkeys, closely related and sympatric folivores. Guerezas generally live in monogamous groups distributed in small territories; red colobus live in big groups with large, overlapping home ranges. Clutton-Brock suggested that these differences in social organization were a consequence of differences in the degree of clumping and temporal stability of the two folivores' foods. While red colobus monkeys concentrate on young leaves and shoots from a wide range of species, guerezas eat more mature leaves and more fruit and select food items from only a handful of species.

One problem was raised by both Altmann and Clutton-Brock, as well as by others before them (Struhsaker 1969; Chalmers and Rowell 1971). This was the assumption that a primate infant can be molded like putty by the environment. But primates are not puttylike. The phylogenetic history of each species is important in determining the responses of its living representatives to the environment. For example, nocturnal strepsirhines rely on their small size and general inconspicuousness to avoid detection by predators. Diurnal, forest-living monkeys, bigger and gregarious, are unlikely to escape notice, and they respond to predators by fleeing or by joining in mobbing actions (noisy bluff attacks calculated to confuse and discourage the predator). In short, different species may react differently to similar circumstances.

In 1977, Clutton-Brock and Harvey (1977a) completed a new investigation of the relationship between ecology and social organization. Instead of making a broad, qualitative classification, they had two goals: (1) to seek statistically significant correlations between pairs from a whole spectrum of behavioral variables to which numerical values could be assigned and (2) to look for associations between ecological features and single behavioral variables rather than general types of social organization. They based their analysis on 100 primate species, classified as nocturnal or diurnal; arboreal or terrestrial; and insectivorous, frugivorous, or folivorous. Estimates of 13 so-called behavioral parameters (Table 9.3) were then made for each species except when congenerics were placed in the same ecological category. In those cases, estimates for each species were lumped into an average for the genus. The authors then looked for consistent associations between pairs of variables and tried to explain the biological and evolutionary causes underlying those that turned out to be statistically significant. Their analysis is one of the most recent and complete of its kind, so let us summarize their results and how they account for them.

1.   Diurnal primates weigh more than nocturnal ones, terrestrial primates more than arboreal ones, and folivores more than frugivores. Nocturnal primates need to be small because they live in the middle and lower levels of the forest, where many of their pathways are fine twigs, and because they rely on being inconspicuous to avoid predators. A large body is advantageous to a terrestrial primate because it permits the animal to range

further each day in search of food and also acts as a deterrent against potential predators. Folivores have larger bodies than frugivores because the former have a lower-quality diet and must process food in great bulk in order to derive enough nutrition.

Table 9.3   Variables used in four studies to describe primate ecology and social organization

Variable	Crook and Gartlan 1966[a]	Eisenberg et al. 1972[b]	S. Altmann 1974[c]	Clutton-Brock and Harvey 1977[d]
ECOLOGY				
Activity cycle	x			x
Substrate use		x		x
Diet	x	x		x
Predator pressure	x		x	
Habitat type	x		x	
Resource distribution			x	
SOCIAL ORGANIZATION				
Body weight				x
Sexual dimorphism	x			x
Feeding group size	x		x	x
Feeding group weight				x
Home range size			x	x
Day range length			x	x
Foraging formation			x	
Intergroup spacing	x		x	
Breeding group size				x
Breeding group sex ratio		x		x
Breeding group type	x	x		x
Male mobility between groups	x			
Population group size[e]				x
Population group weight[e]				x
Population density				x
Population biomass				x

[a] From Crook and Gartlan's Table 1.
[b] From Eisenberg et al.'s Table 1.
[c] From Altmann's 11 principles.
[d] From page 5 of Clutton-Brock and Harvey's article.
[e] Population group refers to animals that regularly associate together and share a common home range. (This is the same as the feeding group or the breeding group in most species, but not in all.)

2.   Nocturnal primates live in small groups; among diurnal species, terrestrial forms live in larger groups than arboreal forms. Among related species, frugivores tend to live in larger groups than folivores. Finally, large-bodied species tend to live in larger groups than small-bodied species. Nocturnal primates live in small groups for the same reasons they have small bodies. Most terrestrial forms live in open country, where the presence of predators and clumped, widely dispersed foods probably favor large groups. No simple explanation is apparent for the distribution of group size among arboreal folivores and frugivores, although Clutton-Brock and Harvey (1977a) suggest two reasons for the trend they found. First, a large group must move further each day than a small group to feed all its members; among animals feeding on an energy-poor food such as leaves, this may be energetically impossible. Second, folivores that feed on a small number of food species throughout the year may be able to defend their home range against neighbors if those food species are spatially clumped. Small group size, then, permits territorial defense.

3.   Nocturnal species live in smaller home ranges than diurnal species, terrestrial species in larger ones than arboreal species, and frugivores in larger ones than folivores. These associations reflect differences in the density and distribution of food supplies. Although data on territoriality are not presented, the authors suggest that it tends to occur when food supplies are relatively dense and evenly distributed in space.

4.   Nocturnal species live at lower densities than diurnal ones, and arboreal species at higher densities than terrestrial ones. The highest densities occur among arboreal folivores. Within the folivore and frugivore categories, population density decreases with increasing body weight. These distinctions are probably related to variation in food density.

5.   Most primates live in social groups in which females outnumber males, but why some live in one-male groups and others in multimale groups is little understood. Clutton-Brock and Harvey (1977a) discuss monogamy at some length, noting that it occurs in territorial species living at low densities. They suggest that monogamy evolves when there is strong selection for territoriality and when the defensible area can only support one female and her offspring.

6.   Sexual dimorphism tends to be greatest in large-bodied species. It may also be more common in species where females heavily outnumber males in the social group. The implication is that in such species, intermale competition for mates is intense and favors bigger, stronger males.

At the end of their paper, Clutton-Brock and Harvey (1977a) looked at the distribution of nine of their social organization variables across ecological categories and found that the distributions were generally similar (Table 9.4). This supports the idea that general types of social organization can usefully be assigned to broad ecological categories.

Table 9.4   Distribution across ecological categories of nine social organization variables

Variable	Nocturnal arboreal insectivore	Nocturnal arboreal frugivore	Nocturnal arboreal folivore	Diurnal arboreal frugivore	Diurnal arboreal folivore	Diurnal terrestrial frugivore	Diurnal terrestrial folivore
Body weight	1	2	3	4	5	6	7
Feeding group size	1.5	1.5	3	4	5	6	7
Feeding group weight	1	2	3	4	5	6	7
Home range	2	3	1	5	4	6	7
Day range	—	2	—	3	1	5	4
Population density	1	5	7	4	6	2	3
Biomass	—	1	4	3	6	2	5
Sex ratio	—	1	—	2	4	3	5
Sexual dimorphism	2	1	—	4	3	5	6

Source: Adapted from Clutton-Brock and Harvey 1977a.
Note: The data indicate rank order of values for each variable in the seven ecological groups.

For example, nocturnal arboreal insectivores tend to be small and solitary and to live in small home ranges while diurnal, terrestrial folivores tend to be big and to live in large groups with a large home range. The authors remained dubious of the value of a broad approach, however, pointing out that some ecological categories are more cohesive than others and that some variables show more marked differences between ecological categories than others. Their final conclusion was that the functional significance of different forms of social organization is better studied by looking at the distribution of single variables rather than at types of social organization or whole sets of variables, as Crook and Gartlan (1966) had done.

9.3   Problems Old and New

Hand in hand with the development of socioecology went a commentary on its weaknesses. Some researchers raised practical problems and tried to solve them, while others attacked the whole approach. The criticisms merge into two, closely related issues. One is the conceptual framework upon which the work was based, the other the interpretation of correlations.

Several researchers pointed out that some or all of the social and ecological categories were too generally defined, of the wrong sort, or of little use in any form. Revisions and refinements were proposed, but fundamental muddles remained. Look at the features used by various authors to characterize social organization and ecology (Table 9.3). The term *ecology* is used in two ways, referring both to attributes of the environment and to attributes of the animals. Social organization likewise subsumes an assortment of qualitatively different variables. There was little initial discussion of the choice of variables or why a particular aspect of the environment was likely to influence a particular aspect of social organization. The terminological and conceptual confusion made it easy to investigate qualitatively different propositions simultaneously without clearly distinguishing between them. Under the banner of research on social organization and ecology, systematic relationships were sought between (1) social organization and the environment; (2) the morphology (e.g., body size and degree of sexual dimorphism) and behavior (e.g., food choices, substrate use, and cycle of activity) of species; and (3) these morphological and behavioral attributes on the one hand and social organization or the environment on the other.

The lack of an adequate framework was recognized by Crook, Ellis, and Goss-Custard (1976). Instead of starting with data and attempting to make sense of them, they began by enumerating the steps necessary to construct a general model of mammalian social systems. The first step, they suggested, is to describe the social system itself, the external environmental variables that shape it, and also the biological attributes of species insofar as they affect the flexibility of social structure and dynamics. The second is to analyze how the social system functions, particularly with reference to how its members exploit resources, avoid predators, mate, and raise their young. The third is to develop models suggesting how, with respect to these "vital functions," characteristics of the environment and of the animals themselves interact to determine the structure and dynamics of the society. The complexity of the task proposed by Crook and his colleagues is great, and we do not have the means to come near to completing it. Nevertheless, their model, incomplete as it was, marked an important milestone, because it formally recognized the biological attributes of species as distinct from social organization and the environment.

The second issue, the interpretation of correlations, presents many difficulties (Smith 1980), of which we here consider two. One is the problem of confounding variables. All gibbon species are frugivorous, and all live in monogamous pairs with their dependent young (Figure 9.3). Is frugivory directly associated with monogamy, or are all gibbons monogamous by virtue of having shared a common monogamous ancestor? One way of trying to resolve this question is to examine the relationship between frugivory and monogamy across a range of unrelated species, including other mammals (Clutton-Brock and Harvey 1977a). Even then the correlation is not conclusive evidence of a causal relationship for these species may share some other unrecognized attribute that is independently associated with frugivory and monogamy.

A well-documented correlation suggests that something is going on, but it does not tell us what that something is. It does not reach the heart of the problem, assigning cause and effect. In most studies, evolutionary cause and effect are inferred from correlations between characteristics of the environment and the biological attributes or social organization of species. The relationship is often described in terms of the *function* of particular traits. For instance, it has been argued that scattered resources and predator pressure have favored the evolution of a large body in savanna-living, terrestrial primates and other mammals. The function of large size, then, is to permit animals to range further each day and to make them less vulnerable to predators. A causal relationship of a different sort is invoked in the interpretation of correlations between features of an animal's biology and behavior. For example, Clutton-Brock and Harvey (1977b:561) comment that "within contemporary primate populations, more stable variables with a strong genetic component (such as body size or digestive anatomy) are likely to constrain more flexible ones, such as home-range size or the proportion of day-time spent feeding. . . . We have assumed that the former usually represents causes and the latter effects." In this context, cause and effect refer to the way in which an animal's morphology constrains its behavior from day to day.

Inferences about evolutionary causes and effects are problematic. First, they assume that single features evolve in response to single environmental causes. But the close interdependence and resulting strong correlations among the attributes of organisms

Figure 9.3   Wherever they have been studied, gibbons (*Hylobates* spp.) have been found to live in monogamous pairs with their dependent offspring. (Courtesy of A. Mitchell.)

make it practically impossible to determine which element is acted on by natural selection and subsequently constrains the others. Moreover, it is unlikely that natural selection often acts specifically on single elements rather than the functional complexes to which they belong (Lewontin 1978). Even if and when it does act in this way, an element that changes and subsequently affects others may itself undergo further change as a result of feedback.

A second assumption implicit in assigning evolutionary cause and effect to pairs of environmental and social organization variables is that social organization changes fast enough to keep pace with environmental changes, so that it is always adapted to the environment. If social group size, for example, is finely tuned to the environment, variation in group size within a particular environment should be less than the variation between environments. Yet at Amboseli alone, the size of baboon social groups ranges from 13 to 200, and there is little evidence of systematic variation in group size among the many different types of environment occupied by baboons. The recent history of Amboseli also illustrates the speed with which an environment may change (Subsection 7.2G).

All evolutionary biologists espouse the view that processes in the world today are much the same as they were in the past. For instance, we believe that the mechanisms responsible for evolutionary change operated in the past as they do in the present. This belief tends to apply to the structure of the natural world as well. However, this application is unwarranted. World climates and vegetation have undergone major transformations in the past 50 million years. While the primates were evolving, so were the animals that preyed upon or competed with them and so were the plants upon which they fed. This process has not stopped. Indeed, human activities are changing the natural world faster than ever.

A single example illustrates the pitfalls for evolutionary reconstruction that result from this constant change. The diet of the brown lemur, *Lemur fulvus,* is mostly mature leaves, while that of the ring-tailed lemur, *L. catta,* contains a much higher proportion of flowers, fruit, and young leaves (Figure 9.4). Mature leaves contain high levels of cellulose, which is difficult to digest, and animals with a high cellulose content in their diet usually have teeth specialized for shearing and a digestive tract specialized for breaking down this complex carbohydrate (Section 5.4). Consequently, we might predict that the brown lemurs should have more specialized teeth and guts than do ringtails. Yet experimental evidence has shown that brown lemurs actually grind up and digest vegetable foods less efficiently than ring-tailed lemurs (Sheine 1979).

How are we to explain these paradoxical results? While there is no clear answer, a look at the paleoecology of Madagascar may provide a clue. Two thousand years ago, there were almost twice as many lemur species as there are today, and Madagascar was also home to a variety of other animals now extinct. The lemurs alive today inhabit for-

Figure 9.4   Brown and ring-tailed lemurs *(Lemur fulvus* and *L. catta)* are sympatric in south-west Madagascar, but their diets overlap little. Dietary differences in the wild are not predictable from differences in the teeth and guts of these species. (Photography by A. Richard.)

ests from which many species, some of them probably competitors, have vanished. It is not difficult to imagine that the surviving lemurs have expanded their life-style to include foods and perhaps whole habitats from which they were once excluded by competitors. It does not matter if they do not make very good use of these new resources, so long as they do not have to compete with more efficient animals. In short, there is no reason to suppose that the distribution, feeding habits, and social organization of lemurs today are the results of a long, slow evolutionary process, each species finely tuned to make the best of its environment. More likely, what we see are animals getting by and making ecological experiments after two thousand years of rapid change.

The final problem is that the concept of social organization is qualitatively different from that of the biological and behavioral attributes of species. The latter characterize each individual belonging to a species; the former characterizes the biological populations comprising a species. If natural selection acts on social organization, the mode of selection is different from that operating on the individual, and it needs to be clearly specified.

What is left? Should we abandon correlational studies of primates in the face of all these problems? I do not think so, for they suggest broad patterns and provide the only means of investigating certain evolutionary issues. However, correlations must be interpreted with great caution, qualitative differences between different kinds of correlations must be made clearer, and a better framework for studying current and evolutionary relationships is needed. Socioecology is a misnomer for an endeavor that ought, as Crook and his colleagues (1976) proposed, to embrace relations between the environment, social organization, and the biological attributes of species. The nature of these relations is explored in Section 9.4.

9.4   New Directions

Social organization refers to a set of spacing and social behaviors that characterize a population of animals of known age and sex (Section 8.1). Broad inferences about the evolution and adaptiveness of social organization need to be complemented by studies of (1) the immediate determinants of these behaviors in individual animals, (2) the implications of these behaviors for other aspects of the animals' lives, and (3) how changes in these behaviors produce changes in social organization. In short, as Stuart Altmann (1974) urged, we need studies of the mechanisms whereby evolutionary changes in social organization may take place. A number of researchers have begun to take this approach, and in this section we review their findings within the framework of Figure 9.5.

Figure 9.5 modifies a model presented by Altmann and Altmann (1979). The purpose of the Altmanns' model was to show relationships between social behavior, demography, and group size and composition. Specifically, "the size and composition of social

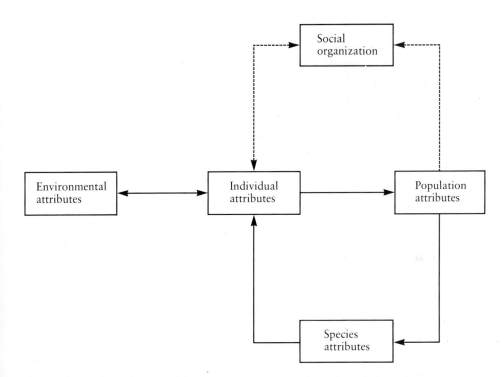

Figure 9.5   A framework within which to consider relationships between social organization and the attributes of the individual, the population, the species, and the environment.

groups, in terms of age, sex and kinship, affect behavior and social relationships. . . . Demographic processes provide delayed feedback on behavior because they affect group size and composition and are altered, in turn, by the effects of behavior on demographic parameters" (p. 48). Figure 9.5 is a more general portrayal of these links. It is hierarchical, its components being the individual, the population, and the species, and its primary purpose is to make useful distinctions among the heterogeneous variables that have been discussed in association with or as possible determinants of social organization. It is clearly not a complete account of the determinants of social organization. After a brief general description of the model, this section considers studies illustrating its components in more detail.

The model traces pathways whereby social organization influences and is influenced by attributes of the individual, population, species, and environment. Attributes of the individual are limited by virtue of the individual's membership in a particular species: the individual is subject to species-specific genetic constraints. These constraints are themselves determined by the phylogenetic history of the species. The degree to which and the way in which such constraints limit or determine behavioral attributes are poorly understood. However, the main goal of this discussion is to examine nongenetic causes of variation in social organization within species, and the role of genetic factors will be only briefly considered.

The following features of the model should be noted: (1) some links are bidirectional, the direction of causation depending on the kind of relationship in question. For instance, animals may change their environment by dispersing the seeds of some trees and destroying those of others (Section 11.4). Alternatively, we may view the environment as an important determinant of individual attributes such as growth rate and longevity. (2) The links portrayed can be investigated within the space of a single generation of animals: we are concerned with biological time, not evolutionary time. (3) Each box in the model represents a set of diverse attributes. For example, those characterizing the individual are behavioral, morphological, physiological, and so on. Links between the attributes within a set are as much a focus of socioecological research as links between attributes in different sets. Both types of links will be considered here. (4) The analytic danger of making social organization its own cause is avoided by distinguishing clearly between patterns of behavior and the behavior of individuals: an animal's behavior may change its position within a social organization without necessarily changing the social organization itself. For instance, an individual may rise in social rank within a social group without changing the description of the social organization as comprising social groups with members arranged in a dominance hierarchy.

Links will be discussed independently of one another wherever possible in this section. However, although each section is titled according to the primary subject matter within it, some studies crosscut these implied categories. Also, not all links will be considered in equal detail. Attention will be focused on those that have been subjects of the most research.

9.4A   Environment and the Individual

The influence of primates on their environment is examined in Chapter 11; here we consider how the environment influences primates. Following Slobodkin and Rapoport (1974), we assume that an individual's initial attempt to cope with an environmental change is behavioral, and only if the behavior proves inadequate is a physiological response elicited. In other words, behavior is the individual attribute most quickly affected by an event in the environment. As we shall see in Subsection 9.4B, changes in *physiology* (i.e., life functions of the individual) and possibly *morphology* (i.e., the structural form of the individual) may follow.

Common sense might suggest that a sudden, severe, and lasting food shortage would increase the level of aggressive competition for food, but this does not appear to be the case. Faced with such conditions, primates avoid conflict and, indeed, interactions of any kind. Rhesus and toque macaques (*Macaca mulatta* and *M. sinica*), chacma baboons (*Papio cynocephalus ursinus*), and vervet monkeys (*Cercopithecus aethiops*) become sluggish and preoccupied with the search for food, and the frequency of aggression drops off sharply while the distance between individuals increases (Loy 1970; Dittus 1977a, 1979, 1980; Hall 1963; Gartlan and Brain 1968).

The behavioral response of primates to normal seasonal fluctuations in their food supply is consistent within populations but varies markedly among species. In the Himalayan winter, for instance, food is scarce and langurs (*Presbytis entellus*) range further each day, whereas the day range of macaques (*Macaca fuscata*) in the temperate Japanese winter is shorter. In strongly seasonal environments, fluctuations are so great that major changes in the food supply are relatively easy to monitor. In less seasonal environments, changes in the food supply are harder to detect. Claims for an association between the behavior of primates and food availability are often tenuous because most primates live in environments with little seasonality, and their food supply has rarely been measured (see Clutton-Brock and Harvey 1977b). Wrangham (1977) has suggested that major changes in spacing behavior among chimpanzees are not directly or uniquely attributable to changes in the distribution and availability of food. We consider his work here to show just how difficult it is to demonstrate a clear effect of the environment on behavior.

Chimpanzees are frugivores (Figure 9.6). They live in large communities, although assemblies of a whole community are rare. More often, individuals forage alone or in small subgroups of varying size and composition. When a male finds a food source, he sometimes gives loud food calls and animals nearby respond by joining him to feed. Since chimpanzees often feed in small groups, why do they ever feed in large ones? Wrangham (1977) outlined two possible explanations. Under certain conditions, it may be more efficient to feed in a large group: a large group may help regulate the management of a food source; in addition, when food is scarce, an individual may increase its chances of finding food by joining others. However, it is also possible that abundant food permits large groups to form for reasons not directly related to feeding. Wrangham set out to explore these ideas with data from the Gombe Stream population.

Wet and dry seasons at Gombe differ with respect to food availability, weather, the density of parasites, and predators. Because Wrangham (1977) was interested in the effect of food availability, he tried to control for other variables by comparing observations made in dry season months in different years. He assumed that the only major difference between them was food related. Most plants produce flowers and fruits on a regular cycle at Gombe, but crop failures do occur, and in the dry season months of 1972 (D_1), a species failed to fruit that provided a major food source for the chimpanzees a year later (D_2). Wrangham concluded that food was more abundant in D_2. The average party size during D_2 also was bigger, and food calling was more frequent in D_2

Figure 9.6   Adult male chimpanzee *(Pan troglodytes)* eating figs. (J. Moore/Anthro-Photo.)

than D_1. In general, the percentage of individuals that succeeded in feeding at a site was higher in a small party than in a large one, suggesting that competition for food is more intense in large groups.

Wrangham (1977) used these findings to reject the idea that large group size increases the efficiency of foraging. If large parties reduce an individual's risk of not finding food, he argued, they should have formed most often in D_1, when food was least available. While lacking the data to reject the idea that large groups manage resources more efficiently than small ones, Wrangham suggested that there was in any case a simpler expla-

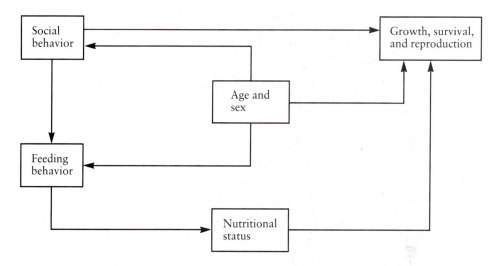

Figure 9.7   A schematic overview of links between attributes of the individual.

nation of his results. Food-calling frequencies provide the clue. Calling was less frequent in D_1, when food was scarce. Yet this, presumably, was the time when learning the whereabouts of food sources would be of greatest value to community members. The meaning of these apparently paradoxical results, Wrangham reasoned, is that while abundant food makes large parties possible without producing excessive competition, the primary reason a male calls is to give himself the opportunity to show off to potential mates. Moreover, a community that often forms large parties is likely to be more successful in intercommunity conflicts, since the outcome of these conflicts is largely determined by the size of the participating groups.

Wrangham's (1977) argument is weakened by his assertion that the lower frequency of food calls in D_1 means that their primary function is not to convey information about the location of food. When food is scarce, males presumably find it more rarely and consequently should call more rarely. The frequency with which males call once they have found a resource must be measured in order to determine the purposes of these calls and of variations in party size.

9.4B   Links between Attributes of the Individual

Central to arguments about the adaptiveness of behavior is the idea that behavior strongly influences physiology with diverse and often persistent consequences. Thus, while an environmental change is likely to elicit a behavioral response first, it may have pervasive indirect effects on other attributes of the individual. In this section we consider research investigating links between various individual attributes (Figure 9.7). Fig-

Figure 9.8   Juvenile baboon *(Papio c. cynocephalus)* cowers as adult male approaches, in the Amboseli National Park, Kenya. (Photography by A. Richard.)

ure 9.7 shows only the relationships considered here, and it should be stressed again (see Section 9.3) that cause and effect are often difficult to distinguish. For example, an animal may be well nourished because it has high social status and hence priority of access to resources, or it may have high social status because it is well nourished and is hence physically more capable of winning fights. It should also be emphasized that we shall review only a few of the recent studies that have investigated the relationships in question.

Post, Hausfater, and McCuskey (1980) compared age, sex, and dominance rank with food intake among the Amboseli baboons (Figure 9.8). The feeding behavior of all but the six youngest infants in a group of about 40 animals was sampled. Adult males and females spent about the same amount of time feeding each day. Since males were almost twice as big as females, males actually spent less time feeding per unit of body weight than females. Time spent feeding might not be nutritionally significant if the bite rate of animals varied, but no systematic differences were found. Post and colleagues ranked age–sex classes according to social status. The feeding bouts of members of low-ranking age–sex classes were more frequently interrupted by aggression than those of high-ranking classes, and frequently interrupted animals fed for shorter bouts. Yet within age–sex classes, there was no relationship between rank and interruption rate. Thus, social rank within the group as a whole, essentially a corollary of age and sex, did indeed confer some immunity from feeding-bout interruption. But within an age–sex class, lower-ranking individuals did not fare worse than higher-ranking ones. Post and colleagues examined dietary overlap between and within age–sex classes. With one exception, overlap within a class was greater than overlap between classes. Among adult males there were significant dietary differences. This may have been because adult males generally maintained a considerable distance between themselves and may therefore have encountered foods at different rates.

Post and colleagues (1980) sat on the fence in their conclusions: "While the relationship between dominance rank (or age or gender) and food intake was not entirely straightforward or simple, neither was it negligible or lacking in potential nutritional importance." However, they went on to suggest that low-ranking individuals may be able to avoid or at least substantially reduce the interference of high-ranking individuals while at the same time assuring adequate food intake. For instance, an animal may choose to feed at sites some distance from the main body of the group or feed longer on poorer-quality food. In short, if social status mediates diet and, by implication, the nutritional state of members of a baboon social group, it does so in a manner subtler than has yet been uncovered.

Relationships between social status and dietary intake have also been studied in toque macaques (*Macaca sinica*) in Sri Lanka (Dittus 1977a, 1979). In this study, further correlations were sought between behavior, body weight, and female reproductive rate. Weight, obtained by coaxing animals onto a scale, was used as an index of nutritional status. The problem with this technique is that many factors, including genetic variation and differences in individual development, contribute to the adult weight range in a population; consequently, a light animal is not necessarily poorly nourished (Figure 9.9). A variety of measures of nutritional status have been developed for humans, but they are difficult to apply to wild nonhuman primates because they all involve close physical examination of the subjects. Capture–release programs for primates are time and energy consuming, and the necessary resources are rarely available (but see Burns in prep.). Dittus's conclusions about social status and dietary intake were based on observations of all members of two social groups, a total of 55 animals, for at least one full day per animal.

Dittus's (1977a, 1979) data showed that adult males received the fewest threats relative to the frequency with which they threatened, subadult males the next fewest, followed in order by adult females, juvenile males, juvenile females, and then infant males and females. Outside the mating season, over 80% of the threats occurred while animals were feeding. Animals used threats while they were foraging to prevent lower-ranking

Figure 9.9   When food is scarce, the mortality rate increases among immature tocque macaques *(Macaca sinica)* like this one. (Courtesy of R. Dewar.)

individuals from approaching and also to displace lower-ranking individuals from feeding sites. Dittus suggested that, as a result, members of lower-ranking age and sex classes had to spend much more of their time foraging each day and were forced to eat a higher proportion of poor-quality foods than higher-ranking animals. Dominant animals consistently fed in the parts of fruiting trees where the fruit was ripest and most abundant and excluded subordinates from these sites. Subordinates generally avoided such locations, feeding instead in poorer areas or on different foods. This was not a true preference: if low-ranking animals were the first to reach a food tree, they fed in the richer areas, vacating them only as more dominant animals arrived.

Dittus supported these qualitative assessments with data on the diets of members of different age and sex classes (Table 9.5). He also showed that among adult males social

Table 9.5   Composition of the diets of toque macaques *(Macaca sinica)*

		% dry weight of total food intake			
Age/sex class	Number of animals observed	Fruits and seeds	Leaf shoots and herbs	Grass	Inverte-brates
Adults Males	2	47	52	1	0.2
Females	1	24	65	1	11
Juveniles Males	1	26	70	3	1
Females	2	13	81	6	0.4

Source: Adapted from Dittus 1977a.

status was positively correlated with weight and, by implication, nutritional state (Table 9.6) and suggested that these behavioral and nutritional differences are eventually reflected in the mortality rate and reproductive success of animals of different social status.

Dittus's research (1977a, 1979) poses important questions, but his results are difficult to interpret for four reasons. First, the sample is very small. The findings in Table 9.5 come from six animals, each watched for one day. Second, the reproductive state of the adult females was not specified, although we would expect females in different states to have different nutritional requirements (Section 4.3; Altmann 1980). Third, the results do not necessarily imply that lower-ranking animals had poorer diets than high-ranking ones. Dittus (1977a:306) himself pointed out that the data are "too sparse to eliminate the possibility that different animals chose foods containing different essential nutrients." A higher proportion of fruit in the diet of males and of leaf shoots and herbs in the diet of females has been reported in several primates (Clutton-Brock 1977c). The difference may reflect differential access to preferred resources, the different needs of

Table 9.6   Weights of adult male toque macaques *(Macaca sinica)* of different social rank

Rank of social status	Rank of weight	Number of adult males	Average weight of males (kg)
1	1	13	5.84
2	2	10	5.22
3	3	7	5.09
4	4	6	4.75

Source: Adapted from Dittus 1977a.
Note: Spearman rank correlation coefficient $r_s = 1.00$, $N = 4$, $p < .05$ one-tailed test.

Table 9.7   **Reproductive success of adult female toque macaques** *(Macaca sinica)* **by social rank**

Social rank of mother	Number of mothers	Annual birth rate per rank	Number of offspring surviving to reproduce		
			Daughters	Sons	Total
High	15	0.750	15	4	19
Middle	15	0.600	2	2	4
Low	14	0.642	4	0	4

Source: From Dittus 1979.
Note: High-ranking mothers have significantly more daughters (*p* < .01) and total offspring (*p* < .01) surviving to reproduce than middle- or low-ranking ones, and offspring survivorship is not proportional (*p* < .03) to differences in birth rates per maternal rank (Kolmogorov-Smirnov one-sample two-tailed test). Data are from nine troops observed for periods of three to nine years.

males and females, or possibly different ways of meeting similar needs. Judging diets containing a high percentage of leaf shoots and herbs as poor compared with diets high in fruit is questionable, since fruit usually provides little protein and leaf shoots are generally protein rich. In short, without a better understanding of the nutritional requirements of animals differing in age, sex, and reproductive state and of the nutrients supplied by different kinds of food, Table 9.5 must be interpreted with great caution.

The fourth and most important problem is that in most of the analysis, the effects of age and sex are not separated from those of social status. It is probable that the small size and high energy requirements of juveniles make them more likely to starve when only poor-quality food is available. Evidence is needed of weight in relation to diet and social status *within* each age and sex class. The positive correlation between weight and social rank among males suggests that the effect may indeed be present within each class, but without data on dietary variation among adult males, it is difficult to draw conclusions concerning the causes of the weight range.

Dittus (1979) also examined the association between social rank and reproductive rate in females. He found that the annual birth rate of high-ranking mothers was higher than that of low-ranking mothers and that the offspring of high-ranking mothers were more likely to survive to reproductive age (Table 9.7). He attributed these differences to the greater ability of high-ranking mothers and their offspring to gain access to preferred food resources. Sade and colleagues (1976) reported that the total reproductive output of high-ranking matrilineages was greater than that of low-ranking ones in the rhesus monkey population on Cayo Santiago (Figure 9.10). This was due to the earlier age at which some young, high-ranking females gave birth for the first time (Drickamer 1974; Sade 1980; Sade et al. 1976). The authors argued that since all animals were provisioned, the discrepancy could not be explained by differences in access to food and corresponding differences in growth rates. They suggested, rather, that animals of dif-

Figure 9.10   Rhesus macaques *(Macaca mulatta)* crowd around a food hopper on Cayo Santiago Island. (Courtesy of D. Sade.)

ferent social status experienced and responded to stress differently under conditions of extreme crowding on the island. Recently, however, Burns (in prep.) has found that low-ranking animals may actually have to work harder for less food, because they are persistently chased away from the provisioning area by members of high-ranking matri-lineage.

Changes in the provisioning regime of various populations of Japanese macaques *(Macaca fuscata)* (Figure 9.11) in conjunction with longitudinal demographic and behavioral records afford an excellent opportunity to examine how differences in the food supply influence the nutritional status and life history of animals and how differences in social status mediate this influence (e.g., Sugiyama and Ohsawa 1982; Mori 1979). Sugiyama and Ohsawa studied animals living in deciduous forest in central Honshu. Between 1969 and 1973 (Stage I) they were heavily provisioned, and between 1973 and 1980 (Stage II) they were not provisioned at all. In late 1973 the original group (Ryozen A) split into two groups (S and P); in 1978 members of the P group dispersed and were excluded from the analysis thereafter. During Stage I, an assortment of foods (wheat, peanuts, soybeans, sweet potatoes, and fruit) was dumped every day at a fixed feeding

Figure 9.11   Japanese macaque *(Macaca fuscata)* at Takasakiyama, Japan. (Courtesy of H. Simon.)

site about 3 by 20 m. If animals crowded together they could all feed at the same time, but sometimes spatially peripheral (and usually low-ranking) individuals were excluded by spatially central (and usually high-ranking) individuals already feeding at the site.

Table 9.8 summarizes some of the authors' findings. During Stage I, when the food supply was abundant but clumped, on average animals grew faster and bigger, survived longer, and gave birth to more offspring. However, high-ranking animals exhibited these trends much more strongly than low-ranking animals. During Stage II, when animals fed only upon foods growing in the wild, on average adult body weight was lower, maturation slower, survival lower, and the interval between births longer, so that females bore fewer offspring. These trends were seen in all social ranks.

Sugiyama and Ohsawa (1982) concluded that the high density and intense competition induced by provisioning in a small area accentuate differences in social status and, by extension, nutritional and reproductive status. They suggest that in the wild the ef-

Table 9.8   Demographic parameters and effects of artificial feeding on Japanese macaques (*Macaca fuscata*)

Demographic parameter	Parameter value under conditions of		Difference by social class	
	Natural feeding	Artificial feeding	Natural feeding	Artificial feeding
Body weight (difference)	1	1.228	Little	Little
Primiparous age (years)	6.74	5.21	$H = 6.69$ $L = 6.77$	$H = 4.82$ $L = 5.54$
Birth rate in high reproductive age (%)	33.58[a]	59.26[b]	$H = 31.67$ $L = 35.14$	$H = 69.77$ $L = 47.37$
Infant mortality (under 2 years) (%)	27.66	14.58	$H = 22.73$ $L = 32.00$	$H = 10.34$ $L = 21.05$
Annual survivorship after 2 years (%)	97.40	99.19	—	—
Number of offspring produced by a female	1.42	3.05	—	—

Source: Sugiyama and Ohsawa 1982.
Note: H = high-ranking animals; L = low-ranking animals.
[a] 6 to 19 years of age.
[b] 5 to 19 years of age.

fects of social differentiation are likely to be less marked because resources are less clumped. C. Burns (pers. com. 1983) is reaching similar conclusions in a comparative study of the Dunga Gali and Cayo Santiago rhesus macaques.

9.4C   Links between Population Attributes

Many individual attributes are found in all members of a population and are in that sense attributes of the population. Here, however, we are concerned with attributes not of each and every member but of the population as a whole. These attributes include density, fertility, mortality, age structure, and sex ratio. Chapter 7 considered some of the links between these attributes, including how a population's age structure and sex ratio influence its fertility and mortality rates. Here we consider the opposite effect through simulations done by Dunbar (1979). His goal was to examine the influence of fertility on the age structure and sex ratio of a population. His findings have major implications for social organization, because they illustrate the importance of population dynamics in determining the age and sex of animals with whom an individual may interact.

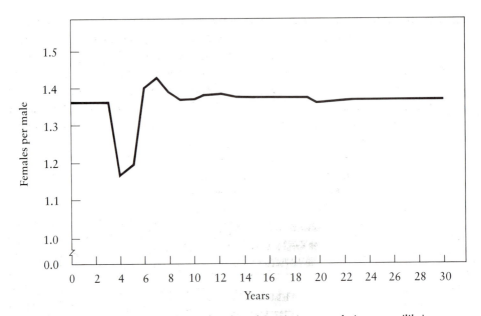

Figure 9.12   Changes in the adult sex (female:male) ratio in a population at equilibrium following a year (year 0) in which no infants were born. (From Dunbar 1979.)

Dunbar (1979) first showed that it would take as many as 65 years for a population of 199 four-year-old females to reach a stable age distribution, given six initial assumptions:

1. Each female gives birth to one infant every 2 years.
2. Half of all females breed each year.
3. Females give birth for the first time at 4 years of age.
4. The fertility rate is constant across all ages.
5. The sex ratio at birth is exactly 1:1.
6. Mortality between birth and age 20 is negligible.

In most environments in the real world, it is extremely unlikely that the six assumptions hold good for anywhere nearly as long as 65 years. This means that there will not be enough time to dampen the effects of one environmental disturbance on a population's age structure before the next disturbance occurs. As a result, Dunbar (1979) argued, it is unlikely that primate populations in the wild ever reach equilibrium.

In a second simulation, Dunbar (1979) added a seventh assumption, namely that adult males reach sexual maturity at the age of 6, and he relaxed the second. Specifically, he started the simulation in a year when no infants were born and found that this "zero year" had profound and long-lasting consequences for the adult sex ratio (Figure 9.12).

Since females mature faster than males, the impact of the zero year is first registered in the female cohort 4 years later, when the number of adult females per male drops sharply. This fall is followed in 2 years by a rise, as the effects of the zero year are felt in the adult male cohort. Three more years pass before the sex ratio approaches stability, and there is one more minor fluctuation as the zero year leaves the population at the end of 20 years, when all animals are assumed to die. Once again, it is shown that a single demographic event can have strong and persistent effects on the composition of a population.

9.4D Links of Species Attributes with Individual and Population Attributes

Just as an attribute possessed by all the members of a population can be called a population attribute, so an attribute possessed by all the populations making up a species can be called a species attribute. In the context of the model presented here, however, the species attribute of importance is its gene pool, and this is not possessed by each population. Rather, all the populations of a species together make up its gene pool. By definition, then, changes in the genetic composition of one of these populations produce changes in the gene pool of the species. How such changes bring about evolutionary change is much debated (e.g., Eldredge and Gould 1972), but these matters are beyond the scope of this discussion.

The extent to which the behavior of individuals is constrained by species-wide genetic predispositions is hotly contested (see, for example, papers in Leeds and Dusek 1981). Among the few species in which such constraints have been tentatively identified are hamadryas and anubis baboons (*Papio hamadryas* and *P. cynocephalus anubis*), although the evidence is open to more than one interpretation. Along the Awash River in Ethiopia, the geographic ranges of these baboons overlap, forming a hybrid zone. Members of hybrid social groups have a pronounced tendency to form harems, a characteristic of hamadryas social organization, but most of the harems are smaller and/or less stable in composition than those typical of hamadryas. These differences are apparently due to the behavior of the hybrid females, who "disobey" their harem leaders more often than do hamadryas females. Both hamadryas and anubis females rapidly modify their behavior in response to different males: female hamadryas soon stop following a particular male when released into a group of olive (anubis) males who do not herd them; conversely, anubis females quickly learn to follow one male when released into a hamadryas band. It thus appears to be the behavior of the hybrid males that facilitates the disobedience of females, and there is some evidence that they are actually less effective herders than their hamadryas counterparts. Olive baboons living in a range of habitats at the species border show several behavioral adaptations to local conditions, but the general features of their social system remain the same. In contrast, the social system of five hybrid groups living in a similar habitat differed in accordance with their genetic makeup, as indicated by their external appearance.

These observations suggest that differences between the social organizations of hamadryas and olive baboons are determined largely by the genetic substrate controlling the behavior of males of the two species. However, determinations about the immediate ancestry of males as indicated by their external appearance have yet to be confirmed by serological analysis. It is still not possible, then, to reject the hypothesis that an appropriate social environment may prompt some individuals to fix their interest persistently on others and that the social possessiveness of hamadryas males and, to a less marked extent, of hybrid males, may be a learned behavior (Gabow 1975; Nagel 1971, 1973; Kummer, Goetz, and Angst 1970; Sugawara 1979).

9.4E   Social Organization, the Population, and the Individual

In Chapter 8 we said that a given social organization could be characterized by a set of spatial and behavioral patterns within a population of animals of specified age and sex. Population processes are thus of central importance for the study of social organization, because they determine the number, age, and sex of potential interactants. The number, age, and sex of animals with whom an individual can interact sometimes influence the behavior of that individual in far-reaching ways. In this sense, social organization can be said to influence individual attributes (Altmann and Altmann 1979). This section is primarily concerned with studies examining the causal pathway from population to social organization and from social organization to the behavior of individuals.

The Altmanns (1979) estimated the number of surviving kin that an individual of a particular age and sex could expect to have in its social group under differing demographic conditions. They found that the number of relatives of any age available to a given group member depends strongly on the recent demographic history of the group. When births greatly exceed deaths and animals born into a group rarely emigrate, an individual will grow up surrounded by close kin. This is not true of a group of stable or declining size. Data from the Amboseli baboons suggest that when the population is stationary, most adult females will not live long enough to become grandmothers.

Structural features of this kind have important behavioral implications. The longer animals live, the greater are the opportunities for social relations to span more than one generation (Figure 9.13). For example, it becomes possible for members of the older generation to help young kin establish their social status in the group (Schulman and Chapais 1980; Chapais and Schulman 1980). It also makes mating between a parent and its offspring a possibility. At Amboseli, in contrast, mating between a baboon mother and her son would be unlikely even if there were no male emigration, because a female

Figure 9.13   Three generations of rhesus macaques *(Macaca mulatta)* on Cayo Santiago Island: a female grooms her infant while two of her daughters watch. One of the daughters also has an infant. (Courtesy of D. Sade.)

will probably not live long enough to see her eldest son reach full adult status (Altmann and Altmann 1979). Similarly, at Amboseli an infant's playmates will usually not include siblings, nieces, or nephews; indeed, it is likely to have few peers of any kind (Figure 9.14). The Altmanns examined the probability of an infant baboon having an opposite-sexed infant within three months of its own age present in the same social group. In a group of about 50 animals, about 6 out of every 100 infants are likely to have such a playmate, and almost a third of all individuals born within six months of one another have no peers of the same sex.

Fluctuations in the proportions of male and female infants that are born and survive each year have important social consequences. For example, none of the females born at Amboseli in 1971 or 1972 survived. In 1973, six out of seven surviving infants were female. Between 1970 and 1975, the dominance hierarchy among adult females was stable, but the following year the 1973 cohort of females reached the age of three years and began challenging adult females to whom they had previously been subordinate. Since then, several changes in dominance rank have occurred, and relations among females are much less peaceful than they were during the years when none were maturing

Figure 9.14   In most primate species, immature animals spend much of their time playing, if playmates are available. Here, two young baboons *(Papio cynocephalus anubis)* play tag through a waterhole in Mikumi National Park, Tanzania. (Courtesy of R. Johnson.)

into the adult hierarchy (Altmann and Altmann 1979; Hausfater, Altmann, and Altmann 1982).

The effects on social organization and on individual behavior of changes in the sex ratio have been examined in another context by Dunbar (1977a, 1979). Every eight years or so, local tribesmen shoot most of the adult males in the gelada baboon *(Theropithecus gelada)* population in the Bole Valley, Ethiopia. The large number of leaderless groups of surviving females join up to form huge harems under the control of the few remaining males. As animals mature over the next few years, the adult sex ratio starts shifting back toward unity. Competition increases among males for access to the big harems, and young males start stealing females from them. Big harems are easily raided and attacks on them are frequent; as a result females are eventually distributed over a greater number of smaller harems. Small harems are easier to control and more difficult to raid, and attacks on them are rare. Thus, when the adult sex ratio approaches one-to-one, harems are usually small and there is little social disruption. When there are many more females, males control large harems and there is little overt antagonism among them. It is when the sex ratio is changing in favor of males that competition is most apparent among the males.

While the theoretical literature confidently predicts that males will compete more intensely for females when the number of females declines (Williams 1975; Orians 1969; Trivers 1972; Emlen and Oring 1977), there is less certainty about the response of females to a decline in the number of males. However, recent field data suggest that females, too, compete for proximity to members of the opposite sex and that a shift in the adult sex ratio affects the intensity of this competition (Seyfarth 1976). Dunbar (1979) argued that while male gelada baboons compete for the control of females, once that control has been established, the females compete for access to the breeding male. The greater the number of females, the more intense is the competition. He also suggested that declining reproductive success in females, often associated with increasing group size (Downhower and Armitage 1971; Rowell 1969), may actually have more to do with a shifting sex ratio than with group size as such. As the number of females in a gelada baboon harem increases, the reproductive success of low-ranking animals decreases, probably because of all the aggression directed at them from their higher-ranking peers.

9.4F Commentary

Figure 9.7 is descriptive rather than predictive. I believe that it reflects a feedback system that exists in the real world, but it can be viewed simply as a way of organizing a disparate array of studies. In this section we have tried to illustrate the links between different sets of attributes as well as some of the relationships between attributes within the same set. These illustrations were drawn from the growing number of studies within this area of research. Some of these studies test predictions derived from a conceptual or theoretical framework. Post, Hausfater, and McCuskey (1980), for example, examined the prediction that high social rank confers priority of access to preferred foods and hence a more nutritious diet. This idea is deeply entrenched in theories of the evolution of social behavior (e.g., Alexander 1974). Taking a different approach, Dunbar (1979) used field data to set the parameters of simulation models designed to explore the long-term consequences of short-term demographic fluctuations on population structure. Like most of the studies reviewed in this section, these two present neither simple *synchronic* (i.e., without time depth) descriptions nor grand evolutionary scenarios. Rather, they are testing and developing, respectively, what I call middle-range theories of social organization (with apologies to Binford 1977). These theories are the general principles that determine relationships between and within the attribute sets portrayed in Figure 9.7. Much narrower in their scope than theories about the evolution of social organization, collectively these principles provide insight into the year-to-year determinants and dynamics of social organization. Interest in middle-range theories is increasing with our awareness of social organization not only as an evolutionary end product but also as a biologically complex and poorly understood system. Further development of these theories is an important task facing students of behavioral ecology.

This chapter has considered the influences of the environment on social organization, but a strongly emerging theme is that evolutionary models should encompass relations between animals as well as relations between animals and their environment. In particular, several authors stressed the importance of reproductive competition for the evolution of social organization (e.g., Goss-Custard, Dunbar, and Aldrich-Blake 1972; Alexander 1974; Crook et al. 1976). The effects of the interaction between environmental constraints and male and female reproductive interests on the evolution of social organization is currently receiving increased attention (e.g., Bradbury and Vehrencamp 1977; Emlen and Oring 1977; Greenwood 1980; Stacey 1982; Wittenberger 1980). We cannot review all of this work, but let us conclude by discussing a model elaborated by Wrangham (1980), referring specifically to primates. It is fitting to end by showing that Zuckerman's (1932) claim for the predominance of sex in social evolution, with which we began, may yet be upheld in spirit if not in specifics.

The logic underlying Wrangham's model has been developed by Wittenberger (1980). Mammalian young need considerable parental care in order to survive, and usually females provide more of this care than do males. From this we can infer that natural selection will favor females who distribute themselves so as to exploit crucial environmental resources efficiently. The distribution of males, in contrast, will be determined by their ability to defend single females or groups of females against competing males. In other words, the environment determines the grouping pattern of females, and this pattern in turn determines the distribution of males. It follows that males and females will have different patterns of dispersal (Greenwood 1980).

Turning to primates, we find that males and females of most species follow the typical mammalian pattern. Males leave their natal group as they approach maturity, whereas females remain in their natal group for life. Males spend their adult lives with few if any kin around them unless or until they father offspring or move into a group in which they have kin. Females, in contrast, are usually surrounded by close relatives and, Wrangham argues, develop tightly knit cooperative social networks under certain conditions. Conditions favoring large, stable groups of female primates, he suggests, are those in which females benefit from jointly defending resources distributed in large, scattered patches. When resources are found in small patches and provide food for only a few individuals, smaller and more territorial social groups are formed, and the female members of these groups allow only one adult male to associate with them. This gives rise to a one-male group form of social organization instead of the multimale group associated with larger aggregations.

Wrangham's model is particularly significant for two reasons, one substantive and one methodological. First, he proposes a clear demarcation between environmental features and interindividual competition for mates as evolutionary influences on social organization. Second, Wrangham's model proceeds from a general assertion about mammalian biology to a set of derived and more specific propositions about how male and female primates distribute themselves. The opposite procedure was used in most of the evolutionary arguments presented in this chapter, which sought general explanations for specific correlations. We have already considered the hazards of the latter approach (Section 9.3; see also Chalmers 1976). In order to test Wrangham's model and its assumptions rigorously, several kinds of data are needed that are presently lacking and that may be difficult or even impossible to collect. For instance, to avoid tautology, food resources would have to be identified and their quality and distribution measured without reference to the feeding behavior of the animals themselves. It is unclear that this can be done. The difficulty of testing this particular model notwithstanding, it is to be hoped that Wrangham's work will stimulate the development of models more amenable to testing.

PART V

PRIMATES IN COMMUNITIES

Introduction

 WHAT is a community? Putting a boundary around a community is no easier than defining a biome. Like biomes, communities are usually continuous with one another, and different criteria will place the boundaries in different positions (Whittaker 1975). Plant communities, for example, may be defined according to their physical structure, the dominant species present, or the whole floristic composition. Because the criterion used depends on the questions the researcher wants to answer, there is no such thing as a correct definition.

For our purposes, the definition given at the beginning of Chapter 2 remains as appropriate as any: "An assemblage of populations of plants, animals, bacteria, and fungi that live in an environment and interact with one another, forming together a distinctive living system with its own composition, structure, environmental relations, development, and function" (Whittaker 1975:2).

Today, the range of questions that community ecologists try to answer is enormous (Cody and Diamond 1975; Whittaker 1975). Historically, attention first focused on community composition, specifically the abundance and diversity of species in different environments. While generating valuable descriptive information, this research provided little insight into how a community functions or the factors governing its organization. The situation was transformed when Lindeman (1942) expounded the idea of trophic structure in a paper that was to become a milestone in the history of ecology. He argued that communities are organized in a hierarchy of trophic levels, or "feeders and fed upon" (Ricklefs 1973), and energy is transferred from one level of the hierarchy to another in a one-way process. Plants provide the ultimate source of energy, harnessed from the sun. They are fed upon by herbivores, who are eaten by carnivores, who may be eaten by other carnivores, and so on. In reality, of course, the situation is not so simple. Some animals eat both plant and animal foods, and the relations among species within a community resemble a complex web more than a rigid hierarchy, but the idea of trophic structure is still an important point of departure for many modern studies.

Today such studies belong in one of the two broad groups into which research on community ecology can roughly be separated. The first concerns the organization and dynamics of communities considered as whole entities, in other words, the sum of their component parts and processes. Into this group falls research on such questions as: what factors regulate the flow of energy and nutrients through the community? How are nutrients recycled? How can we describe community structure, and how is that structure determined? The second group includes research on the interrelationships of particular species within a community, giving rise to such questions as: how do species compete for resources? What is the resulting pattern of resource allocation among community members?

Important as questions concerning whole communities are, Part V considers them but briefly. Individual species or groups of species figure in such questions only to the extent that they contribute to community structure and dynamics, and primates, it seems, contribute little. Rather, the problems taken up in Part V fall primarily within the second group, focusing on the interrelationships of species within a community.

In Chapter 7, we discussed primate populations as though they lived in isolation from each other. They do not. Members of as many as 16 species may inhabit a single forest, and field-workers have devoted considerable energy to the study of interspecific relations (Charles-Dominique et al. 1980; Struhsaker and Leland 1979; Gautier-Hion 1978; Chivers 1980; Raemaekers 1978; Sussman 1974, 1977; Fleagle 1978a, 1978b; Hladik 1975; Hladik and Hladik 1969, 1972; Hladik et al. 1971; Charles-Dominique 1977; Thorington 1967; Klein and Klein 1975; Rodman 1978; Dunbar and Dunbar 1974a). In Chapter 10, we discuss interspecific relationships between sympatric species of primates. Chapter 11 extends the discussion to include other members of the community.

CHAPTER TEN

Sympatry, Competition, and the Niche

10.1 Introduction

ODUM (1971) has identified eight kinds of functional relationships between sympatric species, or species with overlapping ranges (Table 10.1). These relationships are conventionally illustrated in the biological literature by interactions between pairs of organisms, but in social species they are often found between pairs of social groups (Wilson 1975). Note also that more than one of these relationships may exist between a pair of organisms or groups of organisms. Our concerns in this chapter are with competitive relationships between primates and with spatial associations between members of different primate species that probably signify a range of functional relationships.

In Section 10.2 we review general explanations for associations between species, or *polyspecific associations,* and outline briefly the main axioms of competition theory. Although its role as a framework for the study of primate interspecific relationships is not always clearly stated, it is central to much of the research we shall review. In Section 10.3 we survey a selection of field reports on interspecific relationships among primates, including the various authors' interpretations of the data. Finally, in Section 10.4 we evaluate these interpretations and, more generally, the strengths and weaknesses of theories explaining the functional significance of interspecific relations as they apply to the primates.

10.2 Framework of Research

10.2A Polyspecific Associations

Associations between members of closely related species are common in a wide range of vertebrates including bats (Bradbury 1975), ungulates (Bell 1970), primates (Section 10.3), and birds (Moynihan 1962; Morse 1970; Powell 1974; Munn and Terborgh 1979). Observers have also reported associations between species in different orders

Table 10.1   Analysis of population interactions of two species

Type of interaction	Species A	B	Nature of interaction
1. Neutralism	0	0	Neither population affects the other
2. Competition	−	−	Each population inhibits the other
3. Amensalism	−	0	Population A inhibited, B not affected
4. Parasitism	+	−	Population A, the parasite, exploits B, the host
5. Predation	+	−	Population A, the predator, kills and eats B, the prey
6. Commensalism	+	0	Population A, the commensal, benefits while B, the host, is not affected
7. Protocooperation	+	+	Interaction favorable to both but not obligatory
8. Mutualism	+	+	Interaction favorable to both and obligatory

Source: Adapted from Odum 1971.
Note: 0 indicates no significant interaction.
  + indicates growth, survival, or other population attribute benefited.
  − indicates population growth inhibited.

(Section 11.5). Often polyspecific associations are between social groups, and they have been variously interpreted (see Table 10.1) as examples of social commensalism, social protocooperation, and social parasitism. Compared with research on competition (Subsection 10.2B), the study of these associations tends to be less preoccupied with mathematical models and more concerned with empirical descriptions of and speculation about their ecological function. The evidence often focuses on one or more of four central questions (Diamond 1981): are polyspecific associations chance aggregations or always composed of the same species? Who leads and who follows? Why do different species associate? Why do the species sometimes resemble each other in appearance and vocalizations much more closely than one would expect given their taxonomic relationship?

Ecologists studying mixed-species bird flocks in the tropics provide the following answers to these questions. The core of a mixed-species flock consists of individuals that forage together daily for many years, perhaps for their whole adult lives. Each of up to a dozen species is represented by a pair or family, and these individuals defend shared territorial boundaries. Usually individuals from just one species lead the entire flock, often a species with conspicuous plumage, frequent calls, and nervous movements. Other birds follow the "conspicuous, noisy whirlwind" (Diamond 1981) of the leaders.

Why do species associate? Flocking may serve several purposes simultaneously, and different members of a particular flock may be there for different reasons. At least five such reasons have been proposed, all of them concerned with either predator avoidance or food acquisition. Three suggest a protocooperative basis for the relationship. Note that the first two reasons described below apply equally to flocks made up of single species and mixed-species flocks. First, according to the *selfish-herd explanation,* the more individuals there are in a flock, the more chance there is of detecting a predator, and the lower the chance of a given individual being taken (Hamilton 1971). This idea is supported by the observation that lone birds peer about more and spend less time feeding than do birds in a flock (Powell 1974). Second, the *gang explanation* is that flocks can penetrate a territory whose owner would be able to expel a handful of intruders. Third, according to the *feeding explanation,* all participants in an association may benefit if it improves their feeding efficiency. This may happen for several reasons. Morse (1970) found that when food was scarce, birds were more likely to form mixed-species groups, perhaps because the more individuals there are searching, the more likely it is a good feeding patch will be found; because birds can learn new foraging techniques from one another; or because those with similar diets can keep track of resources that have already been depleted (Cody 1974).

In certain contexts, polyspecific associations appear to involve social parasitism or commensalism rather than protocooperation. Thus, the fourth possibility is the *beater explanation:* in the course of their own foraging, some flock members flush prey that others can capture (Munn and Terborgh 1979), a situation in which one species benefits while the other is unaffected. From this, it is a short step to the fifth reason, the parasitism of the *pirate explanation,* according to which birds seize food that another flock member has caught. Why does the victim tolerate the pirate? Perhaps because it would take too much time and energy to chase the pirate away. Alternatively, the victim may benefit from the pirate's presence in some other way such as added protection from predators.

Flock members from widely different taxonomic groups often resemble each other in looks, calls, or both. Why? Moynihan (1962) has argued that members of a mixed-species flock must be able to understand one another's signals if they are to maintain cohesion, and the convergence of signals may be for the sake of economy. For example, it may be easier to stay with the flock if one only has to look for brown birds rather than keep a dozen different color patterns in mind.

10.2B   Competition

Primate ecologists commonly invoke the principle of competitive exclusion to explain differences in the ecology of sympatric primates, although biologists diverge widely in their opinion of the validity and relevance of the principle to the real world (Winterhalder 1980). Indeed, the evidence for competition is one of the most hotly debated issues in ecology today (e.g., Lewin 1983; Roughgarden 1983; Schoener 1982, 1983;

Simberloff 1983). In this section, we look at the principle in its traditional form, as it seems to be understood by most primate ecologists. The reader is referred to Hutchinson (1978a) for a rich history of this tradition. In Section 10.3 we shall return to consider some of the problems with and extensions of what is generally referred to as competition theory.

"When two species jointly utilize a vital resource that is in short supply, either in abundance or availability to the species, one of the species will eventually eliminate the other from the habitat where their distributions overlap" (Jaeger 1974:33). This is a modern statement of Gause's principle of competitive exclusion, developed over 50 years ago from a modification of the logistic equation for population growth. Early laboratory data (Gause 1934) confirmed that only under exceptional circumstances could species competing for a vital resource persist stably in the same habitat. Note that in the pure form of the competitive exclusion principle, the outcompeted species is eliminated; in other words, it becomes extinct in the area of sympatry.

Many assumptions underlie the competitive exclusion principle in its pure form (Winterhalder 1980; MacArthur 1972), among them the following:

1.   the rate at which a population grows is unaffected by its density;
2.   individuals in the competing populations have identical demographic characteristics;
3.   there are no time lags, or random effects, in any of the relevant interactions between the two populations;
4.   there is no spatial patterning in the use of the environment by the competing species;
5.   there are no temporal fluctuations in the relevant environmental components;
6.   the interaction between species is directly affected by only one limiting resource;
7.   the resource in question is of uniform quality;
8.   no evolutionary or behavioral changes occur in either species.

The concept of an ecological niche is integrally related to the competitive exclusion principle. The modern formulation of the niche concept was developed by Hutchinson (1957), although the general idea predates his work by many years (see Hutchinson 1978a for a detailed history). A niche, he proposed, can be regarded as a set of points, each one of which defines a set of environmental values permitting a species to survive. At its simplest, the niche of a species can be described by points plotted along a single dimension. A planktonic animal, for example, might be limited only by temperature. If oxygen concentration is also critical for its survival, then this can be added as a second dimension along which is plotted that particular species' tolerance limits. A full description of the niche of most plants and animals would involve plotting values along many, many dimensions, hence Hutchinson's idea is often called the concept of the *n*-dimensional niche.

According to the competitive exclusion principle, two species competing for a limiting resource cannot persist stably together. (A *limiting resource* is an essential resource that is in short supply and, consequently, reduces the rate of population growth.) By extension, two species cannot stably occupy identical niches in a single environment. According to the pure form of the principle, competition for one or more limiting resources between the two will ensure that one drives the other to extinction in that environment. By implication, then, we might expect competition to reduce natural communities to simple assemblages of a few, markedly different species (Margalef 1975). This does not in fact happen. Why not? Part of the answer lies in the fact that at least one of the assumptions underlying the competitive exclusion principle is incorrect. Specifically, if we change assumption 8, the situation changes dramatically. If species can change, (and they do both behaviorally and morphologically), then extinction is not the inevitable consequence of competition.

At this point, it is useful to distinguish between a species' *fundamental niche* and its *realized niche* (Hutchinson 1957). The fundamental niche represents the total range of conditions under which a species can exist. If its niche overlaps with that of another species and the two species compete, then the area of overlap is either incorporated into the niche of one species and the other becomes extinct, or else the area is divided between the two, producing the realized niche of each. The realized niche results as one or both species undergo an ecological shift. This shift may involve *character displacement,* a morphological change that enables the two species to coexist sympatrically (Brown and Wilson 1956; Grant 1972).

Research on interspecific relations, including much primate field research, has sought evidence of niche separation as support for the importance of competition in organizing natural communities (reviewed by Pianka 1976; Schoener 1974). Recently a few researchers have tried to document not just the outcome of competition but competition itself. In general, observational methods are used in the first kind of study, and experimental methods in the second (Schoener 1974).

The simple observation of niche differences between sympatric species is not necessarily evidence of competition, for differences would exist even if niches were arranged randomly with respect to one another. The observational approach becomes more potent, Schoener (1974) argues, when it is used to examine hypotheses elaborated from the basic idea of the competitive exclusion principle. Schoener used data from 81 studies to build a case for the importance of competition and for the value of observational data in its investigation. The framework of this argument, for which he offers evidence at each step, is as follows. Competition should result in an overdispersion of niches in niche space. In this context, *overdispersion* means that niches are regularly and widely spaced over one or more dimensions. Overdispersion can be achieved in at least three ways. First, species may be regularly spaced along a single dimension. For example, species may differ by a constant ratio in the mean size of their food, a phenomenon often paralleled by differences in body size or feeding apparatus size (Hutchinson 1959; Brown 1975; Grant 1981). Second, if competition is important, species will eventually have to differ in more and more dimensions in order to minimize resource overlap as the number of species in a community increases. (This assumes that there is a limit to the crowding of species along a single dimension.) Third, if species are similar along a dimension involving a resource likely to be crucial to their survival and limited in supply, then they should differ along another dimension of equal importance. For example, species overlapping in habitat may eat different foods or foods of the same size but of different species, or they may partition the habitat vertically.

It has been argued that observational data can provide direct evidence of competition in two contexts (Diamond 1975, 1978). Diamond (1978) dubs these situations the products of "luck" and "wanderlust," respectively. Lucky is the ecologist who happens to be on the spot to watch one species invade another's geographic range and see the realized niches of one or both contract as coexistence is established. Wanderlust is a trademark of the ecologist who travels to the site of a "natural experiment." Such experiments occur because local populations fluctuate in numbers, occasionally fluctuating out of existence and once in a while becoming restored by immigrants, and because a particular species may happen to colonize one island but not another or to disappear on one island but not on another. For all these reasons, similar habitats in the same general area are likely to contain slightly different sets of species. The natural experiment consists of traveling from habitat to habitat, comparing the niche of a species in places where it shares and does not share the environment with a putatively competing species. Niche contraction in the presence of the second species, argues Diamond, is strong evidence of competition between the two. Conversely, niche expansion by one species in the absence of a second suggests that *competitive release,* or a freeing from competition, has occurred.

Observational approaches to the study of competition in the wild have recently been complemented by experiments involving direct manipulation, although practical considerations have so far limited the choice of subjects to small animals such as rodents (e.g., Wondoleck 1978; Holbrook 1979). Difficult as it is to see how this approach could be implemented with primates, it deserves review here. In a study of competition among desert-living rodents, Munger and Brown (1981) excluded larger species of seed-eating, or granivorous, rodents from experimental plots while allowing smaller species with more diverse diets (called omnivores in their study) as well as smaller granivores to enter. Over a period of eight months, the density of small granivores increased over three and a half times, while that of small omnivores did not change. Munger and Brown argue that this was because the small granivores had previously been limited by competition for food with large granivores, and removal of the latter left them with new resources; in contrast, the food supply of small omnivores was unaffected by the manipulation. The stability of small omnivore numbers suggests that the result was not an artifact of the experiment's design. For instance, if exclusion of the large granivores also meant that predators on small rodents were excluded, then the numbers of small omnivores and small granivores alike would have increased.

10.3   Case Studies

10.3A   Strepsirhines in the Makokou Forests of Gabon

The five sympatric lorisids living in the Makokou rain forests of northeastern Gabon (Figure 10.1) were studied by Pierre Charles-Dominique (1977) for a total of 42 months spread over almost eight years. These species represent the nocturnal component of a primate community numbering 16 species in all. In the introduction to his book summarizing this research, Charles-Dominique notes that in captivity all 5 species thrive on the same basic diet of insects, fruit, and milk, and yet in the wild each has a quite distinctive diet. This, he suggests, is because a species "can only make use of certain categories [of food] in its natural habitat because of ecological competition with other species. The more competing species there are for a given type of food, the more they become differentially specialized for the collection of certain restricted categories of that food type" (Charles-Dominique 1977:26). The primary goal of Charles-Dominique's research was to illustrate this point by demonstrating dietary differences among the 5 species, together with associated differences in the height of the forest stratum each exploited and the kind of supports each used during locomotion.

The diet of each species was ascertained primarily by shooting animals throughout the year and examining the contents of their stomachs. Observations of feeding were used to support conclusions based on stomach contents. Note that this approach permanently removed animals from the population and that each stomach represents only one feeding bout out of thousands performed by an individual during a year. Further, as the author himself points out, identifying the foods in the stomach is difficult because they have already been chewed and, in some instances, partially digested. As a result, most of the interspecific comparisons made in this study concern just seven broad categories of dietary components: fruits, gums, leaves, buds, wood fibers, fungi, and animal prey. Items in the last category were divided according to major taxonomic classes, such as beetles, caterpillars and moths, and grasshoppers and locusts.

Table 10.2 gives averages for stomach contents collected regularly throughout the year and at all times of night from animals living in secondary, or disturbed, forest as well as those living in primary, or mature, forest. At this broad level of analysis, no differences are apparent between the five species. Figure 10.2 takes a closer look at diet

Figure 10.1   Nocturnal strepsirhines in the Makokou Forest, Gabon: (clockwise from left) Allen's bush baby *(Galago alleni)*, potto *(Perodicticus potto)*, angwantibo *(Arctocebus calabarensis)*, Demidoff's bush baby *(Galago demidovii)*, and needle-clawed bush baby *(Euoticus elegantulus elegantulus)*.

Table 10.2   Average fresh weights in grams of principal dietary components found in the stomachs of Makokou strepsirhines

| Dietary component | Galagines | | | Lorisines | |
	Galago demidovii (N = 55)	*G. alleni* (N = 12)	*Euoticus elegantulus* (N = 52)	*Perodicticus potto* (N = 41)	*Arctocebus calabarensis* (N = 14)
Insects	1.16	2.22	1.18	3.40	2.00
Fruits	0.30	9.20	0.25	21.00	0.30
Gums[a]	0.15	Negligible	4.80	7.00	None

Source: Adapted from Charles-Dominique 1977.
[a] Because gums do not remain in the stomach for very long, the weights are calculated from combined contents of the stomach and the cecum.

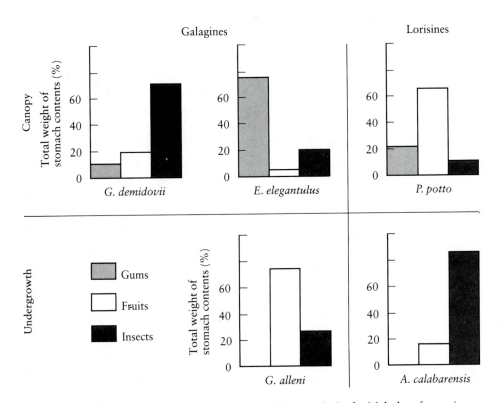

Figure 10.2   Dietary components of the five strepsirhine species in the Makokou forests in Gabon. Horizontal rows distinguish animals of the undergrowth from those of the canopy. (Adapted from Charles-Dominique 1977.)

and also divides the five species according to whether they are usually found in the canopy or undergrowth. In the canopy there is one insectivore *(Galago demidovii)*, one frugivore *(Perodicticus potto)*, and one gummivore *(Euoticus elegantulus)*. In the undergrowth are a frugivore *(Galago alleni)* and an insectivore *(Arctocebus calabarensis)*.

Charles-Dominique (1977) suggests that these differences in vertical distribution in relation to broad categories of diet are important in reducing competition among the five species and that remaining dietary overlaps are further reduced by differences in the choices of items within each broad category. Table 10.3 lists in decreasing order of importance the types of animal prey found in stomach contents, showing that the two lorisines feed largely on "unpalatable" prey ignored by the galagines. This difference in prey type is associated with differences in mode of locomotion and hunting techniques (see Section 5.2). Turning to gums, only the needle-clawed bush baby *(E. elegantulus)* and the potto *(P. potto)* are likely to compete with one another, the author suggests, because the third canopy species, Demidoff's bush baby *(G. demidovii)*, eats so little gum. Gums in the potto's stomach are in large, compact globs, whereas those in the needle-clawed bush baby's stomach consist of fine droplets aggregated in clumps. This is probably because the potto moves slowly, cannot visit many gum sites in a night, and thus has to eat large lumps of rather stale gums. These gums are ignored by the more sprightly needle-clawed bush baby, who spends much of the night dashing from one gum site to another. Charles-Dominique suggests that the latter mode of gum collection is actually uneconomical: more widely spaced visits permit gums to accumulate at points of production, thus reducing the distance that has to be covered to collect a given quantity. He attributes the lack of economy to competitive pressure exerted by the potto: the more often a needle-clawed bush baby visits a gum site in a night, the less chance there is for a potto to discover an accumulation of gum there.

Figure 10.2 indicates a broad division between canopy and undergrowth dwellers. In a more detailed analysis of locomotion, Charles-Dominique (1977) monitored the height of animals above the ground and the type, diameter, and orientation of their support. He found differences in the frequency with which animals were sighted at different heights in the forest and differences in the type of support used at a particular height. These distinctions are summarized in Figure 10.3, a schematic representation of typical pathways followed by the five species in primary forest.

In summary, despite broad similarities in the ecology of sympatric nocturnal strepsirhines in the rain forests of Gabon, particularly similarities in the kinds of food eaten, Charles-Dominique (1977) demonstrates that each species exploits the forest differently. In practice, there is little dietary overlap among them, a separation manifested by differences in their choice of food species and by associated differences in their use of physical space. These distinctions are in turn associated with physiological and morphological differences. Charles-Dominique argues that all these differences are the outcome of competition among nocturnal lorisids over millions of years.

Table 10.3   Animal prey found in the stomachs of the Makokou strepsirhines

	Galagines			Lorisines	
	Galago demidovii (N = 55)	G. alleni (N = 12)	Euoticus elegantulus (N = 52)	Perodicticus potto (N = 41)	Arctocebus calabarensis (N = 41)
	Small beetles: 45% Nocturnal moths: 38% Caterpillars: 10% Hemipterans Orthopterans Centipedes Bugs (homopterans) Pupae	Medium-sized beetles: 25% Snails: 15% Nocturnal moths: 15% Frogs: 8% Ants: 8% Spiders: 8% Orthopterans Termites Centipedes Pupae Caterpillars	Orthopterans: 40% Medium-sized beetles: 25% Caterpillars: 20% Nocturnal moths: 12% Ants Bugs (homopterans) Birds[a]	Hymenopterans (ants): 65% Large beetles: 10% Snails: 10% Caterpillars: 10% Orthopterans Millipedes Spiders Termites Birds[a] Bats[a]	Caterpillars: 65% Beetles: 20% Orthopterans Dipterans Ants

Source: Adapted from Charles-Dominique 1977.
Note: The percentage figures are relative to the total mass of food identified as animal in the stomach contents (fresh weight). For each strepsirhine species, the prey categories are ranked in descending order of importance.
[a] Included not on the basis of stomach content analysis but following direct observations conducted in the forest. (Small vertebrates captured from time to time represent only a minor component of the natural diets.)

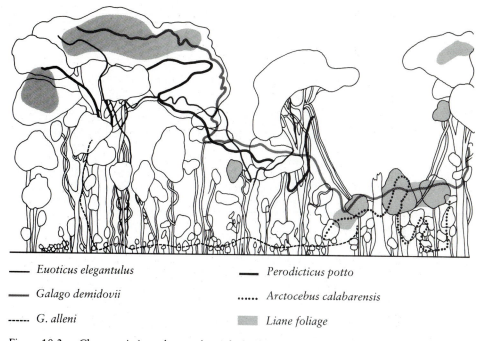

— Euoticus elegantulus   — Perodicticus potto

— Galago demidovii   ...... Arctocebus calabarensis

------ G. alleni   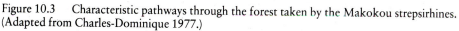 Liane foliage

Figure 10.3   Characteristic pathways through the forest taken by the Makokou strepsirhines. (Adapted from Charles-Dominique 1977.)

10.3B   Monkeys in the Makokou Forests of Gabon

In parallel with Charles-Dominique's (1977) research on the nocturnal primates of the Makokou region, researchers have studied 10 of the 11 diurnal primate species (Table 10.4). Figure 10.4 gives the weights of adult males and females of the cercopithecine members of this array. Of the 11 species, only the mandrill and the crested mangabey are strictly allopatric (Gautier-Hion 1978). Most conclusions about diet in these studies were based on stomach contents. Almost 200 monkeys have now been shot within a radius of about 30 km from the field station. The practical and ethical questions posed by this procedure loom as large as or even larger than they do in the strepsirhine study.

Gautier-Hion (1978) compares and contrasts data on 6 of the 11 species, and we shall focus on her findings. All the diurnal primates range through gallery forest, but for most of them it is a marginal habitat that they use patchily and intermittently. Only three are permanent residents, the crested mangabey *(Cercocebus galeritus)*, the De Brazza monkey *(Cercopithecus neglectus)*, and the talapoin monkey *(Miopithecus talapoin)* (Figure 10.5). Because members of these species never venture far from a river, the home ranges

Table 10.4   Studies of diurnal primates in the Makokou forests[a]

Species	Common name	Study
Pan troglodytes	Chimpanzee	Hladik (1973)
Papio sphinx	Mandrill	Jouventin (1975)
Colobus guereza	Guereza	Gautier-Hion (1978)
Cercocebus albigena	Gray mangabey	Gautier-Hion (1978)
C. galeritus	Crested mangabey	Quris (1975, 1976)
Cercopithecus nictitans	Spot-nosed monkey	Gautier-Hion and Gautier (1974)
		Gautier and Gautier-Hion (1969)
C. pogonias	Crowned monkey	Gautier-Hion (1978)
C. cephus	Moustached monkey	Gautier-Hion (1978)
C. neglectus	De Brazza monkey	Gautier-Hion (1978)
Miopithecus talapoin	Talapoin monkey	Gautier-Hion (1970, 1971, 1973)

Source: Adapted from Gautier-Hion 1978.
[a] The gorilla *(Gorilla gorilla)* is also present but has yet to be studied.

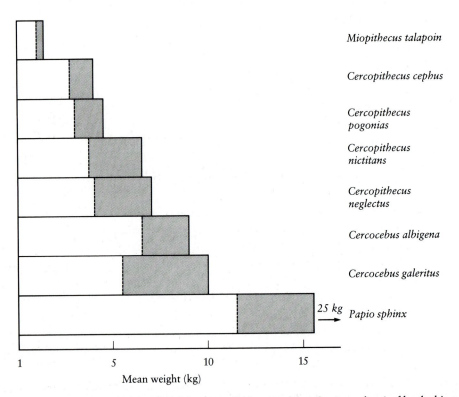

Figure 10.4   Average weights of adult males (solid bars) and females (open bars) of haplorhine primates in the Makokou forests in Gabon. (Adapted from Gautier-Hion 1978.)

Figure 10.5   Three of the haplorhine primates in the Makokou Forest, Gabon: (from the top) De Brazza monkey *(Cercopithecus neglectus)*, talapoin *(Miopithecus talapoin)*, and crested mangabey *(Cercocebus galeritus)*.

of the social groups of all three always include a portion of riverbank. In fact, talapoin monkeys habitually sleep on branches overhanging a river, and if disturbed by a predator, they drop into the river and swim to safety (Rowell 1973)!

Although they use vertical space similarly, the three species vary in their use of horizontal space and in their diet. The talapoin eats a higher proportion of insects than the other two species (Figure 10.6) and ranges in groups of 60 to 80 animals through a wide variety of habitats: groups move away from the riverbank during the day, exploiting primary and secondary growth, old plantations, crops, and even the ponds in which people soak manioc tubers to rid them of toxins (Gautier-Hion 1971). De Brazza monkeys, in contrast, live in small groups with small home ranges and travel slowly and quietly over short distances each day. Animals feed in prolonged bouts in a single tree; thus their daily diet includes few species. The crested mangabey's diet is similar to the De Brazza's, but its ranging pattern is quite different. Crested mangabeys move three to five times further than De Brazza monkeys each day, and the home range of one social group encompasses as many as four home ranges of De Brazza groups. Gautier-Hion (1978) suggests that the low population density of these mangabeys compared with the high density reported for a related subspecies in East Africa (Homewood 1975) results from interspecific competition with the De Brazza monkey. Alternatively, since the limits of its distribution are in Gabon, it may simply be living "at the limits of its adaptation capacities" (see also Quris 1975, 1976).

Guerezas *(Colobus guereza)*, gray mangabeys *(Cercocebus albigena)*, spot-nosed monkeys *(Cercopithecus nictitans)*, moustached monkeys *(C. cephus)*, and crowned monkeys *(C. pogonias)* live primarily in mature rain forest, along with the mandrill *(Papio sphinx)*, gorilla *(Gorilla gorilla)*, and chimpanzee *(Pan troglodytes)*. Guerezas and gray mangabeys, the heaviest of the monkeys, have been largely exterminated in more populated regions by hunters. Like the mandrill and the two great apes, their ecology has yet to be studied in detail in the Makokou region, and Gautier-Hion (1978) limits her comparison mainly to the three *Cercopithecus* species.

A notable characteristic of these three cercopithecines is that they are more likely to be found together in mixed groups than alone (Gautier and Gautier-Hion 1969). Mixed groups included either spot-nosed and moustached monkeys, spot-nosed and crowned monkeys, or all three together, and they were so common that Gautier-Hion (1978:278) comments: "The difficulty is to find one consistently monospecific troop to follow." The composition of at least some mixed-species groups was stable. For example, a group of 13 spot-nosed and 13 crowned monkeys of both sexes was followed for three months during the dry season. The degree of integration of these mixed-species groups varied from minimal, with two social groups simply moving in tandem, to complete spatial mingling. Social interactions were infrequent, however, and agonistic interactions even rarer. Vocal communication, in contrast, was common among members of different species (Gautier-Hion and Gautier 1969). These results may partially reflect

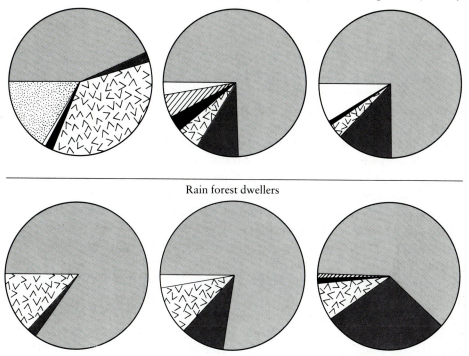

Riverbank forest dwellers

Miopithecus talapoin (N = 9) *Cercopithecus neglectus* (N = 9) *Cercocebus galeritus* (N = 10)

Rain forest dwellers

Cercopithecus pogonias (N = 18) *Cercopithecus cephus* (N = 18) *Cercopithecus nictitans* (N = 38)

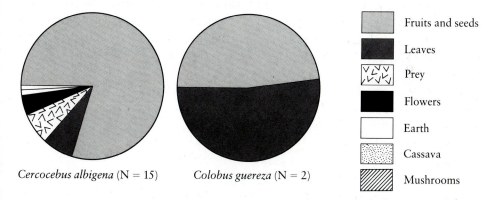

Cercocebus albigena (N = 15) *Colobus guereza* (N = 2)

- Fruits and seeds
- Leaves
- Prey
- Flowers
- Earth
- Cassava
- Mushrooms

Figure 10.6   Diets of the haplorhine primates of the Makokou forests. (Adapted from Gautier-Hion 1978.)

the difficulty observers had in seeing interactions in a forest versus the ease they had in hearing calls.

Despite these close associations among species, certain ecological distinctions are evident. Spot-nosed and crowned monkeys prefer mature forest; moustached monkeys also frequent secondary forest, where they sometimes associate with talapoins. Moustached monkeys also tend to occupy lower strata of the forest than the other two species. Like spot-nosed and crowned monkeys, crested mangabeys preferred the treetops and were commonly found in mixed groups with these species.

According to stomach contents, the three *Cercopithecus* species have different diets (Figure 10.6). The crowned monkey has a less folivorous diet than spot-nosed monkeys and actually eats no leaves at all: the so-called folivorous component of its diet is made up of an assortment of petioles, stems, sapwoods, and bark. Moreover, crowned monkeys eat more insects than spot-nosed monkeys, use different hunting techniques, and catch different prey. Specifically, crowned monkeys capture insects on the wing, while spot-nosed monkeys search for prey on tree trunks and branches. Moustached monkeys occupy an intermediate position between the other two, feeding on leaves as well as fibers and on a wide range of insects.

Gautier-Hion (1978) argues that the data provide ample evidence of mechanisms for avoiding competition. These include the weight gradient among males and females of the three species, coupled with differences in food habits and vertical stratification. Mixed groups gather fruit without interspecific aggression and often simultaneously from the same tree, and Gautier-Hion suggests that fruit is not limiting. Indeed, in view of the hunting pressure on all three species, food may not be limiting at all. Rather than emphasize the effects of competition, Gautier-Hion stresses the probable benefits of polyspecific association: (1) it increases the number of individuals available to watch for predators while not adding to intraspecific competition, and (2) it may enlarge the feeding area and thus provide a more heterogeneous resource base. Neither explanation accounts for the different patterns of associations among species, however, or for anomalies such as the failure of the De Brazza monkey to form interspecific groups at all.

10.3C   Monkeys in the Kibale Forest of Uganda

Five monkey species have been studied for varying lengths of time since 1970 in the Kibale Forest of western Uganda (Table 10.5; Figure 10.7). Struhsaker (1978a) has sum-

Table 10.5   Studies of diurnal primates in the Kibale Forest

Species	Common name	Study
Colobus badius	Red colobus	Struhsaker 1975
C. guereza	Guereza	Oates 1977a, 1977b; Struhsaker and Oates 1975
Cercocebus albigena	Gray mangabey	Waser 1975, 1977a
Cercopithecus mitis	Blue monkey	Rudran 1978a, 1978b
C. ascanius	Red-tailed monkey	Struhsaker 1978a

Source: Adapted from Struhsaker 1978a.

Figure 10.7   Five monkey species in the Kibale Forest, Uganda: (clockwise from left) blue monkey *(Cercopithecus mitis)*, red-tailed monkey *(C. ascanius)*, black mangabey *(Cercocebus albigena)*, guereza *(Colobus guereza)*, and red colobus *(C. badius)*.

marized and compared data on the ranging behavior and food habits of these species in order to identify the nature of niche separation and food competition among them. The methods used in all studies were similar. Feeding data were collected by direct observation of habituated animals, and the figures in summary tables represent the frequencies with which different items were consumed. Remember that this does not necessarily give a good indication of the volume or weight of foods ingested (Section 5.2).

Table 10.6 summarizes the food parts eaten by the five species and the frequency of consumption. Red colobus *(Colobus badius)* and guerezas *(C. guereza)* both feed heavily on young leaf blades, but their diets still differ: red colobus mainly eat young growth, such as buds, flowers, young leaves, and mature petioles from a wide range of plant species, whereas almost half the guereza's food comes from a single species; guerezas eat more fruit than red colobus, and unlike red colobus, guerezas subsist almost entirely on mature leaf blades at certain times of the year. Fruit is the most common item eaten by redtails *(Cercopithecus ascanius),* blue monkeys *(C. mitis),* and mangabeys *(Cercocebus albigena),* with arthropods ranking second. Blues eat about twice as much leafy food as redtails, and mangabeys are distinguished by the almost total absence of leaves and flowers from their diet. Note that these general observations about diet may mask major variations from month to month. For example, the range of the proportion of fruit in the monthly diet of redtails is from 13% to 81%.

Struhsaker (1978a) used two other measures to evaluate the plant component of these primates' diets. First, he ranked diets according to the number of plant species included. Second, he compared the five commonest species-specific plant food items in each diet (Table 10.7). Like Table 10.6, Table 10.7 masks interesting differences in dietary diversity: for example, red colobus feed on a wide variety of species each month, but the identity of these species changes little from one month to another; redtails and blues, in contrast, concentrate each month on one or a few plant species, but these species usually change from month to month. As a result, their diet is less diverse during any one month but more diverse over a year. Table 10.7 shows that only 4 out of the 25 commonest food items are shared by any two monkey species, and none are shared by three or more.

Finally, Struhsaker (1978a) estimated the overall overlap in the plant component of the five species' diets by summing the shared percentages of each species-specific plant food item for each pair of monkey species. Results are presented as a dendrogram in Figure 10.8. Blue monkeys have the greatest dietary overlap within the community of five species, followed by redtails, mangabeys, guerezas, and red colobus. The fruit of one plant species, *Celtis durandii,* accounts for most of the overlap. These fruits are abundant throughout the year at Kibale and are also eaten by many species of birds. Struhsaker considers it unlikely that they are a limiting resource.

Table 10.6   Food habits of five sympatric primates in the Kibale Forest, Uganda

| Food Item | % Total frequency scores | | | | |
	1 Cercopithecus ascanius	2 C. mitis	3 Cercocebus albigena	4 Colobus badius	5 C. guereza
Leaf buds	6.4	2.3	1.1	14.5	4.0
Young leaves	4.5	9.5	3.4	23.7	55.9
Mature leaves	3.3	6.8	0	8.2[a]	12.3
Leaves of undetermined age	0.8	0	0.8	8.0	3.9
Dry leaves	0	0.1	0	0	0
Young leaf petioles	0.5	1.9	0	3.5	0.1
Mature leaf petioles	0		0	15.5	0.1
Stems	0.6	0.3	0.1	1.4	0.04
Bark	0	0	2.5	0	0.3
Flower buds	10.8	1.3	3.4	12.1	1.2
Flowers (includes nectar)	4.5	11.2		3.8	0.9
Fruits	43.6	42.7	58.8	4.2	13.2
Seeds	0.1	2.4		1.4	0
Galls and/or gall-infested, virus- or insect-warped leaves	2.9	0.6	?	0.1	0
Unidentified plant items	0	0.7	3.9	0	6.2
Other plant items	0.5[b]			1.2[c]	1.8
Invertebrates (arthropods)	21.8	19.8	10.9–26.0[d]	2.6	0

Source: Adapted from Struhsaker 1978a.
Note: 1: $N = 1327$ observations, March 1973–June 1974.
   2: $N = 2566$ observations, February 1973–January 1974.
   3: $N = 7670$ observations, April 1972–April 1973.
   4: $N = 2399$ observations, August 1972–March 1975.
   5: $N = 2276$ observations, January–December 1971.
   [a] 3.4% may have been arthropod food.
   [b] Fungus.
   [c] Lichen and moss.
   [d] Arthropod search and ingestion were lumped. In a subsample of 120 scores, Waser (pers. com.) found that only 35% of these scores involved ingestion, giving an estimate of 10.9% compared with 26% for the combination of forage and ingestion.

Table 10.7   Five most common species-specific plant foods of five sympatric primates in Kibale Forest, Uganda

Plant food	Percentage of annual plant diet				
	Colobus badius	C. guereza	Cercocebus albigena	Cercopithecus mitis	C. ascanius
Newtonia leaves	8.8	—	—	—	—
Celtis africana leaf buds	8.4	—	—	—	—
C. africana young leaves	4.6	—	—	—	—
C. africana fruit	—	—	—	8.2	8.1
C. durandii young leaves	—	34.7	—	—	—
C. durandii mature leaves	—	5.3	—	—	—
C. durandii fruit	—	5.9	7.5	—	—
Markhamia young leaf petioles	4.5	—	—	—	—
Markhamia mature leaf petioles	8.0	—	—	—	—
Markhamia young leaves	—	8.0	—	—	—
Ficus exasperata fruit	—	5.9	4.4	—	—
F. brachylepis fruit	—	—	7.2	—	—
Mimusops fruit	—	—	—	—	12.7
Premna leaf buds	—	—	—	—	5.4
Premna fruit	—	—	—	5.3	—
Bosqueia fruit	—	—	—	—	5.3
Millettia flowers and floral buds	—	—	—	—	5.0
Pancovia fruit	—	—	4.9	8.2	—
Teclea fruit	—	—	—	6.5	—
Uvariopsis fruit	—	—	—	6.5	—
Diospyros fruit	—	—	26.9	—	—

Source: Adapted from Struhsaker 1978a.
Note: All data based on a 12-month period in the Kanyawara study area.

Blue monkeys, redtails, and mangabeys eat a lot of arthropods (Table 10.6). Mangabeys differ from the other two because they search for insects in different tree species and because they typically search in the crevices of deadwood and bark and inside hollow sticks, vines, and lianas. Blues and redtails more often scan the surfaces of leaves and bare, live branches and trunks. Differences between blues and redtails are present but harder to see. Blues, for example, take arthropods from a wider variety of microhabitats than redtails and spend more time searching lichen- and moss-covered branches; redtails procure most of their arthropods from mature leaves. Redtails also tend to move around more quickly than blues and may be able to exploit a greater variety of arthropod forms as a result.

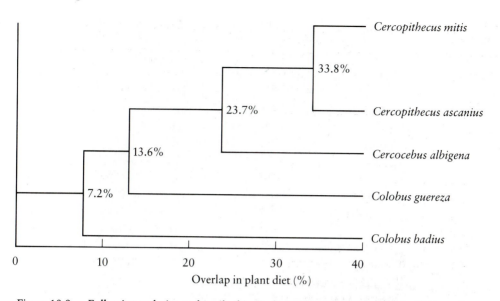

Figure 10.8   Following techniques described in Cody (1974), this community dendrogram for five of the haplorhine primates at Kibale shows similarities in their plant foods. (Adapted from Struhsaker 1978a.)

Pursuing the blue monkey–redtail comparison further, Struhsaker (1978a) notes that most of the dietary overlap between them involves fruit. During 10 months of simultaneous sampling of percentage overlap in the ranging patterns of the two species, the degree of overlap correlated positively with overlap in species-specific plant foods. Struhsaker speculates that this occurred because blue monkey groups intermittently followed redtail groups as an easy way of locating superabundant supplies of fruit.

Vertical stratification does not play an important role in niche separation among these species. Regardless of dietary overlap, there is considerable overlap in feeding heights for all 10 pair combinations of monkeys. However, their horizontal ranging patterns show much more variation. Blue monkeys, redtails, and mangabeys all move further each day than red colobus and guerezas. They also cover a larger proportion of their annual home range in a given five-day period than the other two species, possibly because they are monitoring the status of scattered fruit resources. Among redtails, monthly ranging diversity is positively correlated with the diversity of plant food species and with the proportion of leaves in their diet (Figure 10.9). When fruit is scarce, redtails feed heavily on leaves; this suggests that when forced to feed on foliage, they range more widely to feed from a greater variety of plant species. Blue monkeys and mangabeys used more quarter-hectare quadrats than redtails during the monthly five-day sample periods. Struhsaker suggests that this is related to their greater reliance on superabun-

Figure 10.9   Red-tailed monkey *(Cercopithecus ascanius)* eating bush tomatoes in the Kibale Forest, Uganda. (Courtesy of L. Leland.)

dant fruit and differences in their arthropod-hunting tactics. Of the three species, blue monkeys appear to be the best adapted to a folivorous diet because they do not increase the diversity of food species in their diet or the length of their daily range when they increase their leaf intake. Mangabeys feed very little on foliage.

In short, redtails, mangabeys, and blue monkeys seem to adopt one of two strategies when fruit is scarce. Blues and redtails switch to alternative foods, and mangabeys move to another area where fruit is still abundant. Red colobus and guerezas are different. The large home range of the red colobus seems related to its year-round need for diverse plant species and young growth. The guereza, with the smallest home range of all, relies upon a few abundant, evenly distributed species. Struhsaker (1978a) suggests that much of the dietary overlap among these five primate species at Kibale involves common foods and that competition is more likely to occur over less common items when fruit is scarce. Blue monkeys, with the greatest community overlap of the five, are probably most susceptible to such competition. To support this proposition, Struhsaker compared the abundance of blue monkeys in forests differing in the number of resident primate species and specifically in the presence or absence of red colobus and mangabeys. Based on this comparison, he argues that the absence of mangabeys and red colobus releases blue monkeys from competition, and they achieve higher densities than in the presence of either or both of the former. Corollaries of this argument are that the red colobus outcompetes the blue for young leaves and buds when fruit is scarce and that the mangabey, being bigger, usually displaces the blue at fruiting trees, a crucial advantage when fruit is scarce.

Distinctions in the diets of these primates are mechanisms for coexistence, concludes Struhsaker (1978a), and they go hand in hand with morphological specializations selected by interspecific competition. When food is scarce, animals make further adjustments to reduce competition. Some switch to alternative foods, while others move to areas where preferred foods are still available. Despite these mechanisms, competitive pressures may nonetheless limit the abundance of at least one species, the blue monkey, as evidenced by the higher densities it achieves when postulated competing species are absent.

Struhsaker (1982) also investigated polyspecific associations among the Kibale primates. He found that the frequency of these associations varies with the time of day and between months and years but that there does not seem to be a pattern to the variation. There is also marked variation in the frequency with which species form and initiate associations and in the length of time they remain in an association. Red colobus rarely join other species, for example, but are frequently joined by them, whereas redtails, blue monkeys, and mangabeys frequently join other species. In general, less common species tend to join commoner ones, species with small social groups tend to join species with large ones, and species with large home ranges tend to join species with small ones. A simultaneous study of redtails and blue monkeys showed that their diets were more similar when they were in association than when they were foraging separately, whereas the amount of time redtails associated with red colobus was unrelated to the composition of their diets.

Like Gautier-Hion at Makokou, Struhsaker (1982) concluded that polyspecific associations at Kibale have at least two functions, depending on the species. Redtails, blue monkeys, and mangabeys associate when they are exploiting common resources. In contrast, red colobus and redtails have little dietary overlap and polyspecific associations between them may represent a safety-in-numbers strategy. The large crowned eagle (Stephanoaetus coronatus) is abundant at Kibale and may prey upon monkeys. For the small-bodied redtails in particular, association with a large group of big red colobus monkeys may reduce the likelihood of falling victim to this predator.

10.3D   Monkeys and Apes in the Krau Game Reserve
of Malaysia

In 1968 David Chivers began a detailed study of the ecology and behavior of siamangs
living in forests around the Kuala Lompat Post of the Krau Game Reserve in the middle
of peninsular Malaysia. In subsequent years studies of the siamangs were followed up
and research was extended to include all the permanent diurnal members of the primate
community (Table 10.8). Much of this work has been brought together by Chivers
(1980). We shall first present an overview of niche differentiation among the primates of
Kuala Lompat based on a six-month study carried out by the MacKinnons (1978,
1980b); we shall then look more closely at particular aspects of this differentiation as
reported by other researchers.

The MacKinnons accumulated 1638 hours of observation on each of the five common species, namely the siamang *(Hylobates syndactylus),* lar gibbon *(H. lar),* dusky leaf
monkey *(Presbytis obscura),* banded leaf monkey *(P. melalophos),* and long-tailed macaque *(Macaca fascicularis)* (Figure 10.10). The pigtailed macaque *(M. nemistrina)* was a
rare visitor to the area, and so the MacKinnons inferred its relationship to the other

Table 10.8   Selected studies of primates in the Krau Game Reserve, Malaysia

Hylobates syndactylus	Siamang	Chivers (1974), Fleagle (1976b, 1980b), MacKinnon and MacKinnon (1978, 1980b), Gittins and Raemaekers (1980), Chivers and Raemaekers (1980), Raemaekers and Chivers (1980)
Hylobates lar	Lar gibbon	Gittins and Raemaekers (1980), Fleagle (1980b), MacKinnon and MacKinnon (1978, 1980b), Chivers and Raemaekers (1980), Raemaekers and Chivers (1980)
Presbytis obscura	Dusky leaf monkey	Curtin (1976, 1980), Fleagle (1976a, 1978a,b, 1980b), MacKinnon and MacKinnon (1978, 1980b), Chivers and Raemaekers (1980), Raemaekers and Chivers (1980)
Presbytis melalophos	Banded leaf monkey	Curtin (1976, 1980), Fleagle (1976a, 1978a,b, 1980b), MacKinnon and MacKinnon (1978, 1980b), Chivers and Raemaekers (1980), Raemaekers and Chivers (1980)
Macaca fascicularis	Long-tailed macaque	Aldrich-Blake (1980), Fleagle (1980b), MacKinnon and MacKinnon (1978, 1980b), Chivers and Raemaekers (1980), Raemaekers and Chivers (1980)

Figure 10.10   Monkeys and apes in the Malayan rain forest: (from the top) siamang *(Hylobates syndactylus)*, lar gibbon *(H. lar)*, dusky leaf monkey *(Presbytis obscura)*, banded leaf monkey *(P. melalophos)*, long-tailed macaque *(Macaca fascicularis)*, and pigtailed macaque *(M. nemistrina)*.

species from data collected in northern Sumatra. Although their study was short, it had the advantage that data were assembled on all species almost simultaneously so that "any differences found could be attributed to genuine differences in adaptation and between the species, rather than to anomalies resulting from different weather, locations, food availability or observers" (MacKinnon and MacKinnon 1980b:169).

The diets of all six species contain large proportions of fruit and leaves, and dietary differences are more apparent at the plant species level, where differing preferences divide the six species into three pairs, each made up of congeners. Dietary overlap between different genera is small, and competition for food probably minimal. Dietary similarities between the members of a pair are great, however, and the MacKinnons (1980b) suggest that competition is avoided or reduced by the different foraging strategies that such pairs employ, strategies reflecting differences in body size and/or limb proportions. Thus dusky and banded leaf monkeys both depend heavily on leaves, but morphological differences between the two suit them to travel along different substrates; as a result, they tend to use different feeding sites (see also below). Both hylobatids rely heavily on a single plant genus, the fig (*Ficus*). They eat the smallest range of food types, show the most intensive use of a few plant species, and are by far the least numerous of any primate in the Malaysian community. The MacKinnons suggest that their absence from gallery forests in the area can be attributed to competitive pressure from the pigtailed macaques, which feed heavily on figs but are less dependent on them. Niche separation between the two gibbons is a function of their differing size and foraging strategies. Twice as big as lar gibbons, siamangs travel half as far each day and feed at half as many feeding sites, but do so for twice as long. The lar gibbon, in contrast, exploits smaller, more dispersed food sources (see also below).

The macaques together show dietary preferences between those of the heavily folivorous leaf monkeys and the heavily frugivorous gibbons. Separation between the two macaques is a function of their contrasting morphology, which suits them to different terrain. Pigtailed macaques live in large groups and travel long distances each day on the floor of closed-canopy primary forest in search of large food sources. Long-tailed macaques, in contrast, are usually found near water. They are fine swimmers and cross flooded or swampy areas or even fast-flowing rivers, heavily exploiting a few plant species that are abundant near rivers.

The MacKinnons (1980b:188) pose the question, "Have these complementary foraging strategies evolved mutually as a result of a long history of sympatry, or is it because they were preadapted for coexistence that they were able to coexist so successfully when their respective distributions overlapped?" They suggest that the latter may be the correct answer, arguing that ranging patterns and dietary preferences seem to be based on inherited anatomical features that cannot be rapidly changed to suit new environmental conditions.

The MacKinnons (1980b) broadly outline differences in how the primates of Kuala Lompat use the range of physical substrates afforded by the forest. These differences have been explored in more detail by Fleagle (1976a, 1976b, 1978a, 1978b). Like several other orders of mammals, the primates display a great range of locomotor and postural adaptations. They leap, climb, brachiate, knuckle-walk, and exhibit a variety of forms of arboreal and terrestrial quadrupedalism, and they have been the subjects of many morphological studies and locomotor classifications (e.g., Prost 1965; Napier and Napier 1967; Erikson 1963; Ashton and Oxnard 1964). While hypotheses relating loco-

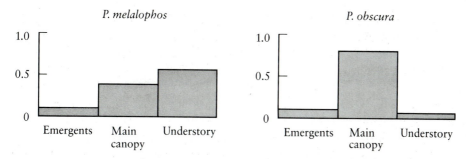

Figure 10.11   Distribution of feeding observations for *Presbytis obscura* and *P. melalophos* in different structural levels at the Krau Game Reserve, Malaysia. (Adapted from Fleagle 1978a.)

motion and posture, or *positional behavior*, to diet and forest use have frequently been proposed and the testing of such hypotheses repeatedly urged (Avis 1962; Napier 1966, 1967; Stern and Oxnard 1973), only recently have researchers begun to collect the kind of data needed to do so (e.g., Sussman 1974; Ward and Sussman 1979; Rose 1974, 1977b; Ripley 1967b; Fleagle and Mittermeier 1980; Napier and Walker 1967; Fleagle in press; Garber 1980). Fleagle's research on the leaf monkeys, combining field observations with laboratory studies of morphology, provides an excellent example of this approach and demonstrates the importance of differences in positional behavior for differences in the ecology of closely related species. Although many of his data were collected at Lima Blas, about 135 km west of Kuala Lompat, comparative evidence indicates that the distinctions found at one site are also present at the other.

What physical substrates are available to the monkeys? Like many rain forests (Section 2.1), Malaysian rain forest can be thought of as comprising three structural levels (Fleagle 1978b). The main canopy is a horizontally continuous stratum made up of a network of the limbs and branches of large trees. It provides a latticework upon which quadrupedal animals can travel horizontally. While the understory may be connected by foliage vertically to the canopy, it is discontinuous horizontally. The trees comprising it are smaller and supports are less stable than those of the canopy. Movement from one part of the understory to another means that an animal must either ascend to the canopy; travel there horizontally and then come back down to the understory; descend to the ground, travel, and come back up; or fly, glide, or leap across the gaps in the understory itself. Emergents form the top layer of the forest. They are huge trees, soaring above the canopy and separated from one another horizontally and often from the continuous canopy. Movement into, out of, or between emergents involves climbing up and down trunks from the main canopy or making long leaps.

How do the leaf monkeys use these strata? The dusky leaf monkey was most often sighted feeding in the canopy and to a lesser extent in emergents, while banded monkeys use all levels of the forest, the understory in particular (Figure 10.11). Duskies usually travel by quadrupedal walking and running along large boughs in the canopy. Bandeds,

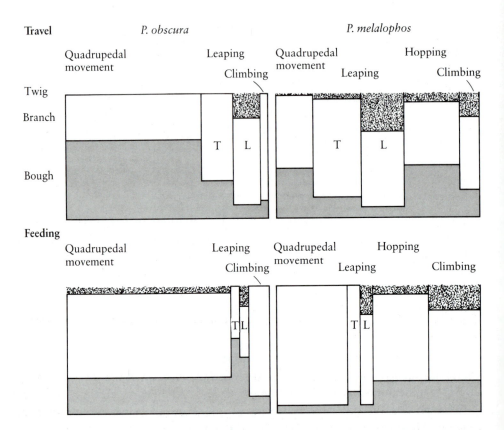

Figure 10.12   Distribution of locomotor patterns on different-sized supports during travel and feeding by *Presbytis obscura* and *P. melalophos*. Horizontal axis shows the contribution of different patterns in each kilometer of progression; vertical axis gives the proportion of that distance that is moved on different types of supports. (Adapted from Fleagle 1978a.)

who feed more in the understory, often travel by leaping and hopping between smaller supports. Movement, support use, and postural behavior are more similar in the two species during feeding than during traveling, but differences between the two activities are consistent with those already mentioned: for instance, duskies move more quadrupedally and more on large supports than do bandeds (Figure 10.12).

The two species show morphological adaptations that can be related to these differences in positional behaviors. These include numerous differences in the distribution of muscle mass, the arrangement of individual muscles, skeletal proportions, joint mobility, and details of skeletal morphology (Fleagle 1976b).

Returning to diet, two studies probe in more detail the differences indicated by the MacKinnons between members of a congeneric pair. Curtin's (1976, 1980) findings on

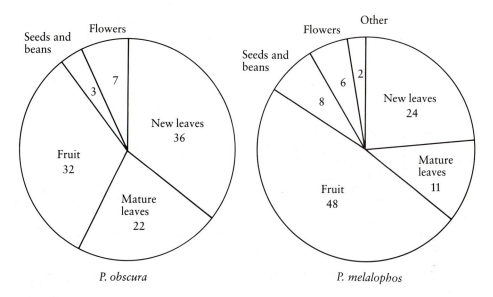

Figure 10.13   Proportions of food types in the diets of *Presbytis obscura* and *P. melalophos*. (Adapted from Curtin 1980.)

leaf monkeys agree with those already reviewed and document additional ways in which the niches of these monkeys differ. Compared with dusky leaf monkeys, banded leaf monkeys have a smaller home range but move further each day, feed from twice as many tree species (197 compared to 87), eat more fruit and fewer young and mature leaves (Figure 10.13), visit more feeding trees in a single day, and spend less time in each. Despite these differences, duskies have in common with bandeds over half their food species, and bandeds share over one-third of theirs with duskies. However, three species upon which both species feed in the same month are large emergents that either are common in the forest or bear conspicuously large fruit crops over longer periods than other tree species. Curtin concludes that the sheer abundance of shared foods mitigates interspecific feeding competition.

The two leaf monkeys weigh about the same, but siamangs are considerably heavier than gibbons. A male siamang weighs from 9.5 to 13 kg, whereas a male gibbon weighs about only 5.5 kg (Napier and Napier 1967). Gittins and Raemaekers (1980) argue that this difference is the key to their ecological separation. Since a gibbon weighs about half as much but its stride is almost as long as a siamang's, the energetic cost of propelling itself over a given distance is less than for a siamang. Lar gibbons travel about twice as far each day as siamangs, seeking out scattered sources of high-energy food in the form of fruit pulp. Siamangs, in contrast, make do with a higher proportion of a less compact energy source, young leaves, which are commoner, thereby relieving themselves of the energetic cost of wide wanderings in search of scattered foods. Siamangs do spend as much time eating figs as gibbons, but figs usually appear in larger clumps than other fruit and yield more energy for the cost of travelling to them.

10.3E   Monkeys in the Manu National Park of Peru

In the forests around Cocha Cashu, a remote Peruvian lake, live 13 sympatric species of primate (Table 10.9). Terborgh's (1983) research focused on 5 of them (Figure 10.14), chosen because they seem to pursue similar lifestyles, seeking out fruit and insects, while differing radically in social organization. His monograph looks in detail at the fine ecological distinctions underlying the broad similarities and at the biological and social concomitants of these distinctions. It also examines the nature and function of the mixed-species associations that the author and his colleagues encountered. Terborgh uses the concept of competition to explain differences in intraspecific relations, but he does not explicitly discuss its importance in establishing and maintaining niche separation among species.

Table 10.9   Primates found at or near Cocha Cashu in the Manu National Park, Peru

Species	Common Name
Ateles paniscus	Black spider monkey
Lagothrix lagotricha	Woolly monkey
Alouatta seniculus	Red howler monkey
Cebus apella[a]	Brown capuchin monkey
Cebus albifrons[a]	White-fronted capuchin
Pithecia monachus	Monk saki
Saimiri sciureus[a]	Squirrel monkey
Aotus trivirgatus	Night monkey
Callicebus moloch	Dusky titi monkey
Callimico goeldii	Goeldi's marmoset
Saguinus imperator[a]	Emperor tamarin
Saguinus fuscicollis[a]	Saddle-backed tamarin
Cebuella pygmaea	Pygmy marmoset

Source: Terborgh 1983.
[a] Studies have been conducted on these species. (See Terborgh 1983).

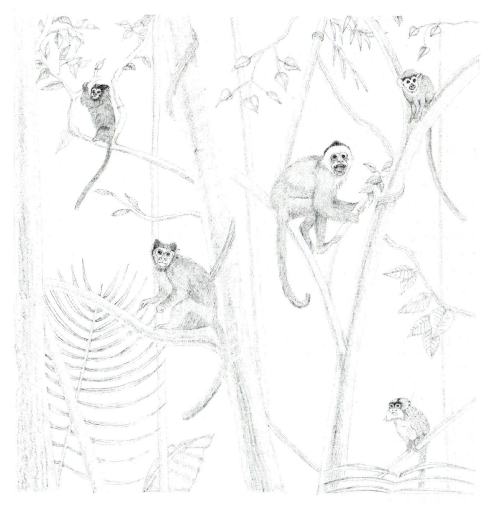

Figure 10.14   Primates studied in the Manu National Park, Peru: (clockwise from left) saddle-backed tamarin *(Saguinus fuscicollis),* brown capuchin *(Cebus apella),* white-fronted capuchin *(C. albifrons),* squirrel monkey *(Saimiri sciureus),* and emperor tamarin *(Saguinus imperator).*

Lake Cocha Cashu is surrounded by tropical vegetation at various stages of succession. On high ground, where trees soar up to 50 or 60 m, the forest at first seems to form a single, endless green ceiling, but closer inspection reveals that it is actually quite variable. Rainfall is highly seasonal, with seven wet and five dry months, and during the latter many trees shed some or all their leaves. Terborgh found that fruit production varies seasonally, with two production peaks in the rainy season and a marked dearth early in the dry season. When fruit production drops, fewer trees fruit and the distance between fruiting trees is greater. Thus, when fruit is scarce it is because fewer trees are fruiting rather than because each tree is producing fewer fruit.

More than 2700 contact hours, most of them over a period of one year, were logged with the five species. The goal was for a team of observers to sample each species for 20 days in each of four seasons, wet, dry, and two transitional periods, but for practical reasons some were sampled in only three of the four.

When soft fruit is abundant, it makes up almost the entire vegetable component of the diets of all animals, and they eat many of the same food species. When the fruit supply dwindles, animals supplement their diets with a wide range of additional items, such as nuts, nectar, and pith. During periods of fruit scarcity, differences among the diets of the five species are most apparent. Brown capuchins *(Cebus apella)* concentrate on palm nuts, a reliable and abundant alternative to soft fruit for animals equipped with the jaws, teeth, and strength to crack them open. Because of the ubiquity of this food, brown capuchins do not have to range more widely when soft fruit is scarce. White-fronted capuchins *(Cebus albifrons)* and squirrel monkeys *(Saimiri sciureus)* rely more and more on figs as other fruit disappears. Since figs are widely spaced and fruit at unpredictable intervals, the ranges of both species increase greatly at this time. Terborgh describes squirrel monkeys in particular as living a seminomadic life in a home range of over 260 ha. To tide themselves over those bleak periods when even the figs run out, white-fronted capuchins slowly and laboriously consume palm nuts, while squirrel monkeys go in for frenetic bouts of insect foraging at a rate of return that probably does not compensate for the energy expended in the hunt. The two tamarin species *(Saguinus imperator* and *S. fuscicollis)* live in tiny territories that contain adequate food supplies year-round. Their principal food species have exceptionally long bearing periods, ripening their crops over a period of many weeks, and these primates choose parts of the forest where such species are common. The result is not only a stable food supply but also one that holds little attraction for most of the larger monkeys, scattered as it is in many small pockets.

Terborgh (1983) argues that the degree to which fruit is concentrated strongly influences the decisions of animals about what to eat. Big monkeys living in big groups cannot profitably exploit resources scattered in tiny packages. Rather, they must find large clumps of food, even if it means traveling long distances to do so. The primates of the Manu National Park vary widely in their dependence on concentrated resources. Squirrel monkeys are at one end of the spectrum, depending on fruit borne on huge fig trees during the dry season. White-fronted capuchins exhibit a similar preference but can make do with palm nuts, which are plentiful but occur at lower concentrations. Brown capuchins are more efficient nutcrackers than white-fronteds and rely heavily on nuts in the dry season. Finally, tamarins live year-round in tiny home ranges and feed on resources that come in tiny, scattered units.

Turning to animal prey, all five species hunt on a variety of substrates, at a variety of heights, and in a variety of habitats, and all five spend more time in this activity when fruit is scarce (Janson in Terborgh 1983). Although insects and insect larvae are the most frequent prey species, capuchins occasionally eat rats and lizards over 20 cm long.

Despite these similarities, there are also clear distinctions among species, for they use different foraging techniques and tend to catch different prey as a result. Saddle-backed tamarins dart up and down tree trunks inspecting cracks and crevices. They rarely catch anything, but when they do, it is often big. The emperor tamarins' behavior is in strong contrast, for they prefer to catch visible prey directly off leaves or branches or to search inside curled-up leaves. Since they usually try to creep up on large, unsuspecting prey, they move slowly and stealthily until the final leap. Squirrel monkeys devote much attention to leaf foraging, but unlike emperor tamarins, they move quickly between foraging spots and eat slow-moving prey such as caterpillars. White-fronted capuchins mix leaf foraging with more strenuous exertions such as breaking open branches and searching the dead bases of palm fronds. Finally, brown capuchins are even more energetic, tearing apart great chunks of deadwood as they search.

Mixed-species associations occur between squirrel monkeys and one or other of the capuchin species and also between the two tamarin species. Terborgh (1983) describes these associations in detail and discusses their value to each species. To this end, he first makes a distinction between two kinds of participation in an association: "When one member of an association receives no benefit, its participation should be passive; active joining, in turn, should indicate some anticipated benefit" (Terborgh 1983:155).

Within the study area, there were almost twice as many social groups of capuchins as of squirrel monkeys, and accordingly, squirrel monkeys were seen more often with capuchins than vice versa. Several lines of evidence indicate that squirrel monkeys are the active participants in associations, the ones who form and break them up. In contrast, capuchins confronted by a milling throng of squirrel monkeys just seem to grin and bear it. Mixed groups persist for a few hours to more than 10 days. Members of the two species mingle extensively but with little physical contact or antagonism. Aggression is most apparent in fruit trees, where capuchins sometimes supplant squirrel monkeys feeding close by or "lose patience and rush pell-mell around a crown expelling all the *Saimiri* in it" (Terborgh 1983:156).

Terborgh suggests that squirrel monkeys benefit in three ways from the association. First, capuchins feed on nuts that are large, tough, and inaccessible to squirrel monkeys; at these times, the latter scavenge partially eaten nuts dropped by the capuchins. More important, squirrel monkeys seem to use capuchins as guides to local fruit sources. Capuchin home ranges are much smaller; thus, a squirrel monkey group coming into an area may use the resident capuchins to lead it to the best fruit trees. Finally, and perhaps most important of all, capuchins apparently have a better predator alarm system than squirrel monkeys. Capuchins do not apparently derive much benefit from the association. In fact, it entails definite disadvantages for both species. Because a mixed group has about twice the biomass of either group alone, members jostle more for position in fruit trees; further, when squirrel monkeys are tagging along, capuchins move about 40% farther each day regardless of season.

Mixed groups containing both capuchin species, with or without squirrel monkeys, occur when fruit production is at a minimum. On these occasions, white-fronted capuchins clearly follow browns, frequently supplanting them at fruit trees and blatantly playing the part of pirates. Like squirrel monkeys, these pirates have much larger home ranges than their victims.

The two tamarin species live in permanent mixed associations consisting of one group of each species; the two groups share and jointly defend a single territory against neighboring mixed groups. Within the territory, they travel separately over parallel paths, coordinating their movements closely when visiting a succession of fruit trees and more loosely at other times. The two species seem to participate equally in maintaining vocal contact and reestablishing close proximity after periods of separation. By chance, the saddle-backed members of one of these tandems disappeared during the study, and Terborgh (1983) was able to watch the remaining group of emperor tamarins alone as well as in association with a neighboring group of saddlebacks, which intermittently moved into the vacant territory for a few days. When moving in tandem, the hourly distance traveled by the emperor tamarins increased by about 30%, but the number of trees visited did not increase.

What advantages do the two partners derive from the association? The answer is unclear. Reciprocal warning of predator attacks is one possibility, cooperative competition (Cody 1974) another. The latter, Terborgh's preferred explanation, refers to a situation whereby species using the same pool of resources profit by exploiting them together and thereby regulating the time of return to particular resources. This maximizes the yield on each visit and minimizes the number of unrewarded visits. The greatest benefit is obtained when resources are scattered, slowly renewing, and little used by other species.

Terborgh (1983) argues that body size is of great importance in structuring the Manu primate community, for it constrains diet, foraging techniques, and strategies against predators. It also determines daily basal metabolic requirements, which in turn influence the size of range needed to support an individual or group. Differences in body size do not, however, tell the whole story, for within the community are several like-sized species. How do they achieve ecological separation? The question remains open in the case of the two tamarin species. Fruit, the most likely critical limiting resource according to Terborgh, is used in seemingly identical fashion by the two species, whose only apparent form of ecological segregation is their insect-hunting techniques. Squirrel monkeys' choice of fruit broadly overlaps that of capuchins, but when hardest pressed in the dry season, squirrel monkeys resort to total insectivory. During these periods the advantage of being smaller than their closest ecological neighbors becomes obvious: the larger species could not subsist on the insects they could catch in a 12-hour day. Thus, facultative insectivory is the card up the squirrel monkeys' sleeve, permitting them to cope with the difficulties of a seasonal environment and at the same time coexist with capuchins. The two capuchin species differ in their ranging behavior, insect-foraging behavior, and use of plant materials, particularly palm nuts and figs. Divergence in their diets and foraging behavior is greatest during the annual period of fruit scarcity.

10.4   Discussion

10.4A   Primate Polyspecific Associations

Polyspecific associations are common among primates. The existence of mixed-species groups was referred to or reported in some detail at three of the five study sites considered in Section 10.3. There is no reference to such groups in the forests of Kuala Lompat. Since nocturnal strepsirhines rarely associate with members of their own species, it is hardly surprising that they have not been seen in association with members of other species. Several field reports confirm the ubiquity of associations among New World monkeys (Thorington 1967; Bernstein 1964; Bernstein et al. 1976; Klein and Klein 1973; Moynihan 1976; Pook and Pook 1982), Old World monkeys (Chalmers 1968a; Gartlan and Struhsaker 1972; Altmann and Altmann 1970), and one study (Bernstein 1967) reports an association between gibbons *(Hylobates lar)* and banded leaf monkeys *(Presbytis melalophos)*.

Let us return to the four questions posed by Diamond (1981). First, are the polyspecific associations seen among primates merely chance aggregations? It has been argued that in some instances they are (Gautier and Gautier-Hion 1969; Gartlan and Struhsaker 1972; Waser 1982) and that in others they are not (Struhsaker 1982; Terborgh 1983).

In most forests, the probability of finding a particular pair of species in association is not the same for all possible pair combinations. In Gabon, for example, De Brazza monkeys form no interspecific associations at all, whereas spot-nosed, moustached, and crowned monkeys range in mixed-species groups most of the time. The problem is to determine whether this nonrandomness reflects real preferences or differences in the density or characteristic group spread of species. Waser (1982) has tried to resolve the problem by developing a model in which groups are treated as though they were gas molecules whose size is equivalent to the average group spread and whose movement is equivalent to the average distance traveled by the group each day. Information about the density of species in the community makes it possible to predict the frequency with which groups will "collide" by chance. Waser's model is the most rigorous attempt yet made to distinguish between random and functionally significant associations, but it fails to take into account the persistence of associations (Struhsaker 1982). For instance, it does not distinguish between two groups from different species that collide and remain in association for five minutes and two that collide and remain in association for five hours. Until modified to do so, the model is probably most useful for analyzing and interpreting brief associations.

The identity of social groups composing associations and their stability through time have not been closely studied because of the practical difficulty of consistently recognizing individual monkeys in the canopy of a rain forest, which is where most associations occur. However, Gautier-Hion's (1978) data suggest that the members of a mixed-species association may be stable over a period of several months, and M. Cords (pers. com. 1983) reports that particular mixed groups of blue and redtail monkeys could repeatedly be seen in particular patches of forest at Kakamega, Kenya. To the extent that the home ranges of species overlap and are of similar size, such consistency is to be expected. Terborgh's observations of tamarin associations suggest a degree of intimacy and stability unparalleled by Old World primates. Pook and Pook (1982) report an equally close relationship between saddle-backed and red-bellied tamarins (*Saguinus fuscicollis* and *S. labiatus,* respectively) in Bolivia. In addition, one of the mixed-species tamarin groups they studied often associated with a group of Goeldi's marmosets *(Callimico goeldii).*

Who leads and who follows? This question has generated considerable controversy in the context of monospecific groups (Altmann 1979; Rhine 1975), and no simple answer emerges from studies of mixed-species groups. Gautier and Gautier-Hion (1969) note that in trispecific associations among spot-nosed monkeys *(Cercopithecus nictitans),* moustached monkeys *(C. cephus),* and mona monkeys *(C. mona),* moustached monkeys are never at the head of the group when it is moving and are always the last to give spacing calls. On different occasions, both spot-nosed and mona monkeys appear to lead progressions. Still, the Gautiers consider the question of leadership an open one, for leadership does not necessarily mean moving at the front of the group. Terborgh's (1983) distinction between active and passive participation in an association brings out this point. In relationships that benefit only one party, the passive participant is usually the leader in that it constrains the movements of the benefiting species. The latter decides whether to remain with the leader, while the leader cannot force the other species to follow or, indeed, to go away.

Terborgh's (1983) discussion of leadership in mixed groups of squirrel monkeys

(Saimiri sciureus) and capuchins (*Cebus* spp.) illustrates the complexity of the situation. Squirrel monkeys are usually in the forefront of progressions, and it initially appeared to observers that they were leading. However, if individuals were randomly distributed throughout the mixed group, a higher frequency of a squirrel monkey being the first individual would be expected simply because there are more squirrel monkeys and they are often more dispersed than capuchins. A closer examination of first-individual records in different contexts showed that this explanation was inadequate. When a group was foraging for insects, a quiet, slow-paced activity during which animals are often widely spaced, squirrel monkeys were in the vanguard of the group about 75% of the time. However, when animals were moving quickly to a fruiting tree or a new foraging area, capuchins were more often in front. One interpretation of this distinction is that individuals are spread out more or less randomly when insect foraging but that capuchins guide major movements between fruit sources. Terborgh and his colleagues sometimes noticed squirrel monkeys running ahead of the capuchins, only to stop and change direction if the capuchins did not follow. When capuchins stopped to feed on palm nuts or some other resource of no interest to squirrel monkeys, the latter waited quietly or spread out to forage in adjacent areas, checking back every half hour or so. Yet capuchins rarely paused when squirrel monkeys stopped to feed, and the latter subsequently had to hurry to catch up.

In contrast to capuchins and squirrel monkeys, neither species of tamarin emerged as a consistent leader. Rather, both participated actively in maintaining their association. Terborgh (1983) concludes that the nature of the functional relationship between associating species is a key to the issue of leadership. When both species benefit from the association, as in the case of the tamarins, there will be no clear leader. When only one species benefits, the other plays the part of willing or unwilling leader.

Why do different primate species associate? As we have already noted, some associations may be chance phenomena that occur because the paths of two social groups belonging to different species happen to cross. A number of advantages have been proposed to explain more persistent associations. These functional explanations resemble those for birds, and, indeed, only the gang explanation (Subsection 10.2A) has yet to be invoked to account for a primate polyspecific association. The following list of possible costs and benefits is distilled from the work of Terborgh (1983), the Gautiers (Gautier and Gautier-Hion 1969, Gautier-Hion and Gautier 1974; Gautier-Hion 1978), Struhsaker (1982), and Gartlan and Struhsaker (1972):

1.   It may be easier to detect and avoid or to dilute the effects of predators in a large, mixed-species group. If so, the relationship is protocooperative, corresponding to the selfish-herd explanation of polyspecific associations among birds given in Subsection 10.2A.

2.   The ability of participants to locate food is increased without increasing intraspecific competition for food or mates. Feeding competition between the participants is minimized by differences in the heights at which they prefer to feed. The relationship, as in point 1, is one of social protocooperation and the associated explanation corresponds to the feeding explanation given in Subsection 10.2A.

3.   Species with similar diets can keep track of resources that have already been depleted, thereby minimizing the number of fruit trees they must visit by maximizing the return provided by each one. The relationship is again one of social protocooperation, and the associated explanation again is feeding efficiency.

4.   In a situation analogous to that of migrating birds from temperate regions that form mixed-species flocks with tropical residents, primates with large home ranges may use members of species with much smaller home ranges to guide them to local resources. This will result in social commensalism if the guiding animals do not experience increased competition for the resources in question, and the underlying explanation is analogous to that of the beater. If competition is increased, then social parasitism occurs, thus recalling the pirate explanation in Subsection 10.2A.

5.   Rates of capturing insects may be increased for one species if its members associate with a second species whose movements flush out prey. As in point 4, the relationship may be commensal or parasitic and thus associated with a beater or pirate explanation, respectively.

The main trouble with all these explanations is that they are hard to evaluate. The context in which associations are most commonly seen may suggest their function, but a strong explanation requires comparative data on the feeding, ranging, and predator-monitoring behavior of animals in and out of mixed groups. The comparison must also control for changes in the environment. If the behavior of a mixed-species group in the wet season differs from that of a monospecific group in the dry season, it is unclear whether the distinction is due to season or to group composition. But forest monkeys are difficult to observe at the best of times, and many species are rarely found in monospecific groups. In those that are so found for extended periods, the switch from single- to mixed-species associations may be correlated with a seasonal change.

To end this section on an optimistic note, it should be added that sometimes the functional significance of polyspecific associations *can* be rigorously examined. At Kibale, for example, changing patterns of association by the animals and persistent teamwork by the observers have made possible a detailed investigation of associations among African rain forest primates, and another study is under way at Kakamega, Kenya (M. Cords, pers. com. 1983). A similar combination of variably associating monkeys and sustained observation in the Manu National Park has facilitated the study of associations among New World primates. At least at certain study sites, then, a better understanding of the function of polyspecific associations is a practical possibility.

10.4B   Evidence for Competition

In most of the studies reviewed in Section 10.3, the idea of competition is invoked in one form or another to explain observed interspecific differences in feeding and ranging behavior. Charles-Dominique (1977) implicitly distinguishes between the fundamental and realized niches of strepsirhines in Gabon to explain why their diets are similar in captivity and different in the wild. Competition over millions of years, he argues, has forced the members of this nocturnal community to become differently specialized in their feeding habits.

Gautier-Hion (1978) suggests that the weight gradient among diurnal primates in Gabon, together with differences in food habits and vertical stratification, show that the competitive exclusion principle has been in operation in that community, too. She also raises two important and unresolved issues: (1) which resource, or combination of resources, is limiting these primates? She suggests that fruit may be superabundant and stresses the need to measure the availability of different foods through time as well as their exploitation by animals other than primates. (2) Since many of the primates in Gabon are heavily hunted, their populations today may be below the level at which competition will occur.

Like Gautier-Hion, Struhsaker (1978a) suggests that competition among the Kibale primates is likely to occur only over rare items during seasonal shortages of fruit, because natural selection has shaped the dietary preferences of sympatric species to minimize competition most of the time. When food is scarce and competition intensifies, animals modify their diet to maintain coexistence. Struhsaker argues that in the blue monkey (Figure 10.15) these adjustments are inadequate and this species suffers from interference competition when in sympatry with red colobus and mangabeys. He presents evidence that blues may be more abundant when these species are not present.

Finally, the MacKinnons (1980b) argue that while the three congeneric pairs of primates they studied in Malaysia are so different that competition for food is probably minimal, similarities between members of each pair are great enough to create a potential for competition. In practice, it is avoided or reduced by different foraging strategies and other behavioral differences reflecting morphological differences. They reject the idea that these distinctions were brought about by competition and character displacement during the evolutionary past. Instead, they suggest that species were preadapted for coexistence before they came into sympatry.

To evaluate these arguments, we must first return to the competitive exclusion principle itself. In Subsection 10.2B we noted that biologists are divided in their opinion of its merits. Let us summarize some of the problems that have been raised (see reviews by Winterhalder 1980, 1981; Whittaker and Levin 1976; Wiens 1977; Schoener 1982; Lewin 1983).

Figure 10.15   A blue monkey *(Cercopithecus mitis)* in the Kibale Forest, Uganda. (Courtesy of L. Leland.)

The hypothesis of the random placement of niches can be rejected for most sympatric species, but this does not rule out other possible causes of species differences (Schoener 1974). For instance, differences in body size may reflect different strategies of predator avoidance rather than an outcome of competition. The idea that factors other than competition may be regulating natural communities becomes increasingly plausible if the assumptions underlying the competitive exclusion principle (Subsection 10.2B) are relaxed. Why relax them? Because they are unrealistic. For instance, contrary to assumption 5, environments do fluctuate over time. A few attempts have been made to take these effects into account (Pianka 1974b; Brown 1975; May and MacArthur 1972), but none come close to approximating the complexity of the real world. According to Wiens (1977), environmental fluctuations make it unlikely that favorable conditions will persist long enough for species to reach the point at which competitive conditions equilibrate. He suggests that populations are often well below the equilibrial size determined by resource availability and are maintained instead at a smaller size by predators or by frequent environmental disturbances. In these circumstances, which characterize temperate biomes in particular, competition will rarely be an important determinant of niche dimensions.

What happens in a variable environment when a similar and potentially competing species moves in? Both the resident and the colonizing species are influenced by the same stresses, but there is evidence that distinctive phenotypes will be favored at times of intense selection. This may reflect a competitive interaction between the two but, alternatively, it may reflect different responses by the two species to the same stresses. One of the strongest arguments for the importance of competition in nature comes from the opposite situation, the absence of a species A from part of the range of another species B with which it typically competes. If B has a broader niche when A is absent, competitive release is said to have occurred. Yet in most such natural experiments, it is initially assumed that the species are competitors and that competition is the sole factor influencing their use of resources. Wiens (1977) points out that if the system is not in an equilibrium determined by resource availability—and this is rarely if ever ascertained—then the species may be responding to different phases of environmental variation in the compared areas.

Finally, some biologists and philosophers of science question not just the assumptions underpinning the principle but the scientific validity of the principle in the first place. If its predictions are logical consequences of the initial premises, then they represent tautologies, or circular reasoning (Peters 1976). The theory predicts that competition will occur among closely related or ecologically similar species and that they must differ in order to coexist. Sure enough, examination of their patterns of resource use reveals differences that are duly labeled "mechanisms" permitting coexistence. But rarely is it established that particular resources really are limiting, and equally rarely is it established that species differences are related directly or solely to competition.

What general conclusions are we to draw? First, even if we are prepared to live with the assumptions of the competitive exclusion principle, it is extremely difficult to demonstrate that patterns of niche separation among sympatric species are an outcome of competition. Second, and more fundamentally, the competitive exclusion principle is a model of a "biologically inconceivable situation" (Winterhalder 1980:60). Does this mean we should abandon it? I do not believe so, for it is a useful starting point in analyzing interactions among species. Hutchinson (1961:143) put it this way:

> "[The competitive exclusion principle] can be used to examine a situation where its main conclusions seem to be empirically false. Just because the theory is analytically true and in a certain sense tautological, we can trust it in the work of trying to find out what has happened to cause its empirical falsification."

With particular reference to the relationships of sympatric primates, we can identify several fruitful directions for research. In the case of all but the two tamarin species, the studies reviewed in Section 10.3 showed niche separation, and the same is true for the other primate studies cited at the beginning of this chapter. Yet none presented a strong case for or against the importance of competition. First, in none were the limiting resources identified, although in some cases an attempt was made to do so and in others the need to do so was strongly emphasized. Careful study of cycles of production in relation to dietary habits is a necessary step in the investigation of competition among primates. It is also feasible. Second, all but one of these studies assumed that populations were maintained at an equilibrium determined by resource availability, but it is far from clear that this is actually so (Chapter 7). Wiens (1977) draws his examples from temperate regions to downplay the importance of competition. We need to know more about primate population dynamics and their determinants before we can reject Wiens's arguments as irrelevant to tropical climes. This too can be achieved, albeit with difficulty. Third, none of these studies revealed an overdispersion of niches, as expounded by Schoener (1974). Unlike the first two problems, it is unclear what kinds of data are needed to solve this one. Overdispersion does not manifest itself in the form of regular differences in body or feeding apparatus size in primates, and the complexity of primate niches is so great that it is difficult to see how to distinguish between a trivial change and a competitively induced addition, subtraction, or compression of niche dimensions. Further thought needs to be given to patterns of separation caused by competition among primates.

Last, all but one of the studies reported here were nonexperimental, although some of the best evidence for competition in other organisms has come from natural or manipulative experiments. Struhsaker (1978a) tried to take advantage of regional differences in primate community composition to examine the competitive relationships of blue monkeys with other arboreal monkeys, but the comparison suffers from a lack of control of the many other differences between the forests being compared, thus leaving the comparison open to exactly the kinds of criticism raised by Wiens. Sussman (1974, 1977) attempted to capitalize on another situation in which two lemur species were distributed both sympatrically and allopatrically. His goal was to study character displacement, but again, the study design was such that a full comparison was not possible. The study of a single primate community is a major undertaking, and it may seem excessive to demand that two or three communities differing with respect to a single primate species be studied simultaneously and in parallel. Excessive or not, such studies would add immeasurably to our understanding of the factors that produce and maintain ecological separation among sympatric species of primates.

CHAPTER ELEVEN

Of Bats, Birds, and Baobabs

11.1 Introduction

THIS book began by emphasizing the distinctiveness of primate biology and biogeography. Here we look once more at primates in a broad context, but this time the emphasis is on their integration into the community. What part do primates play in the overall structure and function of a community (Section 11.2)? What are their relationships with other animals in the community (Section 11.3)? How do their activities affect plant life in the community (Section 11.4)? These questions are about the ecological role of primates today, but they can be asked equally well from an evolutionary perspective.

For example, over a few years primates may repeatedly strip a tree of flowers so that it never bears fruit; over several generations, this action may constitute a selective pressure on members of the tree species to develop appropriate defense mechanisms; the primates, in turn, may evolve specializations rendering these defenses partially or completely ineffective. This process is called *coevolution,* whereby species with close ecological relationships exert selective pressures on one another so that their evolution is interdependent (Ehrlich and Raven 1965). Insight into coevolutionary relationships depends largely upon our understanding of ecological relationships in the present. Only after reviewing the latter, therefore, will we consider the more speculative issue of the coevolution of primates and other forms of life (Section 11.5).

A cautionary note needs to be sounded at the outset of this chapter. Community ecology has been described as a "partially explored wilderness" (Emlen 1973:341); if that be the case, the study of primates in a community context must count as nearly virgin territory. What follows will necessarily be limited in scope, tentative in conclusions, and likely to raise more questions than it answers.

11.2 Primates and Community
Structure and Function

Where do primates fit into the structure of a community? Before looking at this question, we should briefly consider what is meant by *structure.* Community structure can be thought of and measured in several ways (Whittaker 1975). For instance, communities are often described by the growth form of plants (e.g., whether they are trees, shrubs, or herbs) in vertical and horizontal space and by the kinds of animals occupying different levels and patches of vegetation. This we may call the physical structure of a community. Another view considers species diversity, or number of species; a description of structure then involves an enumeration of the species making up variously defined components of the community. For our purposes here, we shall consider community structure primarily from a third perspective, that of the relative contribution of various components to the total biomass of the community. The functional components into which we shall divide the community are trophic levels, described broadly in the preface to Part V as a "hierarchy of feeders and fed upon." The number of links in a food chain varies, although three to five is most common. The links, or levels, are as follows (Whittaker 1975):

1.   *producer,* a photosynthetic plant, which is first organism of the sequence;
2.   *herbivore,* or *primary consumer,* an animal that feeds on plant food;
3.   *first carnivore,* or *secondary consumer,* an animal that feeds on plant-eating animals;
4.   *secondary carnivore,* or *tertiary consumer,* an animal that feeds on first carnivores; and
5.   *tertiary carnivore,* an animal that feeds on secondary carnivores.

Here it is useful to recall the distinction made in Section 2.1 between the net and gross primary production of plants. Gross primary production is the total energy bound by plants, and net primary production is the energy left over after plant respiration. The rate at which animals convert plant material for their own use is called *secondary productivity.* Animals have access only to the net production of plants, and they can harvest only a small fraction of this. Of this fraction, a portion is not assimilated. Of the amount herbivores do assimilate, a portion is used in respiration and is unavailable for harvest by carnivores. For these reasons, there is a sharp decrease in production over successive trophic levels: the production of herbivores, for example, is about one-tenth (or less) that of plants. This relationship between trophic levels can be portrayed visually in the form of a *pyramid of production.* In most communities, both the numbers and biomass of individuals at each trophic level decrease together (Figure 11.1) (see Whittaker 1975 for a clear, extended discussion of trophic structure and function).

We can now make the question of where primates fit into a community's structure more specific: what proportion of a community's biomass is made up of primates, and how is this primate biomass distributed among trophic levels?

Let us begin by noting that animals make up a small proportion of total community biomass. Insects contribute most to the total biomass of animals, and mammals, the

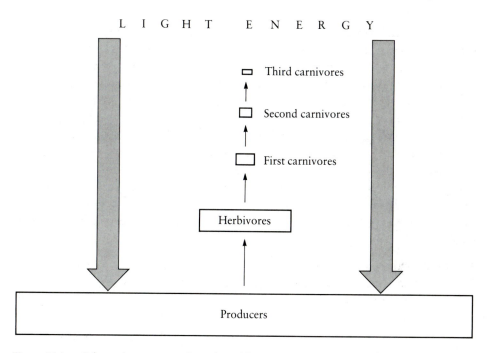

Figure 11.1   Schematic representation of trophic structure as a pyramid of production. Note the sharp decrease in production at each level.

class to which primates belong, contribute proportionately very little. Estimates of biomass in three Central and South American forests are shown in Figure 11.2. Differences between the three are partly attributable to inaccurate measurements and differences in how the measurements were made, but there are certainly real contrasts as well. Such contrasts are to be expected between sites that vary in rainfall, geochemical environment, and degree of human intervention. The impact of such factors on community structure is beyond the scope of this discussion, however, and the data in Table 11.1 are presented primarily to make two simple points. First, in tropical forests the biomass of mammals is an insignificant portion of the total biomass of the community. Second, it is inaccurate and potentially misleading to assume that community structure is the same or even very similar in all tropical forests. While differences between tropical forest and savanna communities and between tropical and temperate-zone communities may be more marked, they should not blind us to contrasts within these ecological categories as well.

What part of the mammalian biomass of a tropical forest is made up of primates? Eisenberg and Thorington (1973) have attempted to answer this question for two New World communities, one a lowland community in Surinam, the other Barro Colorado

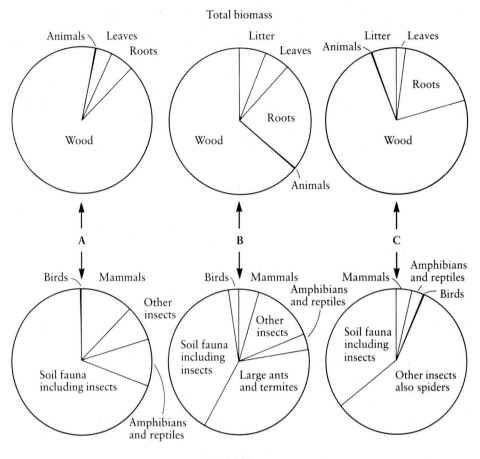

Figure 11.2   Estimated total biomass and animal biomass (percentage of dry weight) in three forests: (**A**) Darien, Panama, (**B**) Central Amazonia, and (**C**) Tabonuco, Puerto Rico. (Data from Golley et al. 1975, Fittkau and Klinge 1973, and Odum et al 1970.)

Island. Estimates of biomass in Surinam are based on records of mammals rescued when the area was flooded during the construction of a dam. Sources of bias in the data are unknown. Eisenberg and Thorington suggest that there is a bias in favor of larger mammals; many of the smaller forms may have escaped unnoticed or drowned. The sample does not include bats, whose biomass has yet to be estimated in any community. It has been suggested that bat biomass may exceed that of all other mammals in New World forests, but Eisenberg and Thorington argue that, although numerous, bats are small and their biomass probably makes up less than 10% of total mammalian biomass. With

Table 11.1   Biomass of selected genera of primates (A) and other mammals (B) on Barro Colorado Island and in Surinam

Genus	Common name	Barro Colorado Island		Surinam	
		Contributed biomass (kg)	% total mammalian biomass[a]	Contributed biomass (kg)	% total mammalian biomass[b]
(A) Primates					
Alouatta	Howler	5,500	6.8	2,634	5.7
Saguinus	Tamarin	40	0.1	58	0.12
Cebus	Capuchin	650	0.8	8	0.1
Total		6,190	7.7	2,700	5.92
(B) Other mammals					
Didelphis	Opossum	1,050	1.3	558	0.59
Bradypus	Three-toed sloth	31,920	40.0	6,732	14.7
Choloepus	Two-toed sloth	9,800	12.3	3,612	7.9
Agouti	Paca	4,000	5.0	1,176	2.5
Dasyprocta	Agouti	2,800	3.5	452	0.98
Proechimys[c]	Spiny rat	2,400	3.0	83	0.18
Tapirus	Tapir	2,300	3.0	9,432	20.6
Tayassu	Peccary	2,300	2.8	1,633	3.5
Mazama	Brocket deer	300	0.4	10,065	22.0
Total		56,970	71.3	33,743	72.95

Source: Adapted from Eisenberg and Thorington 1973.
[a] Percentage calculated from an estimated total of 79,758 kg, which is 20% higher than the known total of 66,465 kg.
[b] Percentage calculated from the recorded total of 45,769 kg.
[c] *Echimys* from Surinam.

the exception of marsupials, small rodents, and bats, the mammals of Barro Colorado Island were directly censused. Estimates for marsupials and rodents were extrapolated from studies on the Panamanian mainland.

The two sets of data (Table 11.1) must be compared with caution, for the area from which the Surinam sample was drawn is not known. The authors themselves assume an area similar in size to Barro Colorado Island, but the basis for this assumption is unclear.

Still, these figures are the best and indeed virtually the only ones we have, and they at least provide a general picture. Primates comprise about 6% of the estimated biomass of nonflying mammals on Surinam and Barro Colorado. From 40% (Surinam) to 70% (Barro Colorado) of mammalian biomass is made up of arboreal forms, and to this the contribution of primates is proportionately somewhat higher, 15% (Surinam) and 14% (Barro Colorado). Sloths (*Bradypus* and *Choloepus* spp.) (Figure 11.3) dominate the biomass in both communities, making up almost a quarter in Surinam and over half on Barro Colorado. Estimates for the tapir (*Tapirus* sp.) and deer (*Mazama* sp.) are different in the two communities. Both are large, terrestrial mammals and are probably over-represented in the Surinam records; in contrast, they are unusually rare on Barro Colorado, perhaps because of poaching in the past or the small size and isolation of the island.

In sum, the contribution of primates to the biomass of nonflying mammals, and even to the arboreal segment of this biomass, is quite small in both these New World communities. By comparison, *Colobus* species have been estimated to form 79% of the nonflying arboreal mammalian biomass in a Ghanian forest, and *Presbytis* species 92% of that in the dry forests of Sri Lanka (Eisenberg, Muckenhirn, and Thorington 1972).

To establish the determinants of differences in trophic structure between ecological communities of broadly similar type, it is essential to examine the ecological relationships of community members. This raises an important and more general point, namely the inextricability of the study of structure and function. If we are to make sense of community structure and specifically the position of primates in that structure, we must understand how communities function. The study of community function, like that of structure, can be approached in several ways. Much effort has been devoted to research on how communities function as whole entities attempting to answer such questions as: how fast and efficiently is energy transferred from one level to another in the trophic structure (Odum 1968)? At what rate are nutrients cycled through and lost from the system (e.g., Bormann and Likens 1970)?

Little is known about the contribution of primates to these broad aspects of community function. What proportion of primary production, for example, is harvested by primates? Only for the leaf portion of primary production, and then only with reference to two primate species, are tentative answers available. Howler monkeys have been estimated to crop 0.87% of total annual leaf production on Barro Colorado Island (Hladik and Hladik 1969); the comparable figure for sloths is 0.63% even though their biomass is higher (Montgomery and Sunquist 1975). Together, vertebrates and invertebrates on Barro Colorado may consume about 15% of the leaves produced each year, although this estimate is probably too high (Leigh and Smythe 1978). At Polonnaruwa, Sri Lanka, the Hladiks (1972) have estimated that groups of purple-faced leaf monkeys (*Presbytis senex*) crop between 5 and 30% of the leaves produced in their territories

Figure 11.3   Three-toed sloth on *Cecropia* tree (top) and two-toed sloth (*Choloepus* spp.) (bottom). (K. Weidmann/Animals Animals; M. R. Stoklos/Animals Animals.)

each year and as much as 15 to 47% of the leaf production of individual *Adina cordifolia* trees. These figures are high, considering that other vertebrates and invertebrates probably crop leaves as well, and further measurements would be useful for confirmation.

The study of community function also includes more focused research. Primates share similar ecological roles and potentially compete with a wide range of mammals, birds, insects, and in the New World even a reptile, the folivorous iguana. The term *guild* is used for species that, regardless of taxonomic position, exploit the same class of environmental resources in a similar way (Root 1967). Guilds are difficult to define rigorously, but the concept is useful because it emphasizes the possible importance of competition between species that are distantly related taxonomically. Historically, evidence of competition has been sought primarily among closely related species, and the guild concept provides a new focus for research. Other animals that may belong in the same guilds as primates will be considered in Section 11.3. The study of primate relationships with other members of the community is not limited, however, to the segment made up by animals. Herbivores rely on plants for food, of course, but many plants rely on herbivores to pollinate their flowers and disperse their seeds. The possibility that special relationships of this kind exist between particular primate and plant species has recently stimulated a wave of research.

In the long run, studies of relationships among community members, be they plants or animals, should provide new insight into the structure and function of the community as a whole. For the immediate future, the first task is to document those relationships. In reviewing the few studies of primates, Sections 11.3 and 11.4 serve mostly to emphasize how much remains to be done. Undaunted, in Section 11.5 we go on to consider the coevolutionary implications of our limited findings.

11.3   Relationships between Primates and Other Herbivores

11.3A   Invertebrates and birds

How much dietary overlap is there between primates and invertebrates that harvest vegetable matter? Do they compete with one another? There are no answers to either of these questions. Dietary overlap has been little studied in this context, and the intensity and, indeed, the very existence of competition between vertebrates and insects, fungi, and bacteria are matters of debate (e.g., Janzen 1977 versus Fleming 1979).

There is one study (Rockwood and Glander 1979) in which the foraging behavior of a primate has been directly compared with that of an insect. Howler monkeys (*Alouatta* spp.) and leaf-cutting ants (*Atta* spp.) (Figure 11.4) are sympatric through most of the

Figure 11.4   Leaf-cutting ants are found in tropical and subtropical America. This individual *(Atta texana)* is clipping a pine needle. (J. Moser/Photo Researchers.)

Neotropics, and both harvest flowers, fruits, and leaves, by preference young ones. The monkeys digest harvested material in their gut with the assistance of microorganisms. The ants carry it back to the nest, where it is broken down by fungi that the ants eat. In a year-long study in Costa Rican forest, howler monkeys and leaf-cutting ants showed broad similarities in foraging behavior but preferred different plant species. Only 7 of the 24 primary food species of each were shared by the other, and they harvested different parts of these 7. Over the year, howler monkeys spent about 40% of foraging time eating young leaves, the remaining 60% more or less evenly distributed among mature leaves, flowers, and fruit. The ants, in contrast, spent almost half their time on mature leaves. Young leaves and flowers were seasonally important, but fruit accounted for less than 2% of the foraging effort. The authors concluded that there was little evidence of dietary overlap or competition between the two.

Pearson (1975) suggested that differences in the biomass of birds at three locations in Amazonian South America were partially due to differences in the biomass of monkeys at the three sites. The site with the highest biomass of monkeys, many of which were presumed to compete with birds for insects and fruit (see also Klein and Klein 1975), had the lowest biomass of birds. An alternative explanation is that the sites differed in the proportion of plants producing bird- or mammal-dispersed fruits. According to this argument, differences in the biomass of birds and monkeys reflect differences in vegetation rather than the outcome of competition (C. Smith, cited in Fleming 1979). The argument rests on proposals made by Snow (1971) and Van der Pijl (1972) that plants have fruit designed to entice birds or bats, but not both. Fruit eaten by birds tends to be red, blue, or black, small, without a strong scent, and clustered at the tips of branches. Fruit eaten by bats tends to be green, large, and strong smelling and to stand out from the foliage. There are grounds for questioning the likelihood that plants produce highly specialized fruit (Section 11.5), and Fleming (1979) himself summarized evidence from the New World suggesting that the diets of primates do overlap significantly with those of bats *and* birds. It is an unanswered question whether this overlap spells competition.

Leighton (1982, pers. com.) examined dietary overlap among frugivorous primates, birds, and bats in Borneo. Well over 100 species of trees and lianas at his study site had fruit that was eaten almost exclusively by the four frugivorous primates present (orangutans, gibbons, and two species of macaque). These fruits had several features in common. They contained a few large seeds surrounded by thin, watery flesh and a 1- to 5-mm thick, indehiscent, inedible rind. This rind was generally yellow-brown, red, or purple. Civets and bats ate some of these fruits, but birds did not. Although birds and primates differed in many of their fruit choices, there were nonetheless instances of overlap that Leighton suggests may have generated intense competition. For instance, Sunda Isles leaf monkeys *(Presbytis aygula)* ate the immature seeds of fruit from several species that, once ripe, were heavily fed upon by hornbills (Figure 11.5).

In summary, the relationship between frugivorous birds and bats has received much more attention than that between frugivorous birds and primates. Leighton's (1982) study provides the only analysis of dietary overlap between the latter. His findings show that many of their dietary preferences differ, but he argues that there may be competition for certain shared resources.

11.3B   Other mammals

At Kuala Lompat, Malaysia, all the primates eat fruit, and for some it is a major dietary item. Agile gibbons, for example, spend over half their feeding time on fruit (Gittins and Raemaekers 1980; Chivers 1980). Primates share the forest with at least 44 other species of mammal and 64 bird species known or believed to be partially or completely frugivorous (Payne 1980). The mammals include a variety of bats, squirrels, rats, civets, and ungulates. Of these, only the tree squirrels are arboreal and diurnal like the anthropoid primates, and efforts to assess the extent to which primates share and perhaps compete for resources with other animals have so far focused on the exploitation of fruit by primates and six species of these squirrels.

A comparison of their feeding patterns (MacKinnon and MacKinnon 1978; Payne 1980) yielded the following major distinctions:

Figure 11.5   With its huge yellow beak and the red, gold, green, and blue of its plumage, this hornbill *(Phyticeros cassidix)* must surely be one of the most striking birds in the forest. (Courtesy of M. Leighton.)

1.   Only three squirrel species habitually fed in the upper layers of the canopy, where all the primates except *Macaca nemestrina* concentrated their feeding activities.

2.   The diet of two of these three species contained a high proportion of seeds embedded in fruit no part of which the observers judged to be edible by gibbons, siamangs, and macaques, who fed primarily on fruit pulp.

Figure 11.6   *Protexerus strangeri,* a large arboreal squirrel found in the Makokou Forest, Gabon.

3.   It is probable that both species of leaf monkey at Kuala Lompat digested the seeds they ate, but neither was seen eating the seeds of species used heavily by the squirrels.
4.   All primates ate a significant quantity of leaves, whereas the squirrels ate almost none.

Payne (1980) concludes that the year-round availability of seeds at Kuala Lompat creates a seed-eating niche for specialists equipped with the gnawing dentition needed to exploit them. Thus, despite the high proportion of fruit in the diets of both tree squirrels and primates, they actually eat few of the same fruits.

In the Makokou Forest of Gabon, Emmons (1980) recorded a decline in the density of a large arboreal squirrel species *(Protexerus strangeri)* (Figure 11.6) between 1971 and 1976. During the same period, the density of another potentially competing squirrel species increased together with that of the *Cercopithecus* monkeys in the area, and Emmons tentatively attributed the decline of *Protexerus* to competition with the monkeys. There are other plausible explanations for the changes. For instance, they may have been caused independently of one another by a fluctuation in the environment.

Sloths and primates are sympatric in many forests in Central and South America. Sloths are specialized leaf eaters with ruminantlike stomachs and digestive processes. Howler monkeys *(Alouatta* spp.), while less specialized, also include a large amount of leaves in their diet. A comparison of the ecology of these similarly sized arboreal folivores is thus of particular interest. On Barro Colorado Island, the primates are sympatric with two-toed and three-toed sloths *(Choloepus hoffmani* and *Bradypus infuscatus,* respectively), and Montgomery and Sunquist (1975) have analyzed their ecological relationships.

Three-toed sloths were rarely observed feeding, and two-toed sloths never. Dietary differences between sloth species and between sloths and primates had to be inferred primarily from the species of trees in which sloths were located (usually doing nothing), although the three-toed sloth's diet was partially reconstructed from stomach contents. The sloth species were found in many of the same tree species: two-toed sloths used 42% of the species used by three-toed sloths, and three-toed sloths used 85% of those used by two-toed sloths. However, there was much less overlap between sloths and primates. Of 28 tree species known to be fed upon by three-toed sloths, only 6 were used as food by primates, and of the 98 tree species in which three-toed sloths were located, only 16 were fed on by primates. Sloths and primates sometimes occupied the same tree at the same time but were never seen interacting. Sloths, it should be said, rarely interact with anything.

The estimated density and biomass of sloths on Barro Colorado Island were much higher than those of howler monkeys. Despite this difference and despite the lower proportion of leaves in the diet of howler monkeys, the primates probably crop about one-third more leaves than sloths. Montgomery and Sunquist (1975) suggest two reasons for the higher leaf-cropping rate of howler monkeys. First, they are morphologically less specialized for digesting leaves and derive less nutritional benefit from a given quantity of leaves than sloths do. Second, the basal metabolic rate of a sloth is 51% lower than that of other mammals of comparable size, and a sloth also lowers its body temperature when the ambient temperature drops, unlike any haplorhine primate. Finally, sloths live in home ranges of 2 ha or less, and they move little or not at all from day to day. An indication of their slothfulness is conveyed by Montgomery and Sunquist's observation that animals were normally captured for radio marking by hoisting them out of the tree in a noose; when this was not feasible, someone climbed the tree and sawed off the branch from which the sloth was hanging! In short, sloths make very efficient use of leaves, do little, and live at high densities, making sympatric howler monkeys look like amateurs in the world of arboreal folivores. It is unclear, however, that the howler monkey population competes with the sloths.

Turning to our last case, that of frugivorous bats, the data are sparse and as equivocal as ever. On Barro Colorado Island, the fruits of a few tree species are eaten by as many as four species of bats, four rodents, three primates, two small carnivores, and the peccary. In general, however, dietary overlap between sympatric frugivorous primates and bats in the New World appears to be low (Fleming 1979).

As we reach the end of this section, our conclusions can be brief. Dietary overlap between primates and distantly related taxa is generally low. This does not rule out the possibility of intense competition between primates and these other taxa, however, for there is no necessary relationship between the degree of dietary overlap and the intensity of competition.

11.4   Primates and Plants

In Chapter 4, we considered the nutritional properties of different plant parts, the distribution of these parts in space and time, and factors other than nutrients affecting a plant's attractiveness as food. In short, we looked at plants from a primate's point of view. Now we take a more balanced view of the relationship and look at the plant's point of view as well. Specifically, are plants ever actually killed by primates? Contrarily, do they receive any benefit from the primates that feed upon them?

11.4A   Primates as Plant Killers

Primates may occasionally starve or die from eating poisonous plant parts. How frequently, if ever, do plants die as a result of attacks by primates? To answer this question best, we first consider the leaves and flowers of trees and then whole plants with a non-woody growth form separately from different kinds of plants (i.e., grasses and other ground cover plants). Some primates digest or crack open tough seed coats with their teeth (Section 5.4), but most make use only of the fleshy part of the fruit; seed predation will not be considered here (but see Subsection 11.4B).

Reports of primates stripping all the leaves from a tree are rare (Figure 11.7), and rarer still are reports of them actually killing trees in this way. Glander (1975) attributed the deaths of two of the four *Cecropia pelatata* trees in a relict riparian forest in Costa Rica to "constant feeding pressure" exerted by a group of howler monkeys *(Alouatta palliata)*. Such chronic overcropping is probably unusual. When clearance for agriculture destroyed much of the natural habitat, this group of monkeys was isolated in a small patch of forest where they probably first exhausted preferred resources before turning, as a last resort, to plant parts that had no nutritional value or were poisonous. On Barro Colorado Island, where the forest is much more extensive, certain *Platypodium elegans* trees are often stripped of new leaf buds at the end of the wet season, but this happens only once a year, and the trees appear to suffer no permanent damage as a result (C. Burns, pers. com. 1981).

If predators consistently eat all the flowers of a tree that reproduces by producing and dispersing seeds, then it will leave no descendants even if the tree itself is not killed. One study has looked at the effect upon subsequent fruit crops of heavy floral predation by primates (Struhsaker 1978b). Struhsaker first noticed that members of a common tree species *(Markhamia platycalyx)* were frequently stripped of their flowers in a sector (K30) of the Kibale Forest, Uganda. A group of about 20 red colobus monkeys, for example, could eat all the flowers in one tree in about an hour (Figure 11.8). The density of primates in K30 was particularly high, and for 24 months the only trees that bore fruit were isolated from the forest or in sectors with low primate densities. In the

Figure 11.7   Ring-tailed lemurs *(Lemur catta)* quickly stripped all the new leaves from this sapling as I watched.

25th month, however, there was an exceptional degree of floral synchrony among *Markhamia* trees (meaning that they all flowered together), and the following months were the first since the beginning of the study in which they bore fruit in K30. It seemed that when trees flowered together, predators were swamped with more flowers than they could eat in the relatively short period of bloom.

Struhsaker (1978b) then began a systematic investigation of the relationship between floral predation, floral synchrony, and subsequent fruit crop size. Of three tree species

Figure 11.8   Red colobus adult female with infant *(Colobus badius)* eating flowers of a dombeya tree, Kibale Forest, Uganda. (Courtesy of L. Leland.)

whose flowers were all heavily fed upon by the Kibale primates, one *(Celtis africana)* showed no significant correlation between degree of floral synchrony and fruit crop size, and the other two *(Strombosia scheffleri* and *Aningeria altissima)* showed significant correlations ($r = 0.67$ and 0.602, respectively).

Struhsaker considered these findings with respect to three hypotheses about the correlation of floral synchrony and fruit crop size:

1.   Floral synchrony will be favored to provide predators with more food than they can eat in a short period of time. Under conditions of heavy predation, therefore, the degree of synchrony and subsequent fruit crop size should be strongly correlated. While this hypothesis explains the correlation found in *S. scheffleri* and *A. altissima,* it does not explain the absence of such a correlation in *C. africana,* which is also subject to heavy floral predation by Kibale primates.

2.   In *xenogamous* species, species dependent upon cross-pollination between individual trees, floral synchrony will increase the chances of fertilization and hence the size of the subsequent fruit crop. In the absence of detailed studies of pollination, this hypothesis could not be evaluated conclusively, but circumstantial evidence suggested that none of the three species was xenogamous.

3.   A positive correlation between floral synchrony and fruit crop size may arise when individual trees respond simultaneously to optimal climatic conditions and appropriate environmental clues. A series of analyses provided limited support for this hypothesis with respect to *S. scheffleri,* but none with respect to *A. altissima.*

In conclusion, Struhsaker (1978b) suggested that the short flowering stage of *C. africana* may minimize the impact of floral predators so that trees do not benefit by flowering simultaneously. The strategy of this species may rather be one of escape. Because trees flower asynchronously and briefly, there is a good chance that a tree will be able to finish flowering before it is found by predators. The best explanation for the positive correlation between floral synchrony and subsequent fruit crop size in *A. altissima* is that synchrony successfully overwhelms predators with prey. In *S. scheffleri,* both pressure from floral predators and optimal climatic conditions may contribute to the correlation. For our purposes we conclude that primates can and do deflower trees given the chance, but that plants use a variety of mechanisms to ensure that the chance is rarely offered.

Perennial ground plants like trees are long-lived and can thus expect repeated assaults by herbivores if they make good eating. The impact of primates on these plant forms has not been studied, and it deserves exploration. Grazing and browsing ungulates stimulate growth in grasses and also help grasses invade new areas by trampling or eating the existing ground cover (McNaughton 1979). In contrast, baboons, the only primates that feed heavily on grass, frequently have a destructive effect because they

Figure 11.9   A baboon *(Papio cynocephalus anubis)* and a cow apparently ignore each other's proximity, at Gilgil, Kenya. (Photography by A. Richard.)

commonly dig up entire plants, roots and all. The magnitude of the destruction thus wrought by savanna baboons is unknown but is probably slight in relation to the total area under grass cover, because the density of baboons is low. Ironically, the destruction may sometimes benefit sympatric humans and their herds. For example, olive baboons at Gilgil dig up, eat, and thereby destroy a grass disdained by cattle browsing in the same area (Figure 11.9). This grass is usually replaced, however, by an invading species that the cattle eat with relish (S. Strum, pers. com. 1981)!

Gelada baboons *(Theropithecus gelada)* exploit grass more efficiently than savanna baboons, and their biomass is much greater. When new growth is available, they behave like proper grazers and eat only the blades, harvesting an estimated 12–15% of the standing crop at any given feeding site (feeding sites being from 2 to 15 m apart) (Wrangham 1976). In the dry season, however, when the grass is dry and coarse, like savanna baboons they resort to digging up and eating whole plants. Horses set loose by the local people probably churn up the ground to obtain grass roots as well, but the impact of the combined efforts of geladas and horses has not been studied.

At Dunga Gali in Pakistan, rhesus monkeys *(Macaca mulatta)* feed primarily on ground vegetation. Clover *(Trifolium* sp.), a perennial ground herb, is uprooted and eaten in large quantities, but it is so abundant compared with the density of monkeys that it seems unlikely that its distribution is significantly affected. Of more interest is the impact of the monkeys upon species with a less vigorous growth habit and upon those that complete their life cycle in one year and depend upon the production of seeds to persist from one year to the next. In most instances, the impact of the monkeys is inseparable from that of the cattle and goats with which they share the forest. One perennial herb, *Oenothera* sp., provides a possible exception. It has a long taproot and only the monkeys exploit this root, pulling it up whenever one is found during the winter months. Indeed, it may be a major winter food (Goldstein 1984). Samples of the ground vegetation show that *Oenothera* plants are rare in the forest. To what extent is this rarity due to the monkeys? Without comparative data on a similar forest from which monkeys have been excluded, we can only guess at the answer, but their impact upon the composition of the ground cover in such temperate forests is an intriguing subject for future study.

11.4B   Primates as Gardeners

Primates can provide three benefits to the plants they eat. These are the stimulation of new growth by pruning, the pollination of flowers, and the dispersal of seeds. Like the negative aspects of the impact of primates upon their food plants, these positive effects have been little studied and our review is correspondingly brief.

Pruning
On Barro Colorado Island, capuchin monkeys *(Cebus capuchinus)* eat the terminal buds of membrillo trees *(Gustavia superba)* during the rainy season. Oppenheimer and Lang (1969) compared the branching pattern of 50 trees of this species on the island with that of 50 individuals of about the same size at another site from which capuchins had been almost or completely eliminated by hunting. At both sites the topography and plant communities were similar, and the vegetation had been undisturbed by people for at least 45 years. Branching was measured by scoring one "fork" each time a branch divided. The number of forks per tree on Barro Colorado Island turned out to be significantly higher than at the other site.

This finding suggests that capuchins on Barro Colorado have an effect on membrillo trees analogous to that produced by browsing ungulates on vegetation close to or on the ground. By removing the terminal buds, animals promote growth in the lateral buds. Moderate browsing of this kind not only increases branching and the amount of food available in the future but also, to the plant's advantage, increases the number of sites available for flowers and fruits the following season. Conversely, of course, the immediate production of flowers and fruits is reduced; thus, if the density of monkeys were high enough, the long-term beneficial effects of intermittent pruning would likely be transformed into a disadvantage for the tree, which would fail to produce any flowers or fruits for years in succession.

It is unclear how widespread effects of this kind are, since this is the only study to my knowledge that has attempted to document pruning by primates.

Pollination

Unlike the major vertebrate pollinators, bats and birds, most primates are much bigger than most flowers. A flower is thus but a mouthful for most primates, whereas it is a large container of a valuable food, nectar (and sometimes the pollen itself), for smaller animals. In general, only when a small primate encounters a particularly big flower does it become practically possible for a primate to act as a pollinator instead of a predator.

Primates believed to act as pollinators include a number of strepsirhines and New World monkeys, in particular the diminutive tamarins (*Saguinus* spp.). It is not easy to demonstrate conclusively that an animal is a pollinator and not just a flower visitor. It must be shown that the suspected pollinator frequently visits flowers of a particular species, does no damage to their reproductive organs, and takes up pollen and transfers it from one plant to another. In few of the cases reviewed below is such conclusive evidence available.

Among the strepsirhines, five species have been seen feeding on flowers without destroying them. Only one study, covering a six-week period, has documented the time spent in this activity, and none has confirmed that pollination takes place, although the evidence certainly points to it.

During the dry season in northwestern Madagascar, mongoose lemurs (*Lemur mongoz*) spent 84% of their feeding time licking floral nectar or eating nectaries of four tree species, the kapok (*Ceiba pentandra*) alone accounting for 80% of the total feeding time (Sussman and Tattersall 1976; Sussman 1978). During the same study, mouse lemurs (*Microcebus murinus*) were also seen licking kapok flowers, and both forked lemurs (*Phaner furcifer*) and dwarf lemurs (*Cheirogaleus medius*) have occasionally been noticed licking flowers in a west coast forest (Petter, Schilling, and Pariente 1975; Charles-Dominique et al. 1980). Finally, during eight nights in a forest in eastern Kenya spent "staking out" baobab trees (*Adansonia digitata*), Coe and Isaac (1965) watched bush babies (*Galago crassicaudatus*) pay repeated visits to these trees and bury their heads in the large, white, pendulous flowers.

In the Manu National Park in Peru, Janson, Terborgh, and Emmons (1981) saw seven species of primates, large as well as small ones, visit the flowers of a canopy liana (*Combretum fruticosum*) (Table 11.2). Monkeys moved systematically from one flower cup to another on the floral spike, lapping the nectar from each. Afterwards, their faces

Table 11.2   Use of flowers of two plant species by five primate species in Manu National Park, Peru

	Plant Species					
	Combretum fruticosum			*Quararibea cordata*		
Primate Species (Year of Sample)	Proportion of Feeding Time[a]	Mean No. Vines Used/Day	Maximum No. Vines Used/Day	Proportion of Feeding Time	Mean No. Trees Used/Day	Maximum No. Trees Used/Day
Saguinus fuscicollis						
(1975)	0	0	0	0.29	2.2	3
(1977)	0.71	6.5	9	—	—	—
Saguinus imperator						
(1975)	0.01	0.1	1	0.29	2.6	4
(1977)	0.70	5.1	7	—	—	—
Saimiri sciureus						
(1977)	0.17	3.4	14	0.16	1.5	6
Cebus albifrons						
(1975)	0.04	0.2	1	0.09	2.0	5
Cebus apella						
(1975)	0.19	2.8	>5	—	—	—
(1977)	0.16	2.9	11	0.35	3.4	5

Source: Adapted from Janson, Terborgh, and Emmons 1981.
[a] Feeding time refers only to the use of vegetable resources. It is the sum of the times spent feeding for all the members of a monkey troop in a given sample.

would be dusted with the pink pollen. Although these feeding bouts damaged the petals, the ovaries were left intact and abundant fruit production afterward indicated that the damage was not critical. The two tamarin species spent almost three-quarters of their feeding time on *Combretum* flowers during a 10-day sample period in 1977. (The authors do not discuss why tamarins spent so much less time on these flowers in 1975.) Flowers of a second species *(Quararibea cordata),* a common tree, were visited more often than the liana by the larger primates and accordingly were more heavily damaged. However, this tree produces large fruit and probably bears more flowers than it can fruit. The flowers' structure is such that animals were always powdered with pollen during a visit; because they visited more than one tree each day, a pollinating role for the large primates is at least possible.

In summary, nectar feeding seems to be quite widespread among the nocturnal primates of Madagascar, and further instances may be uncovered now that attention has been drawn to this behavior. It is likely that pollination occurs during visits to flowers although this is not certain. In South America, a range of primates has been seen feeding on flowers, and at least the smaller species act as pollinators. In the case of the larger species, the evidence is less convincing. Nectar feeding has yet to be reported in any Old World haplorhine. The possibility that certain primates have a coevolutionary relationship with the plant species they pollinate will be discussed in Section 11.5.

Seed Dispersal

Individuals of a given tree species are often widely separated in tropical forests, sometimes by as much as a kilometer, and trees make use of wind, water, and animals to transport their seeds over these distances. In tropical forests, an estimated half or more of the canopy tree species, and perhaps 70% of the lower-stratum species, use birds and mammals for seed dispersal (Longman and Jenik 1974; Van der Pijl 1972). Animals transport seeds in at least three ways. Some species, such as the caviomorph rodents of Central and South America, cache seeds in the ground for later consumption when food is short. Although many of these seeds are eventually eaten, some are missed and consequently germinate (Smythe 1970). Both birds and mammals involuntarily transport seeds stuck to their feathers or fur; these seeds are often equipped with barbs to increase their chance of being dispersed in this way. Finally, frugivores may carry fruit to another tree, eat the soft parts, and drop the seeds, or they may transport seeds some distance in their digestive tract before excreting them, intact and unharmed, in their feces.

Transport of seeds in the gut is called *endozoochory. Exozoochory* refers to the transport of seeds without ingestion. In recent years, increased attention has been paid to the role of primates in endozoochory (Hladik and Hladik 1969; Hladik and Hladik 1967; Howe 1980; Lieberman, Hall, and Swaine 1979; Mittermeier and van Roosmalen 1981; Fleming 1979). Most primates eat fruit, many eat it in large amounts, and in so doing they often ingest large numbers of seeds. Let us consider some of the research into the fate of these seeds.

Lieberman, Hall, and Swaine (1979) found that the feces of baboons *(Papio cynocephalus anubis)* in the Shai Hills, Ghana, contained viable seeds from 29% of the fleshy-fruited and 13% of the dry-fruited plant species known to grow in the area. During their 15-month study, Lieberman, Hall, and Swaine grew 6465 seedlings from baboon feces, representing 59 species. However, baboons probably do not play an important role in dispersing the seeds of all these species, for over 75% of the seedlings were from just 3 species. The germination success of fresh and ingested seeds belonging to 4 species was compared; in 3 of the 4, a significantly higher percentage of seeds germinated from the ingested sample. In the fourth, ingestion had no effect on germination success. Although the relative importance of baboons compared with that of other dispersal agents in the area is unknown, they undoubtedly play some role in seed dispersal for a wide range of plant species and perhaps a central role for a few.

Howe (1980) studied the dispersal of seeds from a canopy tree species *(Tetragastris panamensis)* on Barro Colorado Island. Eight species of mammals, including four primates, and 14 bird species were seen eating or trying to eat the fruit. Germination trials indicated that none of the dispersers influenced the viability of seeds. In assessing the relative importance of each disperser, Howe discounted the number removed by mammals by the mortality expected from competition between germinating seeds concentrated in the same piece of feces. He assumed that there was space for only one of them to reach maturity. Competition with siblings also reduced the probability of survival for seeds dropped under the parent tree.

Table 11.3 traces the fates of an estimated 8592 seeds handled by vertebrates in 19 trees. It shows two striking results. First, most of the dispersal is by mammals, and howler monkeys take 70% of the seeds removed. Of those likely to become established, howler monkeys take 58%, and together with capuchins and raccoonlike animals called coatimundis *(Nasua narica),* howlers account for 75% of the seeds dispersed with a reasonable chance of surviving. Second, an estimated 89% of seeds or seedlings succumbed to sibling competition as a result of being dropped under the parent tree or concentrated in the feces of a mammalian disperser. In other words, dispersers are wasters on a grand scale, too.

The Hladiks (1969) showed that the effect on seeds of passage through a howler monkey's digestive tract is not always as benign as in the case of *Tetragastris*. Germination trials were conducted on seeds of 16 tree species found in the feces of one or more Barro Colorado primates. Slight differences were found in the seeds of 14 species; in most of these cases, seeds passing through the gut germinated faster (Figure 11.10). Major differences were found in 2 species. In one, *Trichilia cipo,* the percentage of seeds that germinated and the rate at which they germinated increased dramatically after

Table 11.3   Fates of estimated 8592 seeds handled by vertebrates during visits recorded in extended watches of *Tetragastris panamensis* on Barro Colorado Island. Numbers in last two columns derived from rates of seed ingestion, drop, and removal, and the incidence of sibling competition in droppings containing many seeds.

Species	Eaten (*n*)	Dropped (*n*)	Removed (*n*)*	Dispersed? (*n*)†	Presumed dead (*n*)‡
Howler monkey	5600	700	4900	544	4356
White-faced monkey	1800	900	900	128	772
Coatimundi	908	304	604	32	572
Slaty-tailed trogon	36	0	36	36	0
Black-throated trogon	2	0	2	2	0
Collared aracari	48	16	32	32	0
Keel-billed toucan	78	26	52	52	0
Swainson's toucan	16	0	16	16	0
Masked tityra	16	0	16	16	0
Purple-throated fruitcrow	88	0	88	88	0

Source: Howe 1980.
* Number removed is equal to number eaten minus number dropped.
† Number potentially dispersed is equal to number removed/seeds per feces. (Birds regurgitate seeds singly.)
‡ Presumed dead due to sibling competition: number presumed dead is equal to number removed minus number potentially dispersed.

passage through a capuchin or spider monkey (Figure 11.10D). Howlers also eat these seeds, but none were obtained from their feces for trials. In one trial with *Ficus insipida* (Figure 11.10E), the percentage of seeds germinating and the speed of germination plummeted after passage through a howler monkey. There was only a slight drop after passage through a spider monkey. Note, however, that results from a second trial were less clear-cut (Figure 11.10G). In a final set of experiments, the Hladiks compared the germination of 100 seeds from the stomach with 100 from the large intestine of a howler monkey. Of those from the stomach, 73 seeds germinated compared with 23 from the large intestine. The authors concluded that fermentation in the howler monkey's large intestine destroyed the seeds of some species.

11.5   Primate Coevolutionary Relationships

In Section 11.1 we gave Ehrlich and Raven's (1965) definition of coevolution as a process whereby "species with close ecological relationships exert reciprocal selective pres-

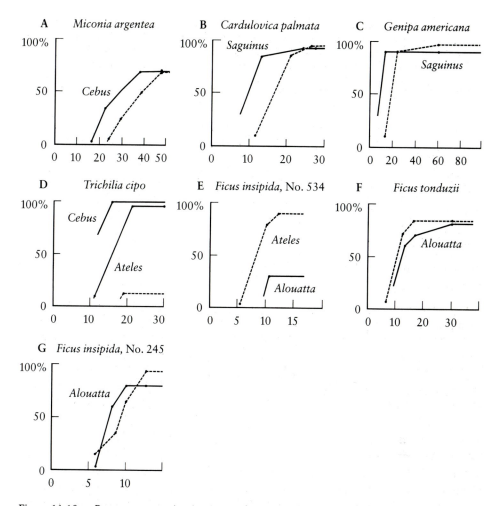

Figure 11.10   Percentage germinating (vertical axis) and number of days taken to germinate (horizontal axis) by seeds collected from primate feces (solid line) and directly from ripe fruit (broken line). (Adapted from Hladik and Hladik 1969.)

sures on one another so that their evolution is interdependent." The place of coevolutionary studies in the history of ecology has been characterized as follows:

> Although it has long been recognized that plants and animals profoundly affect one another's characteristics during the course of evolution, the importance of coevolution as a dynamic process involving such diverse factors as chemical communication, population structure and dynamics, energetics, and the evolution, structure, and functioning of ecosystems has been recognized widely for less than a decade. Indeed, coevolution really represents a point of view about the structure of nature. (Gilbert and Raven 1975:ix)

This chapter has reviewed data on the relationships of primates with plants and other animals, some of which may represent end products of a coevolutionary process. Here we explore this possibility further, paying particular attention to the coevolution of primates with the defense mechanisms, pollination, and seed dispersal systems of plants and to their coevolution with other animals exploiting broadly similar resources.

11.5A   Coevolution of Primates and Plants

The earliest flowering plants were probably visited and pollinated primarily by insects. However, over the last 60 million years, flower-visiting and pollinating vertebrates, notably birds and bats, have evolved as well, and today many plants have flowers specifically adapted for pollination by one of these groups (Baker and Hurd 1968). Insect-pollinated plants produce masses of small flowers, each with a small amount of reward in the form of nectar. Bird-pollinated flowers bloom by day and tend to be brightly colored, usually red, and scentless (Raven 1972; Fenton and Fleming 1976). Except for the specialized flowers pollinated by hummingbirds, many of those visited by birds are also visited by bats, but predominantly bat-pollinated flowers are usually dull-colored, heavily scented, solitary, and borne hanging on the ends of branches, and they bloom at night (Baker 1970). Flowering is often staggered, minimizing the chances of a small bat eating its fill at a single plant (Janson, Terborgh, and Emmons 1981). The kapok, for example, bears bat-adapted flowers (Baker 1973). The tree loses its leaves before flowering, and new foliage grows back only after the seed pods have exploded. This improves seed dispersal as well as access to the flowers by bats. The flowers are strongly scented and produced on terminal branches up in the canopy. Nectar, the bat's primary reward, is produced in large quantities at the base of the flower before it opens; nectar contains sugars and amino acids (Baker and Baker 1975). Flowers open about half an hour after dusk and stay open all night; the next day, the petals drop.

Are there flowers specially adapted for pollination by nonflying mammals? The answer may be yes, although there is no complete agreement concerning their characteristics (Rourke and Wiens 1977; Sussman and Raven 1978; Janson, Terborgh, and Emmons 1981). Some of the disagreement may reflect differences among nonflying mammalian pollinators and corresponding adaptive differences among the flowers with which they have coevolved.

Flowers visited by small, nocturnal, solitary rodents and marsupials in South Africa and Australia have many of the characteristics of bat-pollinated flowers, although they tend to be particularly large and strong and secrete copious amounts of nectar (Rourke and Wiens 1977). Based on these observations and data on nectar feeding among nocturnal primates in Madagascar, Sussman and Raven (1978) suggest that bats and nonflying mammalian pollinators have a reciprocal geographic distribution: pollination systems involving nonflying mammals have arisen primarily in areas like Madagascar and temperate Australia, where pollination by bats takes place rarely or not at all.

A more distinctive set of features may characterize flowers pollinated primarily by diurnal, social, nonflying mammals, particularly primates (Janson, Terborgh, and Emmons 1981). First, in contrast with bat-pollinated flowers, those of the canopy liana visited by primates in the Manu National Park, Peru, were brightly colored rather than strongly scented, providing an obvious target for monkeys heavily dependent on vision to locate food. Second, the flowers of all three species commonly visited by primates were borne not on the very ends of the branches but closer to the central support, and they were shallow-cupped and upright rather than hanging. Third, they opened simultaneously, not one by one, which may increase their attraction for monkeys needing a big enough nectar source to provide substantial nourishment for a large social group. Finally, plant species having flowers with these features also tended to produce a few large fruit from many flowers, thus perhaps reducing the importance of flower loss due to rough handling.

In summary, because there are no nectarivorous bats in Madagascar, nocturnal primates may play an important substitute role as pollinators, but there is little evidence of trees having flowers with features specially adapted for pollination by primates as distinct from bats. In South America, in contrast, there may be a class of flowers specially adapted for pollination by diurnal, social, visually oriented nonflying mammals. This label applies to all New World primates except the owl monkey *(Aotus* spp.*)*, which is nocturnal in most but not all parts of its geographic range (P. Wright 1978, pers. com. 1983).

However, we do not know how many New World primates are in fact important pollinators. Janson, Terborgh, and Emmons (1981) suggest such a role not only for small-bodied species but also for at least two of the larger-bodied forms. Yet the larger monkeys inflict considerable damage on the flowers they supposedly pollinate, and it is not altogether clear whether they are primarily pollinators or floral predators. The data are more convincing for the smaller species. Note that nectarivorous bats are abundant in Central and South America. Thus, bats may not monopolize the role of mammalian pollinator throughout their range as Sussman and Raven (1978) suggest, although they may indeed exclude other small, nocturnal forms. We need more data on the role of New World primates in the pollination systems of their food plants, and we should also explore the intriguing possibility that parallel coevolutionary relationships may occur in the Old World haplorhines, particularly in smaller-bodied species such as talapoin monkeys *(Miopithecus talapoin)*.

All primates except the tarsier eat fruit, and many eat it in large quantities. What limited evidence we have suggests that in so doing, many primate species act as seed dispersers. Have primates coevolved with the plants whose fruit they eat? Parallels between seed and pollen dispersal systems have frequently been drawn (e.g., Van der Pijl 1972; Snow 1971; McKey 1975; Howe 1977, 1979). In both systems the plant provides an incentive to animals, either nectar or fruit, that promotes the plant's reproductive success by dispersing its pollen or seeds.

It has been argued that the marked interdependence of pollinators and plants and the morphological and behavioral specializations characteristic of some pollination systems are likely to be present in certain seed dispersal systems (e.g., McKey 1975). Frugivores are believed to differ in behavioral and ecological traits affecting their suitability as seed dispersers. Plants respond in different ways to the existence of this spectrum of efficiency among dispersers. Some produce many small, nutritionally poor fruits to attract a wide range of "poor-quality" dispersers. Others produce a smaller number of nutritionally rich fruits whose seeds are dispersed by a few "high-quality" dispersers (Snow 1971; McKey 1975). If the theory is correct, then the position of primates in this spectrum deserves a scrutiny it has yet to receive (but see Howe 1980). Recently, however, it has been pointed out that there is little or no evidence of either highly coevolved frugivores or plants adapted to exclude all but a single disperser (Wheelwright and Orians 1982). Only Galapagos tomatoes and tortoises (Rick and Bowman 1961) and a tree species on Mauritius and the extinct dodo (Temple 1977) present possible instances of a close coevolutionary relationship (but see Owadally 1979).

Wheelwright and Orians (1982) have argued that in fact pollen and seed dispersal systems differ in fundamental ways (Table 11.4). Together, these differences suggest that environmental unpredictability and a plant's difficulty in directing its dispersers, even if good germination sites are "knowable," prevent seed dispersal systems from attaining the precision and specialization of some pollen dispersal systems. Plants may actually benefit from having many, not few, dispersers. For example, the more species of dispersers a plant has, the more of its fruits are likely to be carried off (Snow 1971); by using a variety of dispersers, a plant improves the chances that its seeds will be introduced to a range of habitats including new areas for possible colonization (Hamilton and May 1977). From the frugivore's viewpoint, a diet including a range of fruit is also likely to be favored; most frugivores are bigger and longer-lived than most pollinators and cannot compress their annual cycle into the fruiting season of a single plant species.

The role of primates in the seed dispersal systems of living plants demands further investigation, but before we propose coevolutionary relationships between particular primate frugivores and particular plant species, we need a more general understanding of relationships between plants and the whole spectrum of frugivores in a community.

How important have forest primates and other arboreal vertebrates been in shaping the defense mechanisms evolved by the plants upon which they feed? Certainly, we have ample evidence that the toxic effects of certain plant parts constrain the food choices of

Table 11.4   General differences between pollen dispersal and seed dispersal

Characteristic	Pollen dispersal	Seed dispersal
Suitable site for dispersal	Stigma of conspecific flower	Site appropriate for germination and establishment
Characteristics of site as predictors of suitability	Distinctive: color, shape, etc.; often apparent at a distance	Unpredictable: many subtle factors involved; present characteristics often poor indicators of future quality
Temporal pattern of suitable sites	Synchronous with pollen dispersal	Unpredictable: often independent of habitat type or phenology of conspecific plants
Advantage for plant of diet and habitat specificity by animal disperser	High: most pollen lost if visits to other species of plants intervene between visits to conspecifics	Low: presence of adult conspecific plant often an unsuitable site because of density-dependent seed predation or different habitat requirements of seed and tree
Ability of plant to direct animal dispersers to suitable site	High: incentives (nectar, pollen, etc.) provided at suitable site	Low: no incentive for frugivore to deposit seed in favorable site; seed represents heavy and space-consuming ballast that is profitably discarded as quickly as possible

Source: Adapted from Wheelwright and Orians 1982.

primates, yet for various reasons the evolution of plant defense mechanisms is generally attributed more to the destructive potential of herbivorous insects than to vertebrates (Ehrlich and Raven 1965; Price 1975; Freeland and Janzen 1974; Feeny 1975). It has been argued that the coevolutionary role of herbivorous insects is more important simply because they have been around longer than mammals (e.g., Freeland and Janzen 1974). Herbivorous mammals have been a part of forest communities for at least the last 50 million years, however—time enough, in principle, for them to have had some effect on the defense mechanisms of their food species.

Janzen (1978) has written vividly of an evolutionary arms race between plants and insects, with vertebrates largely limited to "making their living through fortuitously picking up what they can handle in a battlefield whose microstructure is determined by the insects and plants" (p. 80). He notes that his metaphor is probably most apt in Neotropical mangrove swamps, where there are virtually no arboreal vertebrate herbivores, and least apt in forests like Kibale, which have a substantial complement of large herbivores. In open grasslands the situation may be reversed entirely, he suggests, with "innocent" insects the benefactors in a war between ground cover plants and mammalian browsers and grazers. Obviously, if certain kinds of herbivore, be they mammals or insects, are absent or rare in a community, they cannot be a major influence on the defense mechanisms of plants.

It is unlikely that there is a simple relationship between biomass and degree of influence, however, because of basic differences in the feeding behavior of herbivorous mammals and many of the herbivorous insects. Specifically, insects show a level of specialization in food choices unequalled by herbivorous mammals. For primates to be major selective agents for chemical defenses in a plant species, they must regularly feed upon many individuals of that species and do more damage than any insect species does (Janzen 1978). While primates may meet the first of these requirements, they are much less likely to meet the second, for there are no primates that come close to matching the degree of specialization shown by insects that attack just one species of plant. These *monophagous* forms have evolved both detoxification systems designed to cope with the chemicals present in their food and chemosensory mechanisms that enable them to locate the particular plants to which they are chemically adapted (Feeny 1975). The onslaught of these insects probably generates stronger selective pressures on the plant than do the foraging patterns of primates, even the most specialized of which include a wide range of plant species and parts in their diets.

In summary, despite the obvious influence of plant biochemistry upon primate dietary habits, it is unlikely that primates have shaped the evolution of chemical defenses in their food species. Still, as with the previous systems discussed, we need to know much more about the depredations of primates compared with insects in the present before we can progress beyond the most tentative conclusions about the past.

11.5B   Coevolution of Primates and Other Animals

In Section 11.3 we discussed the possibility that differences between three Amazonian forest communities in the abundance of frugivorous birds may be due partly to the intensity of competition with primates. We end this chapter with a broader and much more speculative discussion of past competitive relationships between primates and other major taxonomic groups, specifically between the strepsirhine-grade primates of the early Cenozoic and the ancestors of modern bats and birds.

Today, the only region of the world in which nocturnal, solitary primate pollinators are common is Madagascar, where plant-visiting bats are rare. Is this a reciprocal biogeographic relationship resulting from competition in the past? Sussman and Raven (1978) suggest that it is. The virtual extinction of the Paleocene primates, the initial development and later disappearance of the first primates of modern aspect in Europe and North America, and the rise of larger-bodied, diurnal, mainly folivorous–frugivorous primates all took place over about the same period of time that bats evolved the ability to fly and underwent an adaptive radiation during which they spread throughout the Old and New World tropics (cf. Chapter 2).

The earliest strepsirhine-grade primates were all nocturnal. The teeth of several of these Paleocene forms are very similar to those of other nocturnal, flightless mammals living at that time, now thought to be ancestral to the true bats, which appeared in the Eocene. This dentition seems to have been adapted to process a diet composed predominantly of insects and fruit, nectar, or gum. By exploiting the same resources at the same time in the 24-hour cycle, primates may have been thrust into intense competition with bats early in their evolutionary history. For both groups, the consequences would have been profound. Jepson (1970) speculates that competition between primates and bat ancestors precipitated the evolution of flight in the latter. Reciprocally, Sussman and Raven (1978) speculate that competition brought about the extinction of many Paleocene and Eocene primates and perhaps played a role in the evolution of the larger diurnal forms. Small, nocturnal, flower-visiting primates survived only in areas such as Madagascar, where nectarivorous bats are rare.

The possible impact on early primates of competition has been considered even more broadly by Charles-Dominique (1975). Competition between two species eating a similar range of foods is reduced if one is active by day and the other by night, for in general a slightly different array of food items is available at these times (Schoener 1974). Comparing numbers of species of birds and mammals, their body weights, and their activity cycles, Charles-Dominique found that in the Makokou Forest, Gabon, most birds occupy diurnal ecological niches and most mammals nocturnal ones (Figure 11.11). He also found that diurnal mammals tend to be heavier than nocturnal ones. Mammals on Barro Colorado Island, Panama, show similar trends, despite other differences between the two communities. (Note that these findings say nothing about biomass.)

Charles-Dominique (1975) suggests that the division of the 24-hour cycle by birds and mammals is both widespread and ancient. The first arboreal forest mammals were nocturnal climbing forms with a well-developed sense of smell, better suited to negotiate a tangle of branches in poor light than are birds, which depend on vision and are unable to fly along as slowly as caution would dictate in a forest at night. By day, however, flight and visual acuity are advantageous, allowing birds to range over large areas and pick out food sources by sight. Today, birds still rely on vision, whereas most mammals depend more on olfaction to communicate and to locate food.

What about exceptions to the rule of "birds by day and nonflying mammals by night"? Anomalies in the nocturnal world include two groups of birds and all bats, animals that do manage to fly around and make a living at dusk or by night. Goatsuckers

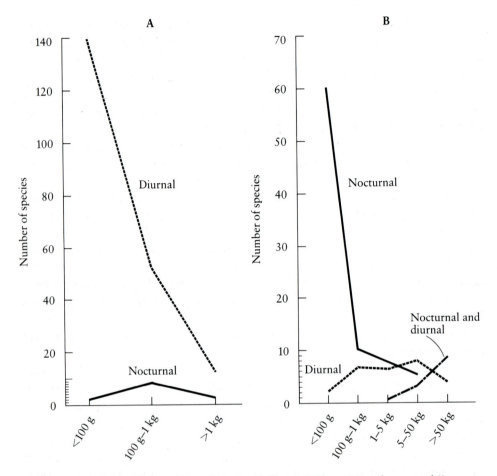

Figure 11.11   (A) Weight categories of sympatric diurnal (208 species) and nocturnal (8 species) birds from the Makokou Forest, Gabon. (B) Weight categories of sympatric nocturnal (83 species), diurnal (26 species) and nocturnal-diurnal (12 species) mammals from the Makokou Forest, Gabon. Note the symmetry between diurnal birds and nocturnal mammals on the one hand, and nocturnal birds and diurnal mammals on the other hand. (Adapted from Charles-Dominique 1975.)

and nightjars hunt at dusk and have a *tapetum lucidem,* a reflecting layer at the back of the eye that helps them see in the twilight. Owls hunt at night but only in relatively open environments. They too have a *tapetum lucidem,* and some species have developed a highly refined system of sound localization. Bats in the suborder Microchiroptera have developed an extraordinarily sensitive sonar mechanism and use echolocation to pinpoint obstacles and prey; members of the Megachiroptera, most of which do not echolocate, are mediocre navigators and usually fly above the forest canopy.

Turning to exceptions in the diurnal world, 20% of the mammals at Makokou are active by day. They include 9 squirrel species, 11 primates, 4 antelopes (out of 15 species), 1 mouse (out of 14 species), and 1 pangolin (out of 3 species). Of these, only the monkeys and squirrels are highly arboreal as well as diurnal and hence in potential competition with birds. In Panama, only 9.5% of the forest mammal species are completely diurnal, and of these, again, only the monkeys and two squirrel species are highly arboreal as well. Charles-Dominique (1975) argues that diurnal, arboreal mammals have all adopted one or more of four "solutions" to the "problem" of competition with birds; these solutions, which permit mammals to exploit resources inaccessible to birds and thereby reduce or eliminate competition, are now described.

First, *an increase in body size,* perhaps the most widespread solution, results in a metabolic economy of scale and provides space for a voluminous digestive tract, so that an animal can exploit foods that are difficult to break down and nutritionally poor and that must be consumed in bulk. Mode of locomotion constrains the size of most birds: the upper limit for body weight compatible with flight appears to be 5 kg. (This limit does not, of course, apply to flightless birds like the ostrich.) Large size has frequently been viewed as a means of escaping the attention of predators, but this explanation probably holds good only for the very biggest mammals such as the elephant, for predators themselves come in many sizes and use various strategies to prey upon large animals as well as small ones.

Why have nocturnal mammals not undergone a comparable radiation of body size? They detect fruit primarily by smell, and this technique is good only over short distances. In the Makokou Forest, trees that were abundant produced few fruit at a time, and those that produced a large fruit crop were rare and widely scattered (Charles-Dominique 1977). Nocturnal species thus tend to find only trees bearing a few fruit. To locate fruit crops plentiful enough to feed large animals or large groups of animals requires extremely good vision over long distances, something possible only by daylight (Figure 11.12). In short, Charles-Dominique suggests that nocturnal mammals are condemned to small size and solitude by virtue of the conditions imposed by nocturnal searching for food.

A second solution to competition with birds is *continuously growing incisors,* which provide certain mammalian species, notably squirrels, with a powerful gnawing mechanism for extracting the contents of tough-coated fruits and nuts that birds cannot open.

Third, *specializations of the claws* enable several groups of mammals to dig out hidden prey more effectively than birds can do with their beaks.

Finally, the *intelligence* of mammals is superior to that of birds and allows them to exploit resources requiring special dexterity and intelligent behavior (Figure 11.13).

Which of these so-called solutions have been adopted by primates? Charles-Dominique (1975) argues for the first and the last. Comparing the weights of nocturnal and

Figure 11.12   Adult female woolly spider monkey with infant *(Brachyteles arachnoidea)* uses all four limbs and tail to move through the periphery of a tree. (Courtesy of A. Young.)

diurnal primates in five forests, he found that 1 kg is the upper limit for nocturnal species except the aye-aye *(Daubentonia madagascariensis)* and the lower limit for diurnal species except the Callitrichidae. The callitrichids have several features that seem to be specializations for feeding on gum (Chapter 5). Gum is not exploited by birds; as a result, large body size is presumably not necessary to avoid competing with them. The aye-aye possesses adaptations that help it to harvest insect prey with particular efficiency, and it may in fact play the part of woodpeckers, which are absent from Madagascar (Cartmill 1974).

Primates at the high end of the size range for diurnal forms avoid competition with birds, argues Charles-Dominique, by eating large quantities of fibrous and nutritionally poor plant parts. Smaller species, still larger than most birds, are also able to exploit a range of foods inaccessible to birds, including large fruits with tough rinds that most

Figure 11.13   Intelligent maybe not, but curious certainly—juvenile rhesus macaques *(Macaca mulatta)* observe the observer on Cayo Santiago Island. (Courtesy of D. Sade.)

birds cannot open. Parrots feed upon nuts and hard fruits, but Charles-Dominique asserts that their beaks, while strong, are less efficient than the haplorhine's combination of hands and teeth.

The diurnal arboreal primates reduce competition with other animals by being bigger than they are. They also make use of dexterity and intelligence to harvest insect prey and fruits largely inaccessible to birds. For example, they rummage efficiently through vegetation, under bark, or in the forest litter in search of insects or small, hidden vertebrates, thanks to their hands, which are specialized for precise and complex movements. Their excellent hand-to-eye coordination allows them to snatch up fast-moving prey. A similar combination of skills allows them to harvest small fruit dangling in a maze of fine twigs on the periphery of trees and to extract the flesh from big, hard-skinned fruit.

Figure 11.14   Silvered leaf monkeys *(Presbytis cristata)* resting in a tree. (Courtesy of K. Wolf.)

In sum, according to these arguments, competition with other animals, bats in particular, drove many primates to extinction early in their evolutionary history. Most of the survivors are diurnal, escaping competition with birds by virtue of their large size, manual dexterity, and intelligent behavior.

In Chapter 1 we saw that a large brain in relation to body size is an identifying characteristic of primates that carries a host of implications, most of them still poorly understood. It is difficult to escape the conclusion drawn by Charles-Dominique (1975) as well as other researchers, that the large brains of primates help them make a living, and it is easy to conjecture why this might be so. However, lacking better comparative information about brain structure and function in primates and other mammals, we shall not pursue these conjectures further. Besides, I prefer to end this work with a chapter that has primates safely ensconced in their natural environment (Figure 11.14), surrounded by birds and bats, pruning trees and dispersing seeds and so forth. Sadly, this is a precarious vision at best. The rate of habitat destruction in the world today is so great that there are few places where primates are safe, and few where their communities are still intact. Now and in the future, the study of these communities must go hand in hand with efforts to conserve what remains.

References

Aldrich-Blake, F. P. G. 1970. Problems of social structure in forest monkeys. In *Social Behaviour in Birds and Mammals*, edited by J. H. Crook, 79–101. London: Academic Press.

———. 1978. Dispersion and food availability in Malaysian forest primates. In *Recent Advances in Primatology*, edited by D. J. Chivers and J. Herbert, Vol. 1, 323–326. New York: Academic Press.

———. 1980. Long-tailed macaques. In *Malayan Forest Primates: Ten Years' Study in Tropical Rain Forest*, edited by D. J. Chivers, 147–162. New York: Plenum.

Aldrich-Blake, F. P. G., T. K. Bunn, R. I. M. Dunbar, and P. M. Headley. 1971. Observations on baboons, *Papio anubis*, in an arid region in Ethiopia. *Folia Primatol.* 15:1–35.

Alexander, R. D. 1974. The evolution of social behavior. *Annu. Rev. Ecol. Syst.* 5:324–382.

Allee, W. C., A. E. Emberson, O. Park, T. Park, and K. P. Schmidt. 1949. *Principles of Animal Ecology*. Philadelphia: Saunders.

Allen, D. L. 1938. Geological studies on the vertebrate fauna of a 500-acre farm in Kalamazoo County, Michigan. *Ecol. Monogr.* 8(3):347–436.

Allen, G. M. 1938. *The Mammals of China and Mongolia*. Part 1. New York: American Museum of Natural History.

Allen, G. M., and A. Loveridge. 1933. Reports on the scientific results of an expedition to the southwestern highlands of Tanganyika Territory. II. Mammals. *Bull. Mus. Comp. Zool. (Harv. Univ.)* 75:45–140.

Allison, J. B. 1964. The nutritive value of dietary protein. In *Mammalian Protein Metabolism*, edited by H. N. Munro and J. B. Allison, vol. 2, 41–134. New York: Academic Press.

Altmann, J. 1974. Observational study of behavior: sampling methods. *Behaviour* 49: 227–267.

———. 1980. *Baboon Mothers and Infants*. Cambridge: Harvard University Press.

Altmann, J., S. A. Altmann, and G. Hausfater. 1978. Primate infant's effects on mother's future reproduction. *Science* 201:1028–1030.

Altmann, J., S. A. Altmann, G. Hausfater, and S. S. McCluskey. 1977. Life history of yellow baboons: infant mortality, physical development, and reproductive parameters. *Primates* 18:315–330.

Altmann, S. A. 1959. Field observations on a howling monkey society. *J. Mammal.* 40:317–330.

———. 1962. A field study of the sociobiology of rhesus monkeys, *Macaca mulatta. Ann. N.Y. Acad. Sci.* 102:338–435.

———. 1965. *Japanese Monkeys: A Collection of Translations.* Atlanta, Ga.: Emory University Press.

———. 1974. Baboons, space, time, and energy. *Am. Zool.* 14:221–248.

———. 1979. Baboon progressions: order or chaos? A study of one-dimensional group geometry. *Anim. Behav.* 27:46–80.

Altmann, S. A., and J. Altmann. 1970. *Baboon Ecology.* Chicago: University of Chicago Press.

———. 1979. Demographic constraints on behavior and social organization. In *Primate Ecology and Human Origins,* edited by I. S. Bernstein and E. O. Smith, 47–62. New York: Garland.

Altmann, S. A., and S. S. Wagner, 1978. A general model of optimal diet. In *Recent Advances in Primatology,* edited by D. J. Chivers and J. Herbert. vol. 1, *Behaviour* 407–414. London: Academic Press.

Amerasinghe, F. P., B. W. B. Van Cuylenberg, and C. M. Hladik. 1971. Comparative histology of the alimentary tract of Ceylon primates in correlation with the diet. *Celon J. Sci. (Biol.) (Colombo)* 9:75–87.

Anand Kumar, T. C. 1974. Oogenesis in adult prosimian primates. In *Reproductive Biology of the Primates,* edited by W. P. Luckett. Vol. 3 of *Contributions to Primatology.,* 82–96. Basel: Karger.

Anderson, R. M., B. D. Turner, and L. R. Taylor, eds. 1979. *Population Dynamics.* New York: Halsted.

Anderson, S., and J. K. Jones, Jr. 1967. *Recent Mammals of the World: A Synopsis of Families.* New York: Ronald Press.

Andrews, P. 1982. Hominoid evolution. *Nature* 295:185–186.

Andrews, P., and J. E. Cronin. 1982. The relationships of *Sivapithecus* and *Ramapithecus* and the evolution of the orangutan. *Nature* 297:541–546.

Andrews, P., and J. Van Couvering. 1975. Palaeoenvironments in the East African Miocene. In *Approaches to Primate Paleobiology,* edited by F. S. Szalay. Vol. 5, 62–103. New York: Karger.

Andriantsiferana, R., and T. Rahandraha. 1973. Variation saisonnière du choix alimentaire spontané chez *Microcebus murinus. C. R. Acad. Sci. (D) (Paris)* 227:2025–2028.

Angst, W., and D. Thommen. 1977. New data and a discussion of infant killing in Old World monkeys and apes. *Folia Primatol.* 27:198–229.

Armstrong, R. A., and M. E. Gilpin. 1977. Evolution in a time-varying environment. *Science* 195:591–592.

Arnold, G. W., and J. L. Hill. 1972. Chemical factors affecting selection of food plants by ruminants. In *Phytochemical Ecology,* edited by J. B. Harborne, 72–101. New York: Academic Press.

Ashton, E. H., and C. E. Oxnard. 1964. Locomotor patterns in primates. *Proc. Zool. Soc. Lond.* 142:1–28.

Avis, V. 1962. Brachiation: the crucial issue for man's ancestry. *Southwest J. Anthropol.* 18:119–148.

Badgley, C., and A. K. Behrensmeyer. 1980. Paleoecology of Middle Siwalik sediments and faunas, northern Pakistan. *Palaeogeogr. Palaeoclimatol. Palaeoecol.* 30:133–155.

Baker, H. G. 1970. Evolution in the tropics. *Biotropica* 2:101–111.

———. 1973. Evolutionary relationships between flowering plants and animals in American and African forests. In *Tropical Forest Ecosystems in Africa and South America: A Comparative Review,* edited by B. J. Meggers, E. S. Ayensu, and W. D. Duckworth, 148–159. Washington, D.C.: Smithsonian Institution Press.

Baker, H. G., and I. Baker. 1975. Studies of nectar-constitution and pollinator-plant coevolution. In *Coevolution of Animals and Plants,* edited by L. E. Gilbert and P. H. Raven, 100–140. Austin: University of Texas Press.

Baker, H. G., and P. D. Hurd. 1968. Intrafloral ecology. *Annu. Rev. Entomol.* 13:385–414.

Baldwin, J. D., and J. I. Baldwin. 1972. Population density and use of space in howling monkeys *(Alouatta villosa)* in southwestern Panama. *Primates* 13:371–379.

———. 1976a. Primate populations in Chiriqui, Panama. In *Neotropical Primates, Field Studies and Conservation,* edited by R. W. Thorington, Jr., and P. G. Heltne, 207–31. Washington, D.C.: National Academy of Sciences.

———. 1976b. Vocalizations of howler monkeys *(Alouatta palliata)* in southwestern Panama. *Folia Primatol.* 26:81–108.

Banfield, A. W. F. 1954. The role of ice in the distribution of mammals. *J. Mammal.* 35:104–107.

Barash, D. P. 1977. *Sociobiology and Behavior.* New York: Elsevier.

Barron, E. J., J. L. Sloan II, and C. G. A. Harrison. 1980. Potential significance of land-sea distribution and surface albedo variations as a climatic forcing factor; 180 m.y. to the present. *Palaeogeogr. Palaeoclimatol. Palaeoecol.* 30:17–40.

Bartholemew, G. A., and W. R. Dawson. 1968. Temperature regulation in desert mammals. In *Desert Biology,* edited by G. W. Brown. Vol. 1, 395–421. New York: Academic Press.

Bate-Smith, E. C. 1972. Attractants and repellants in higher animals. In *Phytochemical Ecology,* edited by J. B. Harborne, 45–56. New York: Academic Press.

Battan, L. J. 1974. *Weather.* Englewood Cliffs, N.J.: Prentice-Hall.

Bauchop, T., and R. W. Martucci. 1968. Ruminant-like digestion of the langur monkey. *Science* 161:698–700.

Bearder, S. K., and G. A. Doyle. 1974. Ecology of bushbabies, *Galago senegalensis* and *Galago crassicaudatus,* with some notes on the behaviour in the field. In *Prosimian Biology,* edited by R. D. Martin, G. A. Doyle, and A. C. Walker, 109–130. London: Duckworth.

Bearder, S. K., and R. D. Martin. 1980. *Acacia* gum and its use by bushbabies, *Galago senegalensis* (Primates: Lorisidae). *Int. J. Primatol.* 1:103–128.

Bekoff, M. 1975. Social behavior and ecology of the African Canidae: a review. In *The Wild Canids,* edited by M. W. Fox, 120–142. New York: Van Nostrand Reinhold.

Bell, R. H. V. 1970. The use of the herb layer by grazing ungulates in the Serengeti. In *Animal Populations in Relation to Their Food Resources,* edited by A. Watson, 111–124. Oxford: Blackwell.

———. 1971. A grazing ecosystem in the Serengeti. *Sci. Amer.* 225(1):86–93.

Berger, M. E. 1972. Population structure of olive baboons *(Papio anubis* [J. P. Fischer]) in the Laikipia District of Kenya. *East Afr. Wildl. J.* 10:159–164.

Berman, C. 1980. Early agonistic experience and rank acquisition among free-ranging infant rhesus monkeys on Cayo Santiago. *Int. J. Primatol.* 1(2):153–170.

Bernstein, I. S. 1964. A field study of the activities of howler monkeys. *Anim. Behav.* 12:92–97.

———. 1967. Intertaxa interactions in a Malayan primate community. *Folia Primatol.* 7:198–207.

Bernstein, I. S., P. Balcaem, L. Dresdale, H. Gouzoules, M. Kavanagh, T. Patterson, and P. Neyman-Warner. 1976. Differential effects of forest degradation on primate populations. *Primates* 17:401–411.

Bernstein, I. S., and E. O. Smith. 1979. *Primate Ecology and Human Origins: Ecological Influences on Social Organization.* New York: Garland STPM Press.

Bertram, B. C. R. 1975. Social factors influencing reproduction in wild lions. *J. Zool. (Lond.)* 177:463–482.

———. 1976. Kin selection in lions and in evolution. In *Growing Points in Ethology,* edited by P. P. G. Bateson and R. A. Hinde, 281–301. Cambridge: Cambridge University Press.

———. 1978. Living in groups: predators and prey. In *Behavioural Ecology,* edited by J. R. Krebs and N. B. Davies, 64–96. Sunderland, Mass.: Sinauer.

———. 1979. Serengeti predators and their social systems. In *Serengeti: Dynamics of an Ecosystem,* edited by A. R. E. Sinclair and M. Norton-Griffiths, 221–248. Chicago: Chicago University Press.

Bigalke, R. C. 1972. The contemporary mammal fauna of Africa. In *Evolution, Mammals, and Southern Continents,* edited by A. Keast, F. C. Erk, and B. Glass, 141–194. Albany: State University of New York Press.

Binford, L. R. 1977. General introduction. In *For Theory Building in Archaeology,* edited by L. R. Binford, 1–10. New York: Academic Press.

Bingham, H. C. 1932. Gorillas in a native habitat. *Carnegie Hist. Wash. Publ.* 426:1–66.

Birchette, M. G. 1982. The postcranial skeleton of *Paracolobus chemeroni.* Ph.D. diss., Harvard University.

Bishop, N. H. 1979. Himalayan langurs: Temperate colobines. *J. Hum. Evol.* 8:251–281.

Boelkins, C. R., and A. P. Wilson. 1972. Intergroup social dynamics of the Cayo Santiago rhesus *(Macaca mulatta)* with special reference to change in group membership by males. *Primates* 13:125–140.

Bogert, L. J., G. M. Briggs, and D. H. Calloway. 1973. *Nutrition and Physical Fitness.* 9h ed. Phila-delphia: Saunders.

Boggess, J. E. 1976. Social behavior of the Himalayan langur *(Presbytis entellus)* in eastern Nepal. Ph.D. diss., University of California, Berkeley.

———. 1980. Intermale relations and troop male membership changes in langurs *(Presbytis en-tellus)* in Nepal. *Int. J. Primatol.* 1:233–274.

Boorman, S. A., and P. R. Levitt. 1980. *The Genetics of Altruism.* New York: Academic Press.

Bormann, F. H., and G. E. Likens. 1970. The nutrient cycles of an ecosystem. *Sci. Amer.* 223(4): 92–101.

———. 1979. Catastrophic disturbance and the steady state in northern hardwood forests. *Am. Sci.* 67:660–669.

Botkin, D. B., P. A. Jordan, A. S. Dominski, H. S. Lowendorf, and G. E. Hutchinson. 1973. So-dium dynamics in a northern ecosystem (moose/wolves/plants). *Proc. Nat. Acad. Sci. USA* 70:2745–2748.

Bourlière, F. 1964. *The Natural History of Mammals.* New York: Knopf.

Bourlière, F., J. J. Petter, and A. Petter-Rousseaux. 1956. Variabilité de la température centrale chez les lémuriens. *Mem. Inst. Rech. Sci. Madagascar Ser. A Biol. Anim.* 10:303–304.

Boyce, M. S. 1979. Seasonality and patterns of natural selection for life histories. *Am. Nat.* 114:569–583.

Boyd, C. E., and C. P. Goodyear. 1971. Nutritive quality of food in ecological systems. *Arch. Hy-drobiol.* 69(2):256–270.

Bradbury, J. W. 1975. Social organization and communication. In *Biology of Bats,* edited by W. Wimsatt. Vol. 3, 1–72. New York: Academic Press.

Bradbury, J. W., and S. L. Vehrencamp. 1977. Social organization and foraging in emballonurid bats. III. Mating systems. *Behav. Ecol. Sociobiol.* 2:1–17.

Brockelman, W. Y., B. A. Ross, and S. Pantuwatana. 1973. Social correlates of reproductive suc-cess in the gibbon colony on Ko Klet Kaeo, Thailand. *Am. J. Phys. Anthropol.* 38:637–640.

———. 1974. Social interactions of adult gibbons *(Hylobates lar)* in an experimental colony. In *Gibbon and Siamang,* vol. 3, edited by D. Rumbaugh, 137–156. Basel: Karger.

Brown, J. H. 1975. Geographical ecology of desert rodents. In *Ecology and Evolution of Commu-nities,* edited by M. L. Cody and J. M. Diamond, 315–341. Cambridge, Mass.: Belknap.

Brown, W. L., and E. O. Wilson. 1956. Character displacement. *Syst. Zool.* 5:49–64.

Budnitz, N., and K. Dainis. 1975. *Lemur catta:* ecology and behavior. In *Lemur Biology,* edited by I. Tattersall and R. W. Sussman, 219–235. New York: Plenum Press.

Buettner-Janusch, J. 1966. *Origins of Man.* New York: Wiley.

Buettner-Janusch, J., and R. J. Andrew. 1962. The use of the incisors by primates in grooming. *Am. J. Phys. Anthropol.* 20:127–130.

Burns, C. In prep. Diet, social status, and reproduction in rhesus macaques, *Macaca mulatta.* Ph.D. diss. Yale University.

Burt, W. H. 1943. Territoriality and home range concepts as applied to mammals. *J. Mammal.* 27:346–352.

Burt, W. H., and R. P. Grossenheider. 1976. *A Field Guide to the Mammals,* 3d ed. Boston: Houghton Mifflin.

Buss, D. H. 1971. Mammary glands and lactation. In *Comparative Reproduction of Non-human Primates,* edited by E. S. E. Hafez, 315–333. Springfield, Ill.: Thomas.

Buss, D. H., and O. M. Reed. 1970. Lactation of baboons fed a low protein maintenance diet. *Lab. Anim. Care* 20: 709–712.

Butler, G. W., and R. W. Bailey, eds. 1973. *Chemistry and Biochemistry of Herbage.* Vol. 1. London: Academic Press.

Butler, H. 1960. Some notes on the breeding cycle of the Senegal galago, *Galago senegalensis senegalensis,* in the Sudan. *Proc. Zool. Soc. Lond.* 135:423–430.

———. 1974. Evolutionary trends in primate sex cycles. In *Reproductive Biology of the Primates,* edited by W. P. Luckett, 2–35. Vol. 3 of *Contributions to Primatology.* Basel: Karger.

Butynski, T. M. 1982. Vertebrate predation by primates: a review of hunting patterns and prey. *J. Hum. Evol.* 11:421–430.

Butzer, K. W. 1977. Environment, culture, and human evolution. *Am. Sci.* 65:572–584.

Bygott, J. D. 1972. Cannibalism among wild chimpanzees. *Nature* 238:410–411.

———. 1979. Agonistic behavior, dominance, and social structure in wild chimpanzees of the Gombe National Park. In *The Great Apes,* edited by D. A. Hamburg and E. R. McCown, 405–428. Menlo Park, Calif.: Benjamin/Cummings.

Carpenter, C. R. 1934. A field study of the behavior and social relations of the howling monkeys *(Alouatta palliata). Comp. Psychol. Monogr.* 10(2):1–168.

———. 1935. Behavior of red spider monkeys in Panama. *J. Mammal.* 16:171–180.

———. 1940. A field study in Siam of the behavior and social relations of the gibbon *(Hylobates lar). Comp. Psychol. Monogr.* 16:1–212.

———. 1942a. Characteristics of social behavior in non-human primates. *Trans. N.Y. Acad. Sci.* 4(8):248–258.

———. 1942b. Societies of monkeys and apes. *Biolog. Symp.* 8:177–204.

———. 1962. Field studies of a primate population. In *Roots of Behavior, Genetics, Instinct and Socialization in Animal Behavior,* edited by E. L. Bliss, 286–294. New York: Harper.

———. 1964. *Naturalistic Behavior of Nonhuman Primates.* University Park, Pa.: Pennsylvania State University Press.

———. 1972. Breeding colonies of macaques and gibbons on Santiago Island, Puerto Rico. In *Breeding Primates,* edited by W. Bereredge, 233–230. Basel: Karger.

Carpenter, J. R. 1940. The grassland biome. *Ecol. Monogr.* 10:617–684.

Cartmill, M. 1972. Arboreal adaptations and the origin of the order Primates. In *The Functional and Evolutionary Biology of Primates*, edited by R. Tuttle, 97–122. Chicago: Aldine/Atherton.

———. 1974. *Daubentonia, Dactylopsila*, woodpeckers, and klinorhynchy. In *Prosimian Biology*, edited by R. D. Martin, G. A. Doyle, and A. C. Walker, 655–670. London: Duckworth.

Casimir, M. J. 1975. Feeding ecology and nutrition of an eastern gorilla group in the Mt. Kahuzi region (Republique du Zaire). *Folia Primatol.* 24:81–136.

Cates, R. G., and G. H. Orians. 1975. Successional status and the palatability of plants to generalized herbivores. *Ecology* 56:410–418.

Caughley, G. 1966. Mortality patterns in mammals. *Ecology* 47:906–918.

———. 1967. Parameters for seasonally breeding populations. *Ecology* 48:834–839.

———. 1977. *Analysis of Vertebrate Populations*. New York: John Wiley.

Cebul, M. S. 1980. The development of social behavior in captive *Saguinus fuscicollis*. Ph.D. diss. Yale University.

Chalmers, A. F. 1976. *What Is This Thing Called Science?* St. Lucia, Australia: University of Queensland Press.

Chalmers, N. R. 1968a. Group composition, ecology and daily activities of free-living mangabeys in Uganda. *Folia Primatol.* 8:247–262.

———. 1968b. The social behavior of free-living mangabeys in Uganda. *Folia Primatol.* 8:263–281.

Chalmers, N. R., and T. E. Rowell. 1971. Behaviour and female reproductive cycles in a captive group of mangabeys. *Folia Primatol.* 14:1–14.

Chance, M. R. A. 1955. The sociability of monkeys. *Man* 55: 162–165.

Chapais, B., and S. R. Schulman. 1980. An evolutionary model of female dominance relations in primates. *J. Theor. Biol.* 82:47–89.

Chapman, G., and W. B. Barker. 1972. *Zoology*, 2d ed. London: Longman.

Chapman, M., and G. Hausfater. 1979. The reproductive consequences of infanticide in langurs: a mathematical model. *Behav. Ecol. Sociobiol.* 5:227–240.

Charles-Dominique, P. 1971. Eco-éthologie des prosimiens du Gabon. *Biol. Gabon.* 7:121–228.

———. 1974. Ecology and feeding behavior of five sympatric lorisids in Gabon. In *Prosimian Biology*, edited by R. D. Martin, G. A. Doyle, and A. C. Walker, 131–150. Pittsburgh: University of Pittsburgh Press.

———. 1975. Nocturnality and diurnality: an ecological interpretation of these two modes of life by an analysis of the higher vertebrate fauna in tropical forest ecosystems. In *Phylogeny of the Primates*, edited by W. P. Luckett and F. S. Szalay, 69–88. New York: Plenum Press.

———. 1977. *Ecology and Behaviour of Nocturnal Primates*. New York: Columbia University Press.

Charles-Dominique, P., H. M. Cooper, A. Hladik, C. M. Hladik, E. Pages, G. F. Pariente, A. Petter-Rousseaux, J. J. Petter, and A. Schilling, eds. 1980. *Nocturnal Malagasy Primates: Ecology, Physiology and Behavior.* New York: Academic Press.

Charles-Dominique, P., and C. M. Hladik. 1971. Le lépilémur du sud de Madagascar: écologie, alimentation, et vie sociale. *Terre Vie* 25:3–66.

Charles-Dominique, P., and R. D. Martin. 1972. Behaviour and ecology of nocturnal prosimians. *Z. Tierpsychol. Suppl.* 9:7–41.

Charles-Dominique, P., and J. J. Petter. 1980. Ecology and social life of *Phaner furcifer.* In *Nocturnal Malagasy Primates: Ecology, Physiology and Behavior,* edited by P. Charles-Dominique et al., 75–96. New York: Academic Press.

Charnov, E. L., and W. M. Schaffer. 1973. Life history consequences of natural selection: Cole's result revisited. *Am. Nat.* 107:791–793.

Cheatum, E. L., and C. W. Severinghaus. 1950. Variation in fertility of white-tailed deer related to range conditions. *Trans. N. Am. Wild. Nat. Resour. Conf.* 15:70–190.

Cheney, D. L. 1977. The acquisition of rank and the development of reciprocal alliances among free-ranging baboons. *Behav. Ecol. Sociobiol.* 2:303–318.

Cheney, D. L., P. C. Lee, and R. M. Seyfarth. 1981. Behavioral correlates of non-random mortality among free-ranging female vervet monkeys. *Behav. Ecol. Sociobiol.* 9:153–161.

Chism, J. 1978. Behavior of group members other than the mother toward captive patas infants. In *Recent Advances in Primatology,* edited by D. J. Chivers and J. Herbert, vol. 1, 173–176. New York: Academic Press.

Chism, J., D. K. Olson, and T. E. Rowell. 1983. Diurnal births and perinatal behavior among wild patas monkeys: Evidence of an adaptive pattern. *Int. J. Primatol.* 4:167–184.

Chism, J., T. Rowell, and D. Olson. 1984. Life history patterns of female patas monkeys. In *Female Primates: Studies by Women Primatologists,* edited by M. Small. New York: Alan Liss.

Chivers, D. J. 1969. On the daily behaviour and spacing of howling monkey groups. *Folia Primatol.* 10:48–102.

———. 1974. The siamang in Malaya: a field study of a primate in tropical rain forest. *Contrib. Primatol.* 4:1–335.

———. 1977. The lesser apes. In *Primate Conservation,* edited by H. S. H. Prince Rainier III and G. H. Bourne, 539–598. New York: Academic Press.

———. 1980. *Malayan Forest Primates: Ten Years' Study in Tropical Rain Forest.* New York: Plenum Press.

Chivers, D. J., and C. M. Hladik. 1980. Morphology of the gastrointestinal tract in primates: comparisons with other mammals in relation to diet. *J. Morphol.* 166:337–386.

Chivers, D. J., and J. J. Raemaekers. 1980. Long-term changes in behaviour. In *Malayan Forest Primates: Ten Years' Study in Tropical Rain Forest,* edited by D. J. Chivers, 109–160. New York: Plenum Press.

Chivers, D. J., J. J. Raemaekers, and F. P. G. Aldrich-Blake. 1975. Long-term observations of sia-mang behaviour. *Folia Primatol.* 23:1–49.

Christian, J. J. 1980. Endocrine factors in population regulation. In *Biosocial Mechanisms of Population Regulation,* edited by M. N. Cohen, R. S. Malpass, and H. G. Klein, 55–115. New Haven: Yale University Press.

Ciochon, R. L., and A. B. Chiarelli, eds. 1980. *Evolutionary Biology of the New World Monkeys and Continental Drift.* New York: Plenum Press.

Clark, C. B. 1977. A preliminary report on weaning among chimpanzees of Gombe National Park, Tanzania. In *Primate Bio-Social Development,* edited by S. Chevalier-Skolnikoff and F. E. Poirier, 235–260. New York: Garland.

Clark, W. E. L. G. 1962. *The Antecedents of Man: An Introduction to the Evolution of the Primates,* 2d ed. Edinburgh: Edinburgh University Press.

Clutton-Brock, T. H. 1974. Primate social organization and ecology. *Nature* 250:539–542.

———. 1977a. Methodology and measurement. In *Primate Ecology,* edited by T. H. Clutton-Brock, 585–590. New York: Academic Press.

———. 1977b. *Primate Ecology.* New York: Academic Press.

———. 1977c. Some aspects of intraspecific variation in feeding and ranging behavior in primates. In *Primate Ecology,* edited by T. H. Clutton-Brock, 539–556. New York: Academic Press.

Clutton-Brock, T. H., and P. H. Harvey. 1977a. Primate ecology and social organization. *J. Zool. (Lond.)* 183:1–39.

———. 1977b. Species differences in feeding and ranging behavior in primates. In *Primate Ecology,* edited by T. H. Clutton-Brock, 557–579. New York: Academic Press.

———. 1980. Primates, brains, and ecology. *J. Zool. (Lond.)* 190:309–323.

Cody, M. L. 1974. *Competition and the Structure of Bird Communities.* Princeton: Princeton University Press.

Cody, M. L., and J. M. Diamond, eds. 1975. *Ecology and Evolution of Communities.* Cambridge: Belknap Press of Harvard University Press.

Coe, M. J., and F. M. Isaac. 1965. Pollination of the baobab *(Adansonia digitata L.)* by the lesser bushbaby *(Galago crassicaudatus* E. Geoffroy). *East Afr. Wildl. J.* 3:123–124.

Coelho, A. M. 1974. Socio-bioenergetics and sexual dimorphism in primates. *Primates* 15:263–269.

Coehlo, A. M., C. A. Bramblett, and L. B. Quick. 1976. Resource availability and population density in primates: a socio-energetic analysis of the energy budgets of Guatemalan howler and spider monkeys. *Primates* 17:63–80.

Cohen, J. E. 1969. Natural primate troops and a stochastic population model. *Am. Nat.* 103:455–477.

——. 1972. Markov population processes as models of primate social and population dynamics. *Theor. Popul. Biol.* 3:119–134.

Cohen, M. N., R. S. Malpass, and H. G. Klein, eds. 1980. *Biosocial Mechanisms of Population Regulation.* New Haven: Yale University Press.

Coimbra-Filho, A. F., and R. A. Mittermeier. 1973. Distribution and ecology of the genus *Leontopithecus* (Lesson 1840) in Brazil. *Primates* 14:47–66.

——. 1978. Tree-gouging, gum-eating, and the "short-tusked" condition in *Callithrix* and *Cebuella.* In *The Biology and Conservation of the Callitrichidae,* edited by D. G. Kleiman, 105–117. Washington, D.C.: Smithsonian Institution Press.

Cole, L. C. 1954. The population consequences of life history phenomena. *Q. Rev. Biol.* 29:103–137.

——.1957. Sketches of general and comparative demography. *Cold Spring Harbor Symp. Quant. Biol.* 22:1–15.

Collias, N., and C. Southwick. 1952. A field study of population density and social organization in howling monkeys. *Proc. Am. Philos. Soc.* 96: 143–156.

Colwell, R. K., and D. J. Futuyuma. 1971. On the measurement of niche breadth and overlap. *Ecology.* 52:567–576.

Conaway, C. H. 1971. Ecological adaptation and mammalian reproduction. *Biol. Reprod.* 4:239–247.

Connell, J. H. 1975. Some mechanisms producing structure in natural communities. In *Ecology and Evolution of Communities,* edited by M. L. Cody and J. M. Diamond, 460–491. Cambridge, Mass.: Belknap.

Cooke, H. B. S. 1978. Africa: the physical setting. In *Evolution of African Mammals,* edited by V. J. Maglio and H. B. S. Cooke. Cambridge: Harvard University Press.

Coppens, Y., V. J. Maglio, C. T. Madden, and M. Beden. 1978. Proboscidea. In *Evolution of African Mammals,* edited by V. J. Maglio and H. B. S. Cooke. Cambridge: Harvard University Press.

Coryndon, S., and R. Savage. 1973. The origin and affinities of African mammal faunas. In *Organisms and Continents through Time,* edited by N. F. Hughes, Vol. 9, 121–135. London: Palaeontology Association, Systematics Association.

Cox, P. R. 1976. *Demography,* 5th ed. Cambridge: Cambridge University Press.

Crook, J. H. 1965. The adaptive significance of avian social organizations. *Symp. Zool. Soc. Lond.* 14:181–218.

——. 1966. Gelada baboon herd structure and movement: a comparative report. *Symp. Zool. Soc. Lond.* 18:237–258.

——. 1970. The socio-ecology of primates. In *Social Behaviour in Birds and Mammals,* edited by J. H. Crook. London: Academic Press.

————. 1972. Sexual selection, dimorphism, and social organization in the primates. In *Sexual Selection and the Descent of Man,* edited by B. G. Campbell, 231–281. Chicago: Aldine.

Crook, J. H., and P. Aldrich-Blake. 1968. Ecological and behavioural contrasts between sympatric ground dwelling primates in Ethiopia. *Folia Primatol.* 8:192–227.

Crook, J. H., J. E. Ellis, and J. D. Goss-Custard. 1976. Mammalian social systems: structure and function. *Anim. Behav.* 24:261–274.

Crook, J. H., and J. S. Gartlan. 1966. Evolution of primate societies. *Nature* 210:1200–1203.

Crowell, J. C., and L. A. Frakes. 1970. Phanerozoic glaciation and the causes of ice ages. *Am. J. Sci.* 268:193–224.

Curtin, R. 1975. The socioecology of the common langur *(Presbytis entellus)* in the Nepal Himalaya. Ph.D. diss. University of California, Berkeley.

Curtin, R., and P. Dolhinow. 1978. Primate social behavior in a changing world. *Am. Sci.* 66:468–475.

Curtin, S. H. 1976. Niche separation in sympatric Malaysian leaf-monkeys *(Presbytis obscura* and *Presbytis melatophos). Yearb. Phys. Anthropol.* 20:421–439.

————. 1980. Dusky and banded leaf-monkeys. In *Malayan Forest Primates: Ten Years' Study in Tropical Rain Forest,* edited by D. J. Chivers, 107–146. New York: Plenum Press.

Darlington, P. J., Jr. 1957. *Zoogeography: The Geographical Distribution of Animals.* New York: Wiley.

Dawson, G. A. 1978. Composition and stability of social groups of the tamarin, *Saguinus oedipus geoffroyi,* in Panama: ecological and behavioral implications. In *The Biology and Conservation of the Callitrichidae,* edited by D. G. Kleiman, 23–38. Washington, D.C.: Smithsonian Institution.

Deag, J. M. 1977. The status of the Barbary macaque *Macaca sylvanus* in captivity and factors influencing its distribution in the wild. In *Primate Conservation,* edited by H. S. H. Prince Rainier III and G. H. Bourne, 268–289. New York: Academic Press.

Deag, J. M., and J. H. Crook. 1971. Social behavior and agonistic buffering in the wild Barbary macaque, *Macaca sylvanus* L. *Folia Primatol.* 15:183–200.

Deevy, E. S. 1947. Life tables for natural populations of animals. *Q. Rev. Biol.* 22:283–314.

Delany, M. J., and D. C. D. Happold. 1979. *Ecology of African Mammals.* London: Longman.

Delson, E. 1975a. Evolutionary history of the Cercopithecidae. In *Approaches to Primate Paleobiology,* edited by F. Szalay. Vol. 5 of *Contributions to Primatology,* 167–217. Basel: Karger.

————. 1975b. Paleoecology and zoogeography of the Old World monkeys. In *Primate Functional Morphology and Evolution,* edited by R. H. Tuttle, 37–64. The Hague: Mouton.

Delson E., and P. Andrews. 1975. Evolution and interrelationships of the catarrhine primates. In *Phylogeny of the Primates: A Multidisciplinary Approach,* edited by W. P. Luckett and F. S. Szalay, 405–446. New York: Plenum Press.

Delson, E., and A. L. Rosenberger. 1980. Phyletic perspectives on platyrrhine origins and anthropoid relations. In *Evolutionary Biology of the New World Monkeys and Continental Drift*, edited by R. L. Ciochon and A. B. Chiarelli, 445–458. New York: Plenum Press.

Demment, M. W. 1982. The scaling of ruminoreticulum size with body weight in East African ungulates. *Afr. J. Ecol.* 20:43–47.

Demment, M. W., and P. J. Van Soest. 1983. Body size, digestive capacity, and feeding strategies of herbivores. Morriston, Ark.: Winrock International Livestock Research Publications.

Denham, W. W. 1971. Energy relations and some basic properties of primate organization. *Am. Anthropol.* 73:77–95.

DeVore, I. 1963. A comparison of the ecology and behavior of monkeys and apes. In *Classification and Human Evolution*, edited by S. L. Washburn, 301–319. Chicago: Aldine.

DeVore, I., ed. 1965. *Primate Behavior: Field Studies of Monkeys and Apes.* New York: Holt.

DeVore, I., and K. R. L. Hall. 1965. Baboon ecology. In *Primate Behavior: Field Studies of Monkeys and Apes,* edited by I. DeVore, 20–52. New York: Holt.

DeVore, I., and S. Washburn. 1963. Baboon ecology and human evolution. In *African Ecology and Human Evolution*, edited by F. C. Howell and F. Bourliere, 335–367. Viking Fund Publications in Anthropology no. 36. Chicago: Aldine/Atherton.

Dewar, R. E. 1984. Recent extinctions in Madagascar: the loss of the subfossil fauna. In *Quaternary Extinctions*, edited by P. S. Martin and R. G. Klein, 57–93. Tucson: University of Arizona Press.

Diamond, J. M. 1975. Assembly of species communities. In *Ecology and Evolution of Communities*, edited by M. L. Cody and J. M. Diamond, 342–444. Cambridge, Mass.: Belknap.

———. 1978. Niche shifts and the rediscovery of interspecific competition. *Am. Sci.* 66:322–331.

———. 1981. Mixed-species foraging groups. *Nature* 292:408–409.

Dittus, W. P. J. 1975. Population dynamics of the toque monkey, *Macaca sinica.* In *Socioecology and Psychology of Primates,* edited by R. H. Tuttle, 125–151. The Hague: Mouton.

———. 1977a. The social regulation of population density and age-sex distribution in the toque monkey. *Behaviour* 63:281–322.

———. 1977b. The socioecological basis for the conservation of the toque monkey *(Macaca sinica)* of Sri Lanka. In *Primate Conservation,* edited by H. S. H. Prince Rainier III and G. H. Bourne, 238–265. New York: Academic Press.

———. 1979. The evolution of behaviors regulating density and age-specific sex ratios in a primate population. *Behaviour* 69: 265–302.

———. 1980. The social regulation of primate populations: a synthesis. In *The Macaques: Studies in Ecology, Behavior and Evolution,* edited by D. G. Lindburg, 263–286. New York: Van Nostrand Reinhold.

Dixson, A. F., and L. George. 1982. Prolactin and parental behaviour in a male New World primate. *Nature* 299:551–552.

Dobzhansky, T. H. 1950. Evolution in the tropics. *Am. Sci.* 38:209–221.

Dolhinow, P. 1977. Normal monkeys? *Am. Sci.* 65:266.

Doutt, R. L. 1964. The historical development of biological control. In *Biological Control of Insect Pests and Weeds,* edited by P. DeBach, 21–42. London: Chapman and Hall.

Dove, W. F. 1935. A study in the individuality in the nutritive instincts and of the causes and effects of variation in the selection of food. *Am. Nat.* 69:469–544.

Downhower, J. F., and K. B. Armitage. 1971. The yellow-bellied marmot and the evolution of polygamy. *Am. Nat.* 105:355–370.

Doyle, G. A., A. Andersson, and S. K. Bearder. 1971. Reproduction in the lesser bushbaby *(Galago senegalensis moholi)* under seminatural conditions. *Folia Primatol.* 14:15–22.

Doyle, G. A., and S. K. Bearder. 1977. The galagines of South Africa. In *Primate Conservation,* edited by H. S. H. Prince Rainier III and G. H. Bourne, 2–37. New York: Academic Press.

Doyle, G. A., A. Pelletier, and T. Bekker. 1967. Courtship, mating, and parturition in the lesser bushbaby *(Galago senegalensis moholi)* under seminatural conditions. *Folia Primatol.* 7:169–197.

Drickamer, L. C. 1974. A ten-year summary of reproductive data for free-ranging *Macaca mulatta. Folia Primatol.* 21:61–80.

Drickmer, L., and S. Vessey. 1973. Group changing in free-ranging male rhesus monkeys. *Primates* 14:359–368.

Dunbar, R. I. M. 1977a. Feeding ecology of gelada baboons: a preliminary report. In *Primate Ecology,* edited by T. H. Clutton-Brock, 251–273. New York: Academic Press.

———.1977b. The gelada baboon: status and conservation. In *Primate Conservation,* edited by H. S. H. Prince Rainier III and G. H. Bourne, 363–384. New York: Academic Press.

———. 1978. Competition and niche separation in a high altitude herbivore community in Ethiopia. *East Afr. Wildl. J.* 16:183–199.

———. 1979. Population demography, social organization, and mating strategies. In *Primate Ecology and Human Origins,* edited by I. S. Bernstein and E. O. Smith, 67–88. New York: Garland.

———. 1980. Demographic and life history variables of a population of gelada baboons *(Theropithecus gelada). J. Anim. Ecol.* 49: 485–506.

Dunbar, R. I. M., and E. P. Dunbar. 1974a. Ecological relations and niche separation between sympatric terrestrial primates in Ethiopia. *Folia Primatol.* 21:36–60.

———. 1974b. Ecology and population dynamics of *Colobus guereza* in Ethiopia. *Folia Primatol.* 21:188–208.

———. 1975. *Social Dynamics of Gelada Baboons. Vol. 6 of Contributions to Primatology,* 1–157. Basel: Karger.

———. 1976. Contrasts in social structure among black-and-white Colobus monkey groups. *Anim. Behav.* 24:84–92.

———. 1977. Dominance and reproductive success among female gelada baboons. *Nature* 266:351–352.

Dunbar, R. I. M., and M. F. Nathan. 1972. Social organization of the Guinea baboon, *Papio papio*. *Folia Primatol.* 17:321–334.

Egerton, F. N., III. 1968a. Ancient sources for animal demography. *Isis* 59:175–189.

———. 1968b. Studies of animal populations from Lamarck to Darwin. *J. Hist. Biol.* 1:225–259.

———. 1969. Richard Bradley's understanding of biological productivity: a study of 18th century ecological ideas. *J. Hist. Biol.* 2:391–410.

Ehrlich, P. R., and L. C. Birch. 1967. The "balance of nature" and "population control." *Am. Nat.* 101:97–107.

Ehrlich, P. R., and P. H. Raven. 1965. Butterflies and plants: a study in coevolution. *Evolution* 8:586–608.

Eisenberg, J. F. 1966. The social organization of mammals. *Handb. Zool. VIII* 10(7):1–92.

———. 1975. Phylogeny, behavior, and ecology in the Mammalia. In *Phylogeny of the Primates: A Multidisciplinary Approach*, edited by W. P. Luckett and F. S. Szalay, 47–68. New York: Plenum Press.

———. 1978. Comparative ecology and reproduction of New World monkeys. In *The Biology and Conservation of the Callitrichidae*, edited by D. G. Kleiman, 13–22. Washington, D. C.: Smithsonian Institution.

———. 1981. *The Mammalian Radiations: An Analysis of Trends in Evolution, Adaptation, and Behavior*. Chicago: Chicago University Press.

Eisenberg, J. F., and R. E. Kuehn. 1966. The behavior of *Ateles geoffroyi* and related species. *Smithson. Misc. Collect.* 151(8): 63.

Eisenberg, J. F., N. A. Muckenhirn, and R. Rudran. 1972. The relation between ecology and social structure in primates. *Science* 176:863–874.

Eisenberg, J. F., and R. W. Thorington, Jr. 1973. A preliminary analysis of a Neotropical mammal fauna. *Biotropica* 5(3):150–161.

Eisenberg, J. F., and D. E. Wilson. 1978. Relative brain size and feeding strategies in the Chiroptera. *Evolution* 32:740–751.

———. 1981. Relative brain size in didelphid marsupials. *Am. Nat.* 118:110–126.

Eldredge, N., and S. J. Gould. 1972. Punctuated equilibria: an alternative to phyletic gradualism. In *Models in Paleobiology*, edited by T. J. M. Schopf, 82–115. San Francisco: Freeman & Cooper.

Ellefson, J. O. 1968. Territorial behaviour in the common white-handed gibbon, *Hylobates lar* Linn. In *Primates: Studies of Adaptation and Variability*, edited by P. Jay, 180–199. New York: Holt, Rinehart & Winston.

———. 1974. A natural history of white-handed gibbons in the Malay Peninsula. In *Gibbon and Siamang,* vol. 3, edited by D. Rumbaugh, 1–136. Basel: Karger.

Elton, C. 1927. *Animal Ecology.* New York: Macmillan.

Emlen, J. M. 1973. *Ecology: An Evolutionary Approach.* Reading, Mass.: Addison-Wesley.

Emlen, S. T., and L. W. Oring. 1977. Ecology, sexual selection, and the evolution of mating systems. *Science* 197: 215–223.

Emmons, L. H. 1980. Ecology and resource partitioning among nine species of African rain forest squirrels. *Ecol. Monogr.* 50:31–54.

Epple, G. 1970. Maintenance, breeding, and development of marmoset monkeys (Callitrichidae) in captivity. *Folia Primatol.* 12:56–76.

———. 1975. The behavior of marmoset monkeys (Callitrichidae). In *Primate Behavior,* vol. 4, edited by L. A. Rosenblum, 195–239. New York: Academic Press.

Erikson, G. E. 1963. Brachiation in New World monkeys and in anthropoid apes. *Symp. Zool. Soc. Lond.* 10:135–164.

Estes, R. D. 1974. Social organization of the African Bovidae. In *The Behavior of Ungulates and Its Relation to Management,* vol. 1, no. 24, edited by V. Geist and F. Walther, 166–205. Morges, Switzerland: International Union for the Conservation of Nature.

Estes, R. D., and J. Goddard. 1967. Prey selection and hunting behavior of the African wild dog. *J. Wildl. Manage.* 31:52–70.

Eudey, A. A. 1980. Pleistocene glacial phenomena and the evolution of Asian macaques. In *The Macaques: Studies in Ecology, Behavior and Evolution,* edited by D. G. Lindburg, 55–83. New York: Van Nostrand Reinhold.

Ewer, R. F. 1968. *Ethology of Mammals.* New York: Plenum Press.

Feeny, P. 1975. Biochemical coevolution between plants and their insect herbivores. In *Coevolution of Animals and Plants,* edited by L. E. Gilbert and P. H. Raven, 3–19. Austin: University of Texas Press.

———. 1976. Plant apparency and chemical defense. In *Biochemical Interaction between Plants and Insects,* edited by J. Wallace and R. Mansell. Vol. 10 of *Recent Advances in Phytochemistry,* 1–40. New York: Plenum Press.

Feeny, P., and H. Bostock. 1968. Seasonal changes in the tannin content of oak leaves. *Phytochemistry (Oxf.)* 7:871–880.

Fenton, M. B., and T. H. Fleming. 1976. Ecological interactions between bats and nocturnal birds. *Biotropica* 8:104–110.

Fittkau, E. J., and H. Klinge. 1973. On biomass and trophic structure of the central Amazonian rain forest ecosystem. *Biotropica* 5:2–14.

Fleagle, J. G. 1976a. Locomotion and posture of the Malayan siamang and implications for hominoid evolution. *Folia Primatol.* 26:245–269.

———. 1976b. Locomotor behavior and skeletal anatomy of sympatric Malaysian leaf-monkeys *(Presbytis obscura* and *Presbytis melalophos). Yearb. Phys. Anthropol.* 20:440–453.

————. 1978a. Locomotion, posture and habitat use of two sympatric leaf-monkeys in West Malaysia. In *Recent Advances in Primatology,* vol. 1, *Behaviour,* edited by D. J. Chivers and J. Herbert, 331–336. New York: Academic Press.

————. 1978b. Locomotion, posture and habitat utilization in two sympatric Malaysian leaf-monkeys *(Presbytis obscura* and *Presbytis melalophos).* In *The Ecology of Arboreal Folivores,* edited by G. G. Montgomery, 243–252. *Symp. Nat. Zool. Park.* Washington, D. C.: Smithsonian Institution Press.

————. 1980a. Locomotion and posture. In *Malayan Forest Primates,* edited by D. J. Chivers, 191–208. New York: Plenum Press.

————. 1980b. Locomotor behavior of the earliest anthropoids: a review of the current evidence. *Z. Morphol. Anthropol.* 71:149–156.

————. In press. Primate locomotion and diet. In *Food Acquisition and Processing in Primates,* edited by D. J. Chivers, B. A. Wood, and A. Bilsborough. New York: Plenum Press.

Fleagle, J. G., and R. A. Mittermeier. 1980. Locomotor behavior, body size and comparative ecology of seven Surinam monkeys. *Am. J. Phys. Anthropol.* 52:301–314.

Fleming, T. H. 1979. Do tropical frugivores compete for food? *Am. Zool.* 19:1157–1172.

Fogden, M. 1974. A preliminary field-study of the western tarsier, *Tarsius bancanus* Horsefield. In *Prosimian Biology,* edited by R. D. Martin, G. A Doyle, and A. C. Walker, 151–165. London: Duckworth.

Folk, G. E., A. Larson, and M. A. Folk. 1976. Physiology of hibernating bears. In *Bears—Their Biology and Management,* edited by M. R. Pelton, J. W. Lentfer, and G. E. Folk. Unipub, n. s., no. 40:373–379. New York.

Fooden, J. 1964. Stomach contents and gastrointestinal proportions of wild-shot Guianan monkeys. *Am. J. Phys. Anthropol.* 22:227–231.

Formozov, A. N. 1946. Snow cover as an integral factor of the environment and its importance in the ecology of mammals and birds. Translated from the Russian by W. Prychodko and W. O. Pruitt. Edmonton: University of Alberta, Occasional Publications No. 1. Originally published in *Zoology (Moscow)* 5(1946), USSR new ser.

Fossey, D. 1974. Observations on the home range of one group of mountain gorillas *(Gorilla gorilla beringei). Anim. Behav.* 22:568–581.

————. 1979. Development of the mountain gorilla *(Gorilla gorilla beringei):* the first thirty-six months. In *The Great Apes,* edited by D. A. Hamburg and E. R. McCown, 139–186. Menlo Park, Calif.: Benjamin/Cummings.

Fossey, D., and A. H. Harcourt. 1977. Feeding ecology of free-ranging mountain gorillas *(Gorilla gorilla beringei).* In *Primate Ecology,* edited by T. H. Clutton-Brock, 415–449. New York: Academic Press.

Frakes, L. A., and E. M. Kemp. 1972. Influence of continental positions on early Tertiary climates. *Nature* 240:97–100.

Freeland, W. J. 1976. Pathogens and the evolution of primate sociality. *Biotropica* 8:12–24.

Freeland, W. J., and D. H. Janzen. 1974. Strategies in herbivory by mammals: the role of plant secondary compounds. *Am. Nat.* 108:269–289.

Freese, C. 1976. Censusing *Alouatta palliata, Ateles geoffroyi,* and *Cebus capuchinus* in Costa Rican dry forest. In *Neotropical Primates, Field Studies and Conservation,* edited by R. W. Thorington, Jr., and P. G. Heltne, 4–9. Washington, D.C.: National Academy of Sciences.

Fretwell, S. D. 1972. *Populations in a Seasonal Environment.* Princeton: Princeton University Press.

———. 1975. The impact of Robert MacArthur on ecology. *Annu. Rev. Ecol. Syst.* 6:1–13.

Frisch, R. E. 1982. Malnutrition and fertility. *Science* 215:1272–1273.

Frisch, R. E., D. M. Hegsted, and K. Yoshinaga. 1975. Body weight and food intake at early estrus of rats on a high fat diet. *Proc. Nat. Acad. Sci. USA* 72:4172–4176.

Frisch, R. E., and R. Revelle. 1970. Height and weight at menarche and a hypothesis of critical body weights and adolescent events. *Science* 169:397–399.

Froelich, J. W., R. W. Thorington, Jr., and J. S. Otis. 1981. The demography of howler monkeys *(Alouatta palliata)* on Barro Colorado Island, Panama. *Int. J. Primatol.* 2:207–236.

Gabow, S. A. 1975. Behavioral stabilization of a baboon hybrid zone. *Am. Nat.* 109:701–712.

Gadgil, M. 1971. Dispersal: population consequences and evolution. *Ecology* 52:253–261.

Gadgil, M., and W. H. Bossert. 1970. Life historical consequences of natural selection. *Am. Nat.* 104:1–24.

Galdikas, B. M. F. 1979. Orangutan adaptation at Tanjung Puting Reserve: mating and ecology. In *The Great Apes,* edited by D. A. Hamburg and E. R. McCown, 195–233. Menlo Park, Calif.: Benjamin/Cummings.

Garber, P. A. 1980. Locomotor behavior and feeding ecology of the Panamanian tamarin *Saguinus oedipus geoffroyi,* Callitrichidae, Primates. *Int. J. Primatol.* 1(2):185–201.

Garber, P. A., L. Moya, and C. Malaga. 1984. A preliminary field study of the moustached tamarin monkey *(Saguinus mystax)* in northeastern Peru: Questions concerned with the evolution of a communal breeding system. *Folia Primatol.*

Garcia, J., and F. R. Ervin. 1968. Gustatory-visceral and telereceptor-cutaneous conditioning-adaptation in internal and external milieus. *Commun. Behav. Biol. Part A Orig. Artic.* 1:389–415.

Gardarsson, A., and R. Moss. 1970. Selection of food by Icelandic ptarmigan in relation to its availability and nutritive value. In *Animal Populations in Relation to Their Food Resources.* Vol. 10 of *Br. Ecol. Soc. Symp.,* edited by A. Watson, 47–72. Oxford: Blackwell.

Gartlan, J. S, and C. K. Brain. 1968. Ecology and social variability in *Cercopithecus aethiops* and *C. mitis.* In *Primates: Studies in Adaptation and Variability,* edited by P. C. Jay, 253–292. New York: Holt.

Gartlan, J. S., and S. C. Gartlan. 1973. Quelques observations sur les groupes exclusivement males chez *Erythrocebus patas. Ann. Fac. Sci. Cameroun* 12:121–144.

Gartlan, J. S., and T. T. Struhsaker. 1972. Polyspecific associations and niche separation of rainforest anthropoids in Cameroon, West Africa. *J. Zool. (Lond.)* 168:221–266.

Gates, D. M. 1962. *Energy Exchange in the Biosphere.* New York: Harper & Row.

Gaulin, S. J. C. 1979. A Jarman/Bell model of primate feeding niches. *Hum. Ecol.* 7:1–19.

Gaulin, S. J. C., and C. K. Gaulin. 1982. Behavioral ecology of *Alouatta seniculus* in Andean cloud forest. *Int. J. Primatol.* 3:1–32.

Gaulin, S. J. C., and M. Konner. 1977. On the natural diet of primates, including humans. In *Nutrition and the Brain,* edited by R. J. Wurtman and J. J. Wurtman, 1–86. New York: Raven Press.

Gause, G. F. 1934. *The Struggle for Existence.* New York: Hafner.

Gautier-Hion, A. 1966. L'Ecologie et l'ethologie du talapoin *Miopithecus talapoin talapoin. Biol. Gabon.* 11:311–329.

———. 1970. L'Organization sociale d'une bande de talapoins *(Miopithecus talapoin)* dans le nord-est du Gabon. *Folia Primatol.* 12:116–141.

———. 1971. L'Ecologie du talapoin du Gabon. *Terre Vie* 25:427–490.

———. 1973. Social and ecological features of talapoin monkey—comparisons with sympatric cercopithecines. In *Comparative Ecology and Behaviour of Primates,* edited by R. P. Michael and J. H. Crook, 148–170. New York: Academic Press.

———. 1978. Food niches and coexistence in sympatric primates in Gabon. In *Recent Advances in Primatology,* vol. 1, *Behaviour,* edited by D. J. Chivers and J. Herbert, 269–286. New York: Academic Press.

———. 1980. Seasonal variations of diet related to species and sex in a community of *Cercopithecus* monkeys. *J. Anim. Ecol.* 49:237–269.

Gautier-Hion, A., and J. P. Gautier. 1974. Les associations polyspécifiques de Cercopithèques du plateau de M'passa (Gabon). *Folia Primatol.* 22:134–177.

Gautier, J. P., and Gautier-Hion, A. 1969. Les associations polyspécifiques chez les Cercopithecidae du Gabon. *Terre Vie* 23:164–201.

Geist, V. 1971. *Mountain sheep: A Study in Behavior and Evolution.* Chicago: University of Chicago Press.

———. 1974a. On the relationship of ecology and behaviour in the evolution of ungulates: theoretical considerations. In *The Behavior of Ungulates and Its Relation to Management,* edited by V. Geist and F. Walther, 235–246. Morges, Switzerland: International Union for the Conservation of Nature Publications, n.s., no. 24.

———. 1974b. On the relationship of social evolution and ecology in ungulates. *Am. Zool.* 14:205–220.

Gengozian, N., J. S. Batson, and T. A. Smith. 1978. Breeding of tamarins *(Saguinus* spp.) in the laboratory. In *The Biology and Conservation of the Callitrichidae,* edited by D. G. Kleiman, 207–214. Washington, D.C: Smithsonian Institution.

Gentry, A. H. 1982. Patterns of neotropical plant species diversity. In *Evolutionary Biology,* vol. 15, edited by M. K. Hecht, B. Wallace, and G. T. Prance, 1–84. New York: Plenum Press.

Gilbert, C., and J. Gillman. 1960. Puberty in the baboon *(P. ursinus)* in relation to age and body weight. *S. Afr. J. Med. Sci.* 25:98–103.

Gilbert, L. E., and P. H. Raven. 1975. *Coevolution of Animals and Plants.* Austin: University of Texas Press.

Gilmore, H. A. 1981. From Radcliffe-Brown to sociobiology: some aspects of the rise of primatology within physical anthropology. *Am. J. Phys. Anthropol.* 56:387–392.

Gittins, S. P., and J. J. Raemaekers. 1980. Siamang, lar, and agile gibbons. In *Malayan Forest Primates: Ten Years' Study in Tropical Rain Forest,* edited by D. J. Chivers, 63–105. New York: Plenum Press.

Glander, K. 1975. Habitat description and resource utilization: a preliminary report on mantled howling monkey ecology. In *Socioecology and Psychology of Primates,* edited by R. H. Tuttle, 37–57. The Hague: Mouton.

———. 1978. Howling monkey feeding behavior and plant secondary compounds: a study of strategies. In *The Ecology of Arboreal Folivores,* edited by G. G. Montgomery, 561–573. Washington, D.C.: Smithsonian Institution Press.

———. 1980. Reproduction and population growth in free-ranging mantled howling monkeys. *Am. J. Phys. Anthropol.* 53:25–36.

———. 1981. Feeding patterns in mantled howling monkeys. In *Foraging Behavior: Ecological, Ethological and Psychological Approaches,* edited by A. Kamil and T. D. Sargent, 231–259. New York: Garland.

———. 1982. The impact of plant secondary compounds on primate feeding behavior. *Yearb. Phys. Anthropol.* 25:1–18.

Goldstein, S. 1984. Ecology of rhesus monkeys, *Macaca mulatta,* in northern Pakistan. Ph.D. diss., Yale University.

Goldstein, S., D. Post, and D. Melnick. 1978. An analysis of Cercopithecoid odontometrics. I. The scaling of the maxillary dentition. *Am. J. Phys. Anthropol.* 49:517–532.

Golley, F. B., J. T. McGinnis, R. G. Clements, G. I. Child, and M. J. Duever. 1975. *Mineral Cycling in a Tropical Moist Forest Ecosystem.* Athens, Ga.: University of Georgia Press.

Goodall, A. G. 1977. Feeding and ranging behavior of a mountain gorilla group *(Gorilla gorilla beringei)* in the Tshibinda-Kahuzi region (Zaire). In *Primate Ecology,* edited by T. H. Clutton-Brock, 450–479. New York: Academic Press.

Goodall, J. 1965. Chimpanzees of the Gombe Stream Reserve. In *Primate Behavior,* edited by I. DeVore, 425–473. New York: Holt, Rinehart & Winston.

———. 1968. The behaviour of free-living chimpanzees in Gombe Stream Reserve. *Anim. Behav. Monogr.* 1(3): 165–311.

———. 1977. Infant killing and cannibalism in free-living chimpanzees. *Folia Primatol.* 28:259–282.

———. 1983. Population dynamics during a 15-year period in one community of free-living chimpanzees in the Gombe National Park, Tanzania. *Z. Tierpsychol.* 61:1–60.

Goodall, J., A. Bandora, E. Bergmann, C. Busse, H. Matama, E. Mpongo, A. Pierce, and D. Riss. 1979. Intercommunity interactions in the chimpanzee population of the Gombe National Park. In *The Great Apes,* edited by D. A. Hamburg and E. R. McCown, 13–54. Menlo Park, Calif.: Benjamin/Cummings.

Gordon, K. D. 1980. Dental attrition in the chimpanzee *(Pan troglodytes verus):* a scanning electron microscope study. Ph.D. diss., Yale University.

Goss-Custard, J. D., R. I. M. Dunbar, and F. P. G. Aldrich-Blake. 1972. Survival, mating, and rearing strategies in the evolution of primate social structure. *Folia Primatol.* 17:1–19.

Gould, S. J. 1971. Geometric similarity in allometric growth: a contribution to the problem of scaling in the evolution of size. *Am. Nat.* 105:113–136.

———. 1975. On the scaling of tooth size in mammals. *Am. Zool.* 15:351–362.

Gould, S. J., and R. C. Lewontin. 1979. The spandrels of San Marco and the Panglossian paradigm: a critique of the adaptationist programme. *Proc. R. Soc. Lond. B Biol. Sci.* 205:581–598.

Graham, C. E. 1970. Reproductive physiology of the chimpanzee. In *The Chimpanzee,* vol. 3, edited by G. Bourne, 183–220. Basel: Karger.

———. 1981a. Menstrual cycle of the Great Apes. In *Reproductive Biology of the Great Apes: Comparative and Biomedical Perspectives,* edited by C. E. Graham, 1–44. New York: Academic Press.

———. 1981b. *Reproductive Biology of the Great Apes: Comparative and Biomedical Perspectives.* New York: Academic Press.

Grant, P. R. 1972. Convergent and divergent character displacement. *Biol. J. Linn. Soc.* 4:39–68.

———. 1981. Speciation and the adaptive radiation of Darwin's finches. *Am. Sci.* 69:653–663.

Greenwood, P. J. 1980. Mating systems, philopatry, and dispersal in birds and mammals. *Anim. Behav.* 28:1140–1162.

Griffin, D. R. 1970. Migrations and homing of bats. In *Biology of Bats,* vol. 1, edited by W. A. Wimsatt, 233–264. New York: Academic Press.

Grime, J. P., G. M. Blythe, and J. D. Thornton. 1970. Food selection by the snail *Cepaca nemoralis* L. In *Animal Populations in Relation to Their Food Resources,* edited by A. Watson. Oxford: Blackwell.

Groves, C. P. 1970. The forgotten leaf-eaters and the phylogeny of the Colobinae. In *Old World Monkeys: Evolution, Systematics and Behavior,* edited by J. R. Napier and P. H. Napier, 555–587. New York: Academic Press.

———. 1972. Systematics and phylogeny of gibbons. In *Gibbon and Siamang,* vol. 1, edited by D. Rumbaugh, 1–89. Basel: Karger.

Gunderson, H. L. 1976. *Mammalogy.* New York: McGraw-Hill.

Gwynne, D. C., and J. M. Boyd. 1970. Relationships between numbers of Soay sheep and pastures at St. Kilda. In *Animal Populations in Relation to Their Food Resources,* edited by A. Watson, 289–302. Oxford: Blackwell.

Hagmeier, E. M., and C. D. Stults. 1964. A numerical analysis of the distributional patterns of North American mammals. *Syst. Zool.* 13(3):125–155.

Hairston, N. G., F. E. Smith, and L. B. Slobodkin. 1960. Community structure, population control, and competition. *Am. Nat.* 94(879):421–425.

Hall, K. R. L. 1962. Numerical data, maintenance activities and biomotion of the wild chacma baboon, *Papio ursinus. Proc. Zool. Soc. Lond.* 139:181–220.

———. 1963. Variations in the ecology of the chacma baboon, *Papio ursinus. Symp. Zool. Soc. Lond.* 10:1–28.

———. 1965. Social organization of the Old World monkeys and apes. *Symp. Zool. Soc. Lond.* 14:265–289.

———. 1966. Behaviour and ecology of the wild patas monkey, *Erythrocebus patas,* in Uganda. *J. Zool. (Lond.)* 148:15–87.

Hall, K. R. L., R. C. Boelkins, and M. J. Goswell. 1965. Behaviour of patas monkeys in captivity with notes on the natural habitat. *Folia Primatol.* 3:22–49.

Hamburg, D. A and E. R. McCown, eds. 1979a. *The Great Apes.* Menlo Park, Calif.: Benjamin/Cummings.

———. 1979b. Introduction to *The Great Apes,* edited by D. A. Hamburg and E. R. McCown. Menlo Park, Calif.: Benjamin/Cummings.

Hamilton, W. D. 1963. The evolution of altruistic behavior. *Am. Nat.* 97:354–356.

———. 1964. The genetical theory of social behavior, I, II. *J. Theoret. Biol.* 7:1–52.

———. 1971. Geometry for the selfish herd. *J. Theor. Biol.* 31:295–311.

Hamilton, W. D., and R. M. May. 1977. Dispersal in stable habitats. *Nature* 269:578–581.

Hamilton, W. J., III, R. E. Buskirk, and W. H. Buskirk. 1978. Omnivory and utilization of food resources by chacma baboons, *Papio ursinus. Am. Nat.* 112:911–924.

Hamilton, W. J., III, and C. D. Busse. 1978. Primate carnivory and its significance to human diets. *Bioscience* 28:761–766.

Hammen, T. van der. 1972. Changes in vegetation and climate in the Amazon basin and surrounding areas during the Pleistocene. *Geol. Mijnbouw.* 51:641–643.

Hammen, T. van der, T. A. Wijmstra, and W. H. Zagwijn. 1971. The floral record of the late Cenozoic of Europe. In *The Late Cenozoic Glacial Ages,* edited by K. K. Turekian, 391–424. New Haven: Yale University Press.

Hampton, S. H., and J. K. Hampton, Jr. 1978. Detection of reproductive cycles and pregnancy in tamarins (*Saguinus* spp.). In *The Biology and Conservation of the Callitrichidae,* edited by D. G. Kleiman. 173–179. Washington, D.C.: Smithsonian Institution.

Happel, R., and T. Cheek. Forthcoming. Evolutionary biology and ecology of *Rhinopithecus.*

Haraway, D. 1978a. Animal sociology and a natural economy of the body politic. Part I: A political physiology of dominance. *Signs* 4(1):21–36.

————. 1978b. Animal sociology and a natural economy of the body politic. Part II: The past is the contested zone: human nature and theories of production and reproduction in primate behavior studies. *Signs* 4(1):37–60.

Harcourt, A. H. 1978. Strategies of emigration and transfer by primates, with particular reference to gorillas. *Z. Tierpsychol.* 48:401–420.

————. 1979. The social relations and group structure of wild mountain gorillas. In *The Great Apes,* edited by D. A. Hamburg and E. R. McCown, 187–192. Menlo Park, Calif.: Benjamin/Cummings.

Harcourt, A. H., K. J. Stewart, and D. Fossey. 1976. Male emigration and female transfer in the wild mountain gorilla. *Nature* 263:226–227.

————. 1981. Gorilla reproduction in the wild. In *Reproductive Biology of the Great Apes: Comparative and Biomedical Perspectives,* edited by C. E. Graham, 265–280. New York: Academic Press.

Harding, R. S. O. 1973. Predation by a troop of olive baboons *(Papio anubis). Am. J. Phys. Anthropol.* 38:587–592.

————. 1975. Meat-eating and hunting in baboons. In *Socioecology and Psychology of Primates,* edited by R. H. Tuttle, 245–257. The Hague: Mouton.

————. 1976. Ranging patterns of a troop of baboons *(Papio anubis)* in Kenya. *Folia Primatol.* 25:143–185.

Harkin, J. M. 1973. Lignin. In *Chemistry and Biochemistry of Herbage,* vol. 1, edited by G. W. Butler and R. W. Bailey, 323–373. New York: Academic Press.

Harrison, D. L. 1964. *The Mammals of Arabia.* Vol. 1. London: Ernest Benn.

Hatley, T., and J. Kappelman. 1980. Bears, pigs and Plio-Pleistocene hominids: a case for the exploitation of belowground food resources. *Hum. Ecol.* 8:371–387.

Hausfater, G. 1972. Intergroup behavior of free-ranging rhesus monkeys *(Macaca mulatta). Folia Primatol.* 18:78–107.

————. 1975. Dominance and reproduction in baboons *(Papio cynocephalus):* a quantitative analysis. In *Contributions to Primatology,* vol. 7. Basel: Karger.

————. 1976. Predatory behavior of yellow baboons. *Behavior* 56(1–2): 44–68.

Hausfater, G., and W. H. Bearce. 1976. Acacia tree exudates: their composition and use as a food source by baboons. *East Afr. Wildl. J.* 14:241–243.

Hausfater, G., J. Altmann, and S. Altmann. 1982. Long-term consistency of dominance relations among female baboons *(Papio cynocephalus). Science* 217:752–755.

Hausfater, G., and D. F. Watson. 1976. Social and reproductive correlates of parasite ova emissions by baboons. *Nature* 262:688–689.

Hazama, N. 1964. Weighing wild Japanese monkeys in Arashiyama. *Primates* 5:81–104.

Hearn, J. P. 1978. The endocrinology of reproduction in the common marmoset *Callithrix jacchus.* In *The Biology and Conservation of the Callitrichidae,* edited by D. G. Kleiman, 163–172. Washington, D.C.: Smithsonian Institution Press.

Hearn, J. P., and S. F. Lunn. 1975. The reproductive biology of the marmoset monkey, *Callithrix jacchus. Lab. Anim. Handb.* 6:191–202.

Heltne, P. G., and R. W. Thorington, Jr. 1976. Problems and potentials for primate biology and conservation in the New World. In *Neotropical Primates: Field Studies and Conservation*, edited by R. W. Thorington, Jr., and P. G. Heltne, 110–124. Washington, D.C.: National Academy of Sciences.

Hendrickx, A. G., and D. C. Kraemer. 1971. Reproduction. In *Embryology of the Baboon*, edited by A. C. Hendrickx, 3–30. Chicago: University of Chicago Press.

Hernandez-Camacho, J., and R. W. Cooper. 1976. The nonhuman primates of Colombia. In *Neotropical Primates: Field Studies and Conservation*, edited by R. W. Thorington and P. G. Heltne, 35–69. Washington, D.C.: National Academy of Sciences.

Hershkovitz, P. 1970. Notes on Tertiary platyrrhine monkeys and description of a new genus from the late Miocene of Colombia. *Folia Primatol.* 12:1–37.

———. 1972. The recent mammals of the neotropical region: a zoogeographic and ecological review. In *Evolution, Mammals, and Southern Continents*, edited by A. Keast, F. C. Erk, and B. Glass, 311–431. Albany: State University of New York Press.

———. 1977. *Living New World Monkeys* (Platyrrhini) *with an Introduction to Primates*. Vol. 1. Chicago: University of Chicago Press.

———. 1982. Supposed squirrel monkey affinities of the late Oligocene *Dolichocebus gaimanensis. Nature* 298:201–202.

Hill, W. C. O. 1953–1966. *Primates: Comparative Anatomy and Taxonomy*. 6 vols. Edinburgh: Edinburgh University Press.

Hinde, R. A. 1974. *Biological Bases of Human Social Behaviour*. New York: McGraw-Hill.

———. 1976. Interactions, relationships and social structure. *Man* 11:1–17.

Hinde, R. A., and M. J. A. Simpson. 1975. Qualities of mother–infant relationships in monkeys. In *The Parent-Infant Interaction*, Ciba Foundation Symposium, no. 33. Amsterdam: Associated Scientific Publishers.

Hinde, R. A., and Y. Spencer-Booth. 1967. The behaviour of socially living rhesus monkeys in their first two and a half years. *Anim. Behav.* 15:169–196.

Hladik, A. 1978. Phenology of leaf production in rain forest of Gabon: distribution and composition of food for folivores. In *The Ecology of Arboreal Folivores*, edited by G. G. Montgomery, 51–71. Washington, D.C.: Smithsonian Institution Press.

Hladik, A., and C. M. Hladik. 1969. Rapports trophiques entre végétation et primates dans la forêt de Barro Colorado (Panama). *Terre Vie* 23:25–117.

Hladik, C. M. 1967. Surface relative du tractus digestif de quelques primates, morphologie des villosités intestinales et correlations avec le régime alimentaire. *Mammalia* 31:120–147.

———. 1973. Alimentation et activité d'un groupe de chimpanzes reintroduits en forest gabonaise. *Terre Vie* 27:343–413.

———. 1975. Ecology, diet and social patterning in Old and New World primates. In *Socioecology and Psychology of Primates,* edited by R. H. Tuttle, 3–35. The Hague: Mouton.

———. 1977a. Chimpanzees of Gabon and chimpanzees of Gombe: some comparative data on the diet. In *Primate Ecology,* edited by T. H. Clutton-Brock, 481–501. New York: Academic Press.

———. 1977b. A comparative study of the feeding strategies of two sympatric species of leaf monkeys: *Presbytis senex* and *Presbytis entellus.* In *Primate Ecology,* edited by T. H. Clutton-Brock, 324–353. New York: Academic Press.

———. 1979. Diet and ecology of prosimians. In *The Study of Prosimian Behavior,* edited by G. A. Doyle and R. D. Martin, 307–357. New York: Academic Press.

Hladik, C. M., P. Charles-Dominique, and J. J. Petter. 1980. Feeding strategies of five nocturnal prosimians in the dry forest of the west coast of Madagascar. In *Nocturnal Malagasy Primates,* edited by P. Charles-Dominique et al, 41–74. New York: Academic Press.

Hladik, C. M., and D. J. Chivers. 1978. Concluding discussion: ecological factors and specific behavioural patterns determining primate diet. In *Recent Advances in Primatology,* vol. 1, edited by D. J. Chivers and J. Herbert, 433–444. New York: Academic Press.

Hladik, C. M., and A. Hladik. 1967. Observations sur le rôle des primates dans la dissémination des végétaux de la forêt Gabonaise. *Biol. Gabon.* 3:43–58.

———. 1972. Disponibilités alimentaires et domaines vitaux des primates à Ceylon. *Terre Vie* 26:149–215.

Hladik, C. M., A. Hladik, J. Bousset, P. Valdebouze, G. Viroben, and J. DeLort-Laval. 1971. Le régime alimentaire des primates de Ile de Barro-Colorado (Panama): résultats des analyses quantitatives. *Folia Primatol.* 16:85–122.

Hladik, C. M., and I. Quequen. 1974. Géophagie et nutrition minérale chez les primates sauvages. *C. R. Acad. Sci. (D) (Paris)* 279:1393–1396.

Hoffman, R. S. 1958. The role of reproduction and mortality in populations of voles *(Microtus).* *Ecol. Monogr.* 28:79–108.

Holbrook, S. 1979. Habitat utilization, competitive interactions, and coexistence of three species of cricetine rodents in east-central Arizona. *Ecology* 60:758–769.

Hollings, C. S. 1966. The strategy of building models of complex ecological systems. In *Systems Analysis in Ecology,* edited by K. E. F. Watt, 195–214. New York: Academic Press.

Homewood, K. 1975. Can the Tana mangabey survive? *Oryx* 13:53–59.

Hook, J. 1960. Seasonal variations in physiological functions of arctic ground squirrels and black bears. In *Mammalian Hibernation,* edited by C. P. Lyman and A. R. Dawe. *Bull. Mus. Comp. Zool. (Harv. Univ.)* 124:155–171.

Hooton, E. A. 1954. The importance of primate studies to anthropology. *Hum. Biol.* 26:179–188.

Horn, H. S. 1978. Optimal tactics of reproduction and life-history. In *Behavioural Ecology: An Evolutionary Approach,* edited by J. R. Krebs and N. B. Davies, 411–429. Sunderland, Mass.: Sinauer.

Horn, H. S., and R. M. May. 1977. Limits to similarity among coexisting competitors. *Nature* 270:660–661.

Howe, H. F. 1977. Bird activity and seed dispersal of a tropical wet forest tree. *Ecology* 58:539–550.

———. 1979. Fear and frugivory. *Am. Nat.* 114:925–931.

———. 1980. Monkey dispersal and waste of a Neotropical fruit. *Ecology* 61:944–959.

Howell, D. J. 1974. Bats and pollen: physiological aspects of the syndrome of chiropterophily. *Comp. Biochem. Physiol.* 48A:263–276.

Hrdy, S. B. 1974. Male–male competition and infanticide among the langurs *(Presbytis entellus)* of Abu, Rajasthan. *Folia Primatol.* 22:19–58.

———. 1977. *The Langurs of Abu: Female and Male Strategies of Reproduction.* Cambridge: Harvard University Press.

———. 1979. Infanticide among animals: a review, classification, and examination of the implications for the reproductive strategies of females. *Ethol. Sociobiol.* 1:13–40.

———. 1980. Primate behavior. *The Great Apes. Science* 207:632–634.

Hrdy, S. B., and D. B. Hrdy. 1976. Hierarchical relations among female Hanuman langurs (Primates: Colobinae, *Presbytis entellus*). *Science* 197:913–915.

Hsu, K. I., W. B. F. Ryan, and M. B. Cita. 1973. Late Miocene desiccation of the Mediterranean. *Nature* 242:240–244.

Humphrey, N. K. 1976. The social function of the intellect. In *Growing Points in Ethnology,* edited by P. P. G. Bateson and R. A. Hinde, 303–317. Cambridge: Cambridge University Press.

Hutchinson, G. E. 1957. Concluding remarks. *Cold Spring Harbor Symp. Quant. Biol.* 22:415–427.

———. 1959. Homage to Santa Rosalia; or why are there so many kinds of animals? *Am. Nat.* 93:145–159.

———. 1961. The paradox of the plankton. *Am. Nat.* 95:137–145.

———. 1978a. *An Introduction to Population Ecology.* New Haven: Yale University Press.

———. 1978b. Zoological iconography in the West after A.D. 1200. *Am. Sci.* 66:675–684.

Hylander, W. L. 1975. Incisor size and diet in anthropoids with special reference to Cercopithecidae. *Science* 189:1095–1098.

Irving, L. 1972. *Arctic Life of Birds and Mammals Including Man.* New York: Springer-Verlag.

Itani, J. 1975. Twenty years with Mt. Takasaki monkeys. In *Primate Utilization and Conservation,* edited by G. Bermant and D. G. Lindburg, 101–125. New York: Wiley.

———. 1979. Distribution and adaptation of chimpanzees in an arid area. In *The Great Apes,* edited by D. A. Hamburg and E. R. McCown, 55–71. Menlo Park, Calif.: Benjamin/Cummings.

Itani, J., and A. Suzuki. 1967. The social unit of wild chimpanzees. *Primates* 8:355–381.

Itani, J., K. Tokuda, Y. Furuya, K. Kano, and Y. Shin. 1963. The social construction of natural troops of Japanese monkeys in Takasakiyama. *Primates* 4:1–42.

Izawa, K. 1970. Unit-groups of chimpanzees and their nomadism in the savannah woodland. *Primates* 11:1–45.

———. 1975. Foods and feeding behavior of monkeys in the Upper Amazon basin. *Primates* 16:295–316.

Jacobsen, N. 1970. Salivary amylase. II: Alpha amylase in salivary glands of the *Macaca irus* monkey, the *Cercopithecus aethiops* monkey, and man. *Caries Res.* 4:200–205.

Jaeger, E. C. 1957. *The North American Desert.* Stanford: Stanford University Press.

Jaeger, R. G. 1974. Competitive exclusion: comments on survival and extinction of species. *Bioscience* 24:33–39.

Janson, C. H., J. Terborgh, and L. H. Emmons. 1981. Non-flying mammals as pollinating agents in the Amazonian forest. *Reprod. Botany* (supplement to *Biotropica*) 13(2):1–6.

Janzen, D. H. 1969. Seed-eaters versus seed size, number, and toxicity and dispersal. *Evolution* 23:1–27.

———. 1971. Escape of juvenile *Dioclea megacarpa* (Leguminosae) vines from predators in a deciduous tropical forest. *Am. Nat.* 105:97–112.

———. 1977. Why fruits rot, seeds mold, and meat spoils. *Am. Nat.* 111:691–713.

———. 1978. Complications in interpreting the chemical defenses of trees against tropical arboreal plant-eating vertebrates. In *The Ecology of Arboreal Folivores,* edited by G. G. Montgomery, 73–84. Washington, D.C.: Smithsonian Institution Press.

———. 1979. New horizons in the biology of plant defenses. In *Herbivores: Their Interaction with Secondary Plant Metabolics,* edited by G. A. Rosenthal and D. H. Janzen, 331–350. New York: Academic Press.

———. 1983. Dispersal of seeds by vertebrate guts. In *Coevolution,* edited by D. J. Futuyuma and M. Slatkin, 232–262. Sunderland, Mass.: Sinauer.

Jarman, P. J. 1974. The social organization of antelope in relation to their ecology. *Behaviour* 48:215–267.

Jay, P. 1965. The common langur of north India. In *Primate Behavior: Field Studies of Monkeys and Apes,* edited by I. DeVore, 197–249. New York: Holt.

Jenness, R. 1974. The composition of milk. In *Lactation: A Comprehensive Treatise,* vol. 3, edited by B. L. Larson and V. R. Smith, 3–107. New York: Academic Press.

Jepson, G. L. 1970. Bat origins and evolution. In *Biology of Bats,* vol. 1, edited by W. A Wimsatt, 1–64. New York: Academic Press.

Jerison, H. J. 1973. *Evolution of the Brain and Intelligence*. New York: Academic Press.

Johnson, G. E. 1931. Hibernation in mammals. *Q. Rev. Biol.* 6:439–461.

Johnson, R. W., and W. D. Raymond. 1965. The chemical composition of some tropical food plants. V. Mango. *Trop. Sci.* 7:156–164.

Jolly, A. 1966a. *Lemur Behavior: A Madagascan Field Study*. Chicago: University of Chicago Press.

———. 1966b. Lemur social intelligence and primate intelligence. *Science* 153:501–506.

———. 1972a. *The Evolution of Primate Behavior*. New York: Macmillan.

———. 1972b. Troop continuity and troop spacing in *Propithecus verreauxi* and *Lemur catta* at Berenty (Madagascar). *Folia Primatol.* 17:335–362.

Jolly, A., H. Gustafson, W. L. R. Oliver, and S. M. O'Connor. 1982a. Population and troop ranges of *Lemur catta* and *Lemur fulvus* at Berenty, Madagascar: 1980 census. *Folia Primatol.* 39:115–123.

———. 1982b. *Propithecus verreauxi* population and ranging at Berenty, Madagascar, 1975 and 1980. *Folia Primatol.* 39:124–144.

Jolly, C. J. 1972. The classification and natural history of *Theropithecus (Simopithecus)* (Andrews 1916) baboons of the African Plio-Pleistocene. *Bull. Br. Mus. (Nat. Hist.) Geol.* 22:1–122.

Jones, C. 1970. Stomach contents and gastrointestinal relationships of monkeys collected in Rio Muni, West Africa. *Mammalia* 34:107–117.

Jones, C. B. 1980. The functions of status in the mantled howler monkey, *Alouatta palliata* Gray: intraspecific competition for group membership in a folivorous neotropical primate. *Primates* 21:389–405.

Jones, C. J., and J. Sabater Pi. 1968. Comparative ecology of *Cercocebus albigena* (Gray) and *Cercocebus torquatus* (Kerr) in Rio Muni, West Africa. *Folia Primatol.* 9:99–113.

Jones, K. C. 1983. Inter-troop transfer of *Lemur catta* males at Berenty, Madagascar. *Folia Primatol.* 40:145–160.

Jordan, P. A., P. C. Shelton, and D. L. Allen. 1967. Numbers, turnover, and social structure of the Isle Royale wolf population. *Am. Zool.* 7:233–252.

Jorde, L. B., and J. N. Spuhler. 1974. A statistical analysis of selected aspects of primate demography, ecology, and social behavior. *J. Anthropol. Res.* 30:199–244.

Jouventin, P. 1975. Observations sur la socio-écologie du mandrill. *Terre Vie* 29:493–532.

Kalabukhov. N. I. 1960. Comparative ecology of hibernating mammals. In *Mammalian Hibernation*, edited by C. P. Lyman and A. R. Dawe. *Bull. Mus. Comp. Zool. (Harv. Univ.)* 124:45–74.

Kamil, A. C., and T. D. Sargent, eds. 1981. *Foraging Behavior: Ecological, Ethological, and Psychological Approaches*. New York: Garland STPM Press.

Kano, T. 1971. The chimpanzees of Filabanga, western Tanzania. *Primates* 12(3–4):229–246.

Kawamura, S. 1958. The dominance hierarchy of the Minoo-B group—a study on the rank system of Japanese macaque. *Primates* 1:149–156.

Kay, R. F. 1975. The functional adaptations of primate molar teeth. *Am. J. Phys. Anthropol.* 43(2):195–215.

———. 1981. The nut-crackers—a new theory of the adaptations of the Ramapithecinae. *Am. J. Phys. Anthropol.* 55(2):141–151.

Kay, R. F., and M. Cartmill. 1977. Cranial morphology and adaptation of *Palaechthon naciamenti* and other Paromomyidae (Plesiadapoidea, Primates) with a description of a new genus and species. *J. Hum. Evol.* 6:19–53.

Kay, R. F., and W. L. Hylander. 1978. The dental structure of mammalian folivores with special reference to primates and Phalangeroids (Marsupiala). In *The Ecology of Arboreal Folivores,* edited by G. G. Montgomery, 173–192. Washington, D.C.: Smithsonian Institution Press.

Kay, R. F., and W. S. Sheine. 1979. On the relationship between chitin particle size and digestibility in the primate *Galago senegalensis. Am. J. Phys. Anthropol.* 50:301–308.

Kay, R. F., J. G. Fleagle, and E. L. Simons. 1981. A revision of the Oligocene apes of the Fayum Province, Egypt. *Am J. Phys. Anthropol.* 55:293–322.

Keast, A. 1972. Introduction: the southern continents as backgrounds for mammalian evolution. In *Evolution, Mammals and Southern Continents,* edited by A. Keast, F. C. Erk and B. Glass, 19–22. Albany: State University of New York Press.

Kendeigh, S. C. 1961. *Animal Ecology.* New Jersey: Prentice-Hall.

Kennedy, G. C., and J. Mitra. 1963. Body weight and food intake as the initiating factors for puberty in the rat. *J. Physiol. (Lond.)* 166:408–418.

Kennedy, G. E. 1980. *Paleoanthropology.* New York: McGraw-Hill.

Kennett, J. P. 1977. Cenozoic evolution of Antarctic glaciation, the Circum-Antarctic Ocean, and their impact on global paleoceanography. *J. Geophys. Res.* 82:3843–3860.

Kerr, G. R. 1972. Nutritional requirements of subhuman primates. *Physiol. Rev.* 52(2):415–467.

Kihlstrom, J. E. 1972. Period of gestation and body weight in some placental mammals. *Comp. Biochem. Physiol.* 43A:673–679.

King, J. A. 1973. The ecology of aggressive behavior. *Annu. Rev. Ecol. Syst.* 4:117–138.

Kinzey, W. G. 1977. Diet and feeding behaviour of *Callicebus torquatus.* In *Primate Ecology,* edited by T. H. Clutton-Brock, 127–151. New York: Academic Press.

———. 1982. Distribution of primates and forest refuges. In *Biological Diversification in the Tropics,* edited by G. T. Prance, 455–482. New York: Columbia University Press.

Kinzey, W. G., A. L. Rosenberger, and M. Ramirez. 1975. Vertical clinging and leaping in a neotropical anthropoid. *Nature* 255:327–328.

Kleiman, D. 1977. Monogamy in mammals. *Q. Rev. Biol.* 52:39–69.

———, ed. 1978a. *The Biology and Conservation of the Callitrichidae.* Washington, D. C.: Smithsonian Institution.

———. 1978b. Characteristics of reproduction and sociosexual interactions in pairs of lion tamarins *(Leontopithecus rosalia)* during the reproductive cycle. In *The Biology and Conservation of the Callitrichidae,* edited by D. G. Kleiman, 181–190. Washington, D. C.: Smithsonian Institution.

Kleiman, D., and J. F. Eisenberg. 1973. Comparisons of canid and felid social systems from an evolutionary perspective. *Anim. Behav.* 21:637–659.

Klein, D. R. 1970. Food selection by North American deer and their response to over-utilization of preferred plant species. In *Animal Populations in Relation to Their Food Resources,* edited by A. Watson, 25–46. Oxford: Blackwell.

Klein, L. 1971. Observations on copulation and seasonal reproduction of two species of spider monkeys, *Ateles belzebuth* and *A. geoffroyi. Folia Primatol.* 15:233–248.

———. 1974. Agonistic behavior in neotropical primates. In *Primate Aggression, Territoriality and Xenophobia: A Comparative Perspective,* edited by R. Holloway, 77–122. New York: Academic Press.

Klein, L. L., and D. J. Klein. 1973. Observations on two types of neotropical primate intertaxa associations. *Am. J. Phys. Anthropol.* 38(2):649–653.

———. 1975. Social and ecological contrasts between four taxa of neotropical primates *(Ateles belzebuth, Alouatta seniculus, Saimiri sciureus, Cebus apella).* In *Socioecology and Psychology of Primates,* edited by R. H. Tuttle, 59–85. The Hague: Mouton.

———. 1977. Feeding behavior of the Colombian spider monkey. In *Primate Ecology,* edited by T. H. Clutton-Brock, 153–181. New York: Academic Press.

Koenigswald, G. H. R. Von. 1962. Potassium-argon dates for the Upper Tertiary. *Proc. K. Ned. Akad. Wet. Ser. B. Phys. Sci.* 65:31–34.

Koering, M. J. 1974. Comparative morphology of the primate ovary. In *Reproductive Biology of the Primates,* edited by W. P. Luckett, 38–81. Vol. 3 of *Contributions to Primatology.* Basel: Karger.

Koford, C. B. 1963. Rank of mothers and sons in bands of rhesus monkeys. *Science* 141:356–357.

———. 1965. Population dynamics of rhesus monkeys on Cayo Santiago. In *Primate Behavior,* edited by I. DeVore, 160–174. New York: Holt, Rinehart & Winston.

———. 1966. Population changes in rhesus monkeys: Cayo Santiago 1960–1964. *Tulane Stud. Zool.* 13:1–7.

Koyama, N., K. Norikoshi, and T. Mano. 1975. Population dynamics of Japanese monkeys at Arashiyama. In *Contemporary Primatology,* 411–417. 57th International Congress of Primatology. Basel: Karger.

Krebs, C. J. 1972. *Ecology: The Experimental Analysis of Distribution and Abundance.* New York: Harper & Row.

Kriewaldt, F. H., and A. G. Hendrickx. 1968. Reproductive parameters of the baboon. *Lab. Anim. Care* 18:361–370.

Kruuk, H. 1972. *The Spotted Hyaena.* Chicago: University of Chicago Press.

Kruuk, H., and M. Turner. 1967. Comparative notes on predation by lion, leopard, cheetah and wild dog in the Serengeti area, East Africa. *Mammalia* 31:1–27.

Kuhme, W. 1965. Communal food distribution and division of labour in African hunting dogs. *Nature* 205:443–444.

Kummer, H. 1968. *Social Organization of Hamadryas Baboons.* Chicago: University of Chicago Press.

Kummer, H., W. Goetz, and W. Angst. 1970. Cross-species modification of social behavior in baboons. In *Old World Monkeys: Evolution, Systematics, and Behavior,* edited by J. R. Napier and P. H. Napier, 351–363. London: Academic Press.

Labov, J. B. 1977. Phytoestrogens and mammalian reproduction. *Comp. Biochem. Physiol.* 57A:3–9.

Lancaster, J. B., and R. B. Lee. 1965. The annual reproductive cycle in monkeys and apes. In *Primate Behavior: Field Studies of Monkeys and Apes,* edited by I. DeVore, 486–513. New York: Holt.

Landau, M., D. Pilbeam, and A. Richard. 1982. Human origins a century after Darwin. *Bioscience* 32(6):507–512.

Laurent, R. F. 1973. A parallel survey of equatorial amphibians and reptiles in Africa and South America. In *Tropical Forest Ecosystems in Africa and South America: A Comparative Review,* edited by B. J. Meggers, E. S. Ayensu, and W. D. Duckworth, 259–266. Washington, D. C.: Smithsonian Institution Press.

Lawick, H. van. 1974. *Solo: The Story of an African Wild Dog.* Boston: Houghton Mifflin.

Lawick, H. van, and J. van Lawick-Goodall. 1970. *Innocent Killers.* London: Collins.

Leakey, R. E., and R. Lewin. 1977. *Origins.* New York: Dutton.

Leeds, A., and V. Dusek, eds. 1981. *Sociobiology: The Debate Evolves.* Boston: The Philosophical Forum XIII (2–3)

Leigh, E. G., Jr., and N. Smythe. 1978. Leaf production, leaf consumption and the regulation of folivory on Barro Colorado Island. In *The Ecology of Arboreal Folivores,* edited by G. G. Montgomery, 33–50. Symposia of the National Zoological Park. Washington, D.C.: Smithsonian Institution Press.

Leighton, M. 1982. Fruit resources and patterns of feeding, spacing, and grouping among sympatric Bornean hornbill (Bucerotidae). Ph.D. diss., University of California, Davis.

Levin, D. A. 1971. Plant phenolics: an ecological perspective. *Am. Nat.* 105:157–181.

———. 1976. Alkaloid-bearing plants: an ecogeographic perspective. *Am. Nat.* 110:261–284.

Levins, R. 1975. Evolution in communities near equilibrium. In *Ecology and Evolution of Communities,* edited by M. L. Cody and J. M. Diamond, 16–50. Cambridge, Mass.: Belknap.

Lewin, R. 1983. Santa Rosalia was a goat. *Science* 221:636–639.

Lewontin, R. C. 1974. *The Genetic Basis of Evolutionary Change.* New York: Columbia University Press.

———. 1978. Fitness, survival and optimality. In *Analysis of Ecological Systems,* edited by D. H. Horn, R. Mitchell, and G. R. Stairs, 3–21. Columbus: Ohio State University Press.

Lieberman, D., J. B. Hall, and M. D. Swaine. 1979. Seed dispersal by baboons in Ghana. *Ecology* 60:65–75.

Lindburg, D. G. 1969. Rhesus monkeys: mating season mobility of adult males. *Science* 166:1176–1178.

———. 1971. The rhesus monkey in North India: an ecological and behavioral study. In *Primate Behavior: Developments in Field and Laboratory Research II,* edited by L. A. Rosenblum, 1–106. New York: Academic Press.

———. 1977. Feeding behaviour and diet of rhesus monkeys *(Macaca mulatta)* in a Siwalik forest in North India. In *Primate Ecology,* edited by T. H. Clutton-Brock, 223–249. New York: Academic Press.

Lindeman, R. L. 1942. The trophic–dynamic aspect of ecology. *Ecology* 23:399–418.

Lock, J. M. 1972. Baboons feeding on *Euphorbia candelabrum. East Afr. Wildl. J.* 10:73–76.

Longman, K. A., and J. Jenik. 1974. *Tropical Forest and Its Environment.* London: Longman.

Loy, J. 1970. Behavioral responses of free-ranging rhesus monkeys to food shortage. *Am. J. Phys. Anthropol.* 33:263–272.

Luckett, W. P. 1974. Comparative development and evolution of the placenta in primates. *Contrib. Primatol.* 3:142–234.

———. 1975. Ontogeny of the fetal membranes and placenta: their bearing on primate phylogeny. In *Phylogeny of the Primates,* edited by W. P. Luckett and F. S. Szalay, 157–182. New York: Plenum Press.

Lydekker, R. 1922. *The Royal Natural History.* Vol. 1. London: Warne.

MacArthur, R. H. 1968. Selection for life tables in periodic environments. *Am. Nat.* 102:381–383.

———. 1972. Coexistence of species. In *Challenging Problems in Biology,* edited by J. A. Behnke, 253–259. New York: Oxford University Press.

MacArthur, R. H., and R. Levins. 1967. The limiting similarity, convergence, and divergence of coexisting species. *Am. Nat.* 101:377–385.

MacArthur, R. H., and E. R. Pianka. 1966. On the optimal use of a patchy environment. *Am. Nat.* 100:603–609.

MacArthur, R. H., and E. O. Wilson. 1967. *The Theory of Island Biogeography.* Princeton: Princeton University Press.

McCance, R. A., and E. M. Widdowson. 1964. Protein metabolism and requirements in the newborn. In *Mammalian Protein Metabolism,* vol. 2, edited by H. N. Munro and J. B. Allison, 226–246. New York: Academic Press.

McCann, C. 1933. Notes on some Indian macaques. *J. Bombay Nat. Hist. Soc.* 36:796–810.

McGinnis, P. R. 1979. Sexual behavior in free-living chimpanzees: consort relationships. In *The Great Apes,* edited by D. A. Hamburg and E. R. McCown, 429–440. Menlo Park, Calif.: Benjamin/Cummings.

McGrew, W. C. 1979. Evolutionary implications of sex differences in chimpanzee predation and tool use. In *The Great Apes,* vol. 5, edited by D. A. Hamburg and E. R. McCown, 441–464. Menlo Park, Calif.: Benjamin/Cummings.

McHugh, P. R., T. H. Moran, and G. N. Barton. 1975. Satiety: a graded behavioral phenomenon regulating caloric intake. *Science* 190:167–169.

McKenna, J. J. 1979. The evolution of allomothering behavior among colobine monkeys: function and opportunism in evolution. *Am. Anthropol.* 81:818–840.

McKey, D. 1974. Adaptive patterns in alkaloid physiology. *Am. Nat.* 108:305–320.

———. 1975. The ecology of coevolved seed dispersal systems. In *Coevolution of Animals and Plants,* edited by L. E. Gilbert and P. H. Raven, 159–191. Austin: University of Texas Press.

———. 1978. Soils, vegetation, and seed-eating by black colobus monkeys. In *The Ecology of Arboreal Folivores,* edited by G. G. Montgomery, 423–437. Washington, D.C.: Smithsonian Institution.

McKey, D., P. G. Waterman, C. N. Mbi, J. S. Gartlan, and T. T. Struhsaker. 1978. Phenolic content of vegetation in two African rain forests: ecological implications. *Science* 202:61–64.

MacKinnon, J. R. 1974. The behaviour and ecology of wild orangutans *(Pongo pygmaeus). Anim. Behav.* 22:3–74.

———. 1976. A comparative ecology of the Asian apes. *Primates* 18:747–772.

———. 1979. Reproductive behavior in wild orangutan populations. In *The Great Apes,* edited by D. A. Hamburg and E. R. McCown, 257–274. Menlo Park, Calif: Benjamin/Cummings.

MacKinnon, J. R., and K. S. MacKinnon. 1978. Comparative feeding ecology of six sympatric primates in West Malaysia. In *Recent Advances in Primatology,* vol. 1, edited by D. J. Chivers and J. Herbert, 305–321. New York: Academic Press.

———. 1980a. The behavior of wild spectral tarsiers. *Int. J. Primatol.* 1:361–379.

———. 1980b. Niche differentiation in a primate community. In *Malayan Forest Primates,* edited by D. J. Chivers, 167–190. New York: Plenum Press.

McNab, B. K. 1971. On the ecological significance of Bergmann's rule. *Ecology* 52:845–854.

———. 1980. Food habits, energetics, and the population biology of mammals. *Am. Nat.* 116:106–124.

McNaughton, S. J. 1979. Grassland-herbivore dynamics. In *Serengeti: Dynamics of an Ecosystem,* edited by A. R. E. Sinclair and M. Norton-Griffiths, 46–81. Chicago: University of Chicago Press.

Manning, T. H. 1948. Notes on the country birds and mammals west of Hudson Bay between Reindeer and Baker Lakes. *Can. Field-Nat.* 62:1–28.

Margalef, R. 1975. Diversity, stability and maturity in natural ecosystems. In *Unifying Concepts in Ecology,* edited by W. H. van Dobben and R. H. Lowe-McConnel, 151–160. The Hague: Junk.

Marler, P. 1969. *Colobus guereza:* territoriality and group composition. *Science* 163:93–95.

Marsh, C. W. 1979a. Comparative aspects of social organization in the Tana River red colobus, *Colobus badius rufomitratus. Z. Tierpsychol.* 51:337–362.

———. 1979b. Female transference and mate choice among Tana River red colobus. *Nature* 281: 568–569.

Marshall, L. G., S. D. Webb, J. J. Sepkoski, Jr., and D. M. Raup. 1982. Mammalian evolution and the Great American interchange. *Science* 215:1351–1357.

Martin, R. D. 1972. Adaptive radiation and behaviour of the Malagasy lemurs. *Philos. Trans. R. Soc. Lond. B Biol. Sci.* 264:295–352.

———. 1975. The bearing of reproductive behavior and ontogeny on strepsirhine phylogeny. In *Phylogeny of the Primates,* edited by W. P. Luckett and F. S. Szalay, 265–297. New York: Plenum Press.

———. 1981. Relative brain size and basal metabolic rate in terrestrial vertebrates. *Nature* 293:57–60.

Martin, R. D., and S. K. Bearder. 1979. Radio bush baby. *Natural History* 88(8):76–81.

Mason, W. A. 1978. Ontogeny of social systems. In *Recent Advances in Primatology,* vol. 1, edited by D. J. Chivers and J. Herbert, 4–14. New York: Academic Press.

Masui, K., Y. Sugiyama, A. Nishimura, and H. Ohsawa. 1975. The life table of Japanese monkeys at Takasakiyama. In *Contemporary Primatology,* edited by S. Kondo, M. Kawai, and A. Ehara, 401–406. Basel: Karger.

May, R. M. 1973. *Stability and Complexity in Model Ecosystems.* Monographs in Population Biology, vol. 6. Princeton: Princeton University Press.

———. 1977. Optimal life-history strategies. *Nature* 267:394–395.

May, R. M., and R. H. MacArthur. 1972. Niche overlap as a function of environmental variability. *Proc. Nat. Acad. Sci. USA* 69:1109–1113.

Maynard Smith, J. 1978. Optimization theory in evolution. *Annu. Rev. Ecol. Syst.* 9:31–56.

Mayr, E. 1963. *Animal Species and Evolution.* Cambridge, Mass.: Belknap.

———. 1970. *Populations, Species, and Evolution.* Cambridge, Mass.: Belknap.

Mech, L. D., D. M. Barnes, and J. R. Tester. 1968. Seasonal weight changes, mortality, and population structure of raccoons in Minnesota. *J. Mammal.* 49:63–73.

Melnick, D. 1981. Microevolution in a population of Himalayan rhesus monkeys *(Macaca mulatta).* Ph.D. diss., Yale University.

Melnick, D. J., and K. K. Kidd. In Press. Genetic and evolutionary relationships among Asian macaques. *Int. J. Primatol.*

Melnick, D. J., M. C. Pearl, and A. F. Richard. In press. Male migration and inbreeding avoidance in wild rhesus monkeys. *Am. J. Primatol.*

Menzel, E. W. 1973. Chimpanzee spatial memory organization. *Science* 182:943–945.

Merrell, M. 1947. Time-specific life tables contrasted with observed survivorship. *Biometrics* 3:129–136.

Mertl-Milhollen, A., H. L. Gustafson, N. Budnitz, K. Dainis, and A. Jolly. 1979. Population and territory stability of the *Lemur catta* at Berenty, Madagascar. *Folia Primatol.* 31:106–122.

Merzenich, M. M., and J. H. Kaas. 1980. Principles of organization of sensory-perceptual systems in mammals. *Prog. Psychobiol. Physiolog. Psych.* 9:1–42.

Meyer, G. E. 1978. Hyracoidea. In *Evolution of African Mammals,* edited by V. J. Maglio and H. B. S. Cooke, 284–314. Cambridge: Harvard University Press.

Miller, R. S. 1957. Observations on the status of ecology. *Ecology* 38:353–354.

———. 1967. Pattern and process in competition. *Adv. Ecol. Res.* 4:1–74.

Milton, K. 1978. Behavioral adaptations to leaf-eating by the mantled howler monkey *(Alouatta palliata).* In *The Ecology of Arboreal Folivores,* edited by G. G. Montgomery, 535–549. Washington, D.C.: Smithsonian Institution.

———. 1979. Factors influencing leaf choice by howler monkeys: a test of some hypotheses of food selection by generalist herbivores. *Am. Nat.* 114:362–378.

———. 1980. *The Foraging Strategy of Howler Monkeys: A Study in Primate Economics.* New York: Columbia University Press.

———. 1981a. Distribution patterns of tropical plant foods as an evolutionary stimulus to primate mental development. *Am. Anth.* 83:534–548.

———. 1981b. Food choice and digestive strategies of two sympatric primate species. *Am. Nat.* 117:496–505.

Milton, K., and M. L. May. 1976. Body weight, diet and home range area in primates. *Nature* 259:459–462.

Missakian, E. 1972. Genealogical and cross-genealogical dominance relations in a group of free-ranging rhesus monkeys *(Macaca mulatta)* on Cayo Santiago. *Primates* 13:169–180.

Mittermeier, A., Jr. 1973. Group activity and population dynamics of the howler monkey on Barro Colorado Island. *Primates* 14:1–19.

Mittermeier, R. A., and A. F. Coimbra-Filho. 1977. Primate conservation in Brazilian Amazonia. In *Primate Conservation,* edited by H.S.H. Prince Rainier III and G. H. Bourne, 117–167. New York: Academic Press.

Mittermeier, R. A., and M. G. M. van Roosmalen. 1981. Preliminary observations on habitat utilization and diet in eight Surinam monkeys. *Folia Primatol.* 36:1–39.

Montgomery, G. G., and M. E. Sunquist. 1975. Impact of sloths on neotropical forest energy flow and nutrient cycling. In *Tropical Ecological Systems: Trends in Terrestrial and Aquatic Research,* edited by F. B. Golley and E. Medino, 69–98. New York: Springer-Verlag.

Moore, J. In press. Female transfer in primates. *Int. J. Primatol.*

Moreau, R. E. 1969. Climatic changes and the distribution of forest vertebrates in West Africa. *J. Zool.* 158:39–61.

Morgan, L. H. 1868. *The American Beaver and His Works.* Philadelphia: Lippincott.

Mori, A. 1979. Analysis of population changes by measurement of body weight in the Koshima troop of Japanese monkeys. *Primates* 20:371–397.

Morse, D. H. 1970. Ecological aspects of some mixed-species foraging flocks of birds. *Ecol. Monogr.* 40:119–168.

Moustegaard, J. 1977. Nutrition and reproduction in domestic animals. In *Reproduction in Domestic Animals,* 3d ed., vol. 2, edited by H. H. Cole and P. T. Cupps, 170–223. New York: Academic Press.

Moynihan, M. 1962. The organization and probable evolution of some mixed species flocks of neotropical birds. *Smithson. Misc. Collect.* 143(7):1–140.

———. 1970. Some behavior patterns of platyrrhine monkeys. II. *Saguinus geoffroyi* and some other tamarins. *Smithson. Contrib. Zool.* 28:1–77.

———. 1976. *The New World Primates.* Princeton: Princeton University Press.

Muckenhirn, N. A., and Eisenberg, J. F. 1973. Home ranges and predation of the Ceylon leopard. In *The World's Cats,* vol. 1, *Ecology and Conservation,* edited by R. L. Eaton, 142–175. Winston, Oreg.: World Wildlife Safari.

Muller, C. H. 1966. Role of chemical inhibition (allelopathy) in vegetational composition. *Bull. Torrey Bot. Club* 93:332–351.

Munger, J. C., and J. H. Brown. 1981. Competition in desert rodents: an experiment with semipermeable exclosures. *Science* 211:510–512.

Munn, C. A., and J. W. Terborgh. 1979. Multi-species territoriality in neotropical foraging flocks. *Condor* 81:338–347.

Munro, H. N. 1964. An introduction to nutritional aspects of protein metabolism. In *Mammalian Protein Metabolism,* vol. 2, edited by H. N. Munro and J. B. Allison, 3–39. New York: Academic Press.

———. 1969. Evolution of protein metabolism in mammals. In *Mammalian Protein Metabolism,* vol. 3, edited by H. N. Munro, 133–182. New York: Academic Press.

Murdoch, W. W. 1966. Community structure, population control, and competition—a critique. *Am. Nat.* 100:219–226.

Murphy, D. A., and J. A. Coates. 1966. Effects of dietary protein on deer. *Trans. N. Am. Wildl. Nat. Resour. Conf.* 31:129–139.

Murphy, G. I. 1968. Pattern in life history and the environment. *Am. Nat.* 102:391–403.

Murray, P. 1975. The role of cheek pouches in cercopithecine monkey adaptive strategy. In *Primate Functional Morphology and Evolution,* edited by R. H. Tuttle. 151–194. The Hague: Mouton.

Nadler, R. D. 1981. Laboratory research on sexual behavior of the great apes. In *Reproductive Biology of the Great Apes: Comparative and Biomedical Perspectives,* edited by C. E. Graham, 192–238. New York: Academic Press.

Nadler, R. D., C. E. Graham, D. C. Collins, and O. R. Kling. 1981. Postpartum amenorrhea and behavior of great apes. In *Reproductive Biology of the Great Apes: Comparative and Biomedical Perspectives,* edited by C. E. Graham, 69–81. New York: Academic Press.

Nagel, U. 1971. Social organization in a baboon hybrid zone. *Proc. Int. Congr. Primatol.* 3:48–57.

———. 1973. A comparison of anubis baboons, hamadryas baboons, and their hybrids at a species border in Ethiopia. *Folia Primatol.* 19:104–165.

Napier, J. 1966. Stratification and primate ecology. *J. Anim. Ecol.* 35:411–412.

———. 1967. Evolutionary aspects of primate locomotion. *Am. J. Phys. Arthropol.* 27:333–341.

———. 1970. Paleoecology and catarrhine evolution. In *Old World Monkeys,* edited by J. R. Napier and P. H. Napier. New York: Academic Press.

———. 1973. *Bigfoot: The Yeti and Sasquatch in Myth and Reality.* New York: Dutton.

Napier, J. R., and P. H. Napier. 1967. *A Handbook of Living Primates.* New York: Academic Press.

Napier, J. R., and A. C. Walker. 1967. Vertical clinging and leaping, a newly recognized category of locomotor behavior among primates. *Folia Primatol.* 6:204–219.

National Research Council, Committee on Nonhuman Primates. 1981. *Techniques for the Study of Primate Population Ecology.* Washington, D. C.: National Academy Press.

Neville, M. K. 1968. Ecology and activity of Himalayan foothill rhesus monkeys *(M. mulatta). Ecology* 49:110–123.

———. 1972. The population structure of red howler monkeys *(Alouatta seniculus)* in Trinidad and Venezuela. *Folia Primatol.* 17:56–86.

Newsome, A. E. 1965. The influence of food on breeding in the red kangaroo in central Australia. *Wildl. Res. Bull.* C.S.I.R.O. 11:187.

Neyman, P. F. 1978. Aspects of the ecology and social organization of free-ranging cotton-top tamarins *(Saguinus oedipus)* and the conservation status of the species. In *The Biology and Conservation of the Callitrichidae,* edited by D. G. Kleiman, 39–72. Washington, D.C.: Smithsonian Institution.

Niemitz, C. 1979. Outline of the behavior of *Tarsius bancanus.* In *The Study of Prosimian Behavior,* edited by G. A. Doyle and R. D. Martin, 631–660. New York: Academic Press.

Nishida, T. 1968. The social group of wild chimpanzees in the Mahali Mountains. *Primates* 9:167–224.

———. 1972. A note on the ecology of the red-colobus monkeys *(Colobus badius vephrosceles)* living in the Mahali Mountains. *Primates* 13:57–64.

————. 1979. The social structure of chimpanzees of the Mahale Mountains. In *The Great Apes,* edited by D. A. Hamburg and E. R. McCown, 73–122. Menlo Park, Calif.: Benjamin/Cummings.

Nissen, H. W. 1931. A field study of the chimpanzee. *Comp. Psychol. Monogr.* 8:1–122.

Noble, G. K. 1939. The role of dominance in the social life of birds. *Auk* 56:263–273.

Noy-Meir, I. 1973. Desert ecosystems: environment and producers. *Annu. Rev. Ecol. Syst.* 4:25–51.

Oates J. F. 1977a. The guereza and its food. In *Primate Ecology,* edited by T. H. Clutton-Brock, 276–321. New York: Academic Press.

————. 1977b. The social life of a black-and-white colobus monkey, *Colobus guereza. Z. Tierpsychol.* 45:1–60.

————. 1978. Water-plant and soil consumption by guereza monkeys *(Colobus guereza):* a relationship with minerals and toxins in the diet? *Biotropica* 10:241–253.

Oates, J. F., T. Swain, and J. Zantovska. 1977. Secondary compounds and food selection by colobus monkeys. *Biochem. Syst. Ecol.* 5:317–321.

Oates, J. F., P. G. Watermann, and G. M. Choo. 1980. Food selection by the South Indian leafmonkey, *Presbytis johnii,* in relation to leaf chemistry. *Oecologia (Berl.)* 45:45–56.

Oberlander, G. T. 1956. Summer fog precipitation on the San Francisco Peninsula. *Ecology* 37:851–852.

Odum, E. P. 1964. The new ecology. *Bioscience* 14(7):14–16.

————. 1968. Energy flow in ecosystems: a historical review. 11–18.

————. 1971. *Fundamentals of Ecology.* 3d ed. Philadelphia: Saunders.

Odum, H. T., W. Abbott, R. W. Selander, F. B. Golley, and R. F. Wilson. 1970. Estimates of chlorophyll and biomass of the Tabonuco forest of Puerto Rico. In *A Tropical Rain Forest,* edited by H. T. Odum and R. F. Pigeon, 13–119. Washington, D.C.: United States Atomic Energy Commission.

Ognev, S. I. 1962. *Mammals of the U.S.S.R. and Adjacent Countries.* Vol. 5, *Rodents.* Translated from the Russian by A. Birron and Z. S. Cole. Israel Program for Scientific Translations, Jerusalem. Washington, D.C.: National Science Foundation.

Ohsawa, H. 1979. The local gelada population and environment of the Gich area. In *Ecological and Sociological Studies of Gelada Baboons,* edited by M. Kawai, 4–45. Basel: Karger.

Ohwaki, K., R. E. Hungate, L. Lotter, R. R. Hofmann, and G. Maloiy. 1974. Stomach fermentation in East African colobus monkeys in their natural state. *Appl. Microbiol.* 27:713–723.

Oppenheimer, J. R. 1977. *Presbytis entellus,* the hanuman langur. In *Primate Conservation,* edited by H.S.H. Prince Rainier III and G. H. Bourne, 469–512. New York: Academic Press.

Oppenheimer, J. R., and G. E. Lang. 1969. *Cebus* monkeys: effect on branching of *Gustavia* trees. *Science* 165:187–188.

Orians, G. H. 1969. On the evolution of mating systems in birds and mammals. *Am. Nat.* 103:589–603.

Osgood, W. H. 1943. The mammals of Chile. *Field Mus. Nat. Hist. Publ. Zool. Ser.* 30:1–268.

Otis, J. S., J. W. Froelich, and R. W. Thorington, Jr. 1981. Season and age-related differential mortality by sex in the mantled howler monkey, *Alouatta palliata*. *Int. J. Primatol.* 2:197–205.

Owadally, A. W. 1979. The dodo and the tambala coque tree. *Science* 203:1363–1364.

Oxnard, C. E. 1966. Vitamin B_{12} nutrition in some primates in captivity. *Folia Primatol.* 4:424–431.

Packer, C. 1975. Male transfer in olive baboons. *Nature* 255:219–220.

———. 1977. Reciprocal altruism in *Papio anubis*. *Nature* 265:441–443.

———. 1979a. Inter-troop transfer and inbreeding avoidance in *Papio anubis*. *Anim. Behav.* 27:1–36.

———. 1979b. Male dominance and reproductive activity in *Papio anubis*. *Anim. Behav.* 27:37–45.

Packer, C., and A. E. Pusey. 1982. Cooperation and competition within coalitions of male lions: kin selection or game theory? *Nature* 296:750–742.

———. 1983. Adaptation of female lions to infanticide by incoming males. *Am. Nat.* 121:716–728.

Parra, R. 1978. Comparison of foregut and hindgut fermentation in herbivores. In *The Ecology of Arboreal Folivores,* edited by G. G. Montgomery, 205–230. Washington, D.C.: Smithsonian Institution.

Passingham, R. 1982. *The Human Primate.* Oxford and New York: W. H. Freeman and Company.

Patterson, B. 1978. Pholidota and Tubulidentata. In *Evolution of African Mammals,* edited by V. J. Maglio and H. B. S. Cooke, 268–278. Cambridge: Harvard University Press.

Patterson, B., and R. Pascual. 1972. The fossil mammal fauna of South America. In *Evolution, Mammals and Southern Continents,* edited by A. Keast, F. C. Erk, and B. Glass, 247–309. Albany: State University of New York Press.

Payne, J. B. 1980. Competitors. In *Malayan Forest Primates: Ten Years' Study in Tropical Rain Forest,* edited by D. J. Chivers, 261–277. New York: Plenum Press.

Pearl, M. C. 1982. Networks of social relations among Himalayan rhesus monkeys. Ph.D. diss., Yale University.

Pearl, M. C., and S. R. Schulman. 1983. Techniques for the analysis of social structure in animal societies. In *Advances in the Study of Behavior,* vol. 13, edited by J. Rosenblatt, 107–146. New York: Academic Press.

Pearson, D. L. 1975. The relation of foliage complexity to ecological diversity of three Amazonian bird communities. *Condor* 77:453–466.

Peters, R. H. 1976. Tautology in evolution and ecology. *Am. Nat.* 110:1–12.

Petter, J.-J. 1962. Recherches sur l'écologie et l'éthologie des lémuriens Malgaches. *Mém. Mus. Natl. Hist. Nat. Sér. A Zool.* 27:1–146.

———. 1977. The aye-aye. In *Primate Conservation,* edited by H.S.H. Prince Rainier III and G. H. Bourne, 38–59. New York: Academic Press.

Petter, J.-J., A. Schilling, and G. Pariente. 1971. Observations éco-éthologiques sur deux lémuriens malgaches nocturnes: *Phaner furcifer* et *Microcebus coquereli*. *Terre Vie* 25:287–327.

———. 1975. Observations on the behavior and ecology of *Phaner furcifer*. In *Lemur Biology*, edited by I. Tattersall and R. W. Sussman, 209–218. New York: Plenum Press.

Petter-Rosseaux, A. 1974. Photoperiod, sexual activity, and body weight variations of *Microcebus murinus* (Miller 1977). In *Prosimian Biology*, edited by R. D. Martin, G. A. Doyle, and A. C. Walker, 365–373. London: Duckworth.

———. 1980. Seasonal activity rhythms, reproduction, and body weight variations in five sympatric nocturnal prosimians, in simulated light and climatic conditions. In *Nocturnal Malagasy Primates,* edited by P. Charles-Dominique et al, 137–152. New York: Academic Press.

Pianka, E. R. 1974a. *Evolutionary Ecology.* New York: Harper & Row.

———. 1974b. Niche overlap and diffuse competition. *Proc. Nat. Acad. Sci. USA* 71:2141–2145.

———. 1976. Competition and niche theory. In *Theoretical Ecology: Principles and Applications,* edited by R. May, 114–141. Philadelphia: Saunders.

Pianka, E. R., and W. S. Parker. 1975. Age-specific reproductive tactics. *Am. Nat.* 109:453–464.

Pickford M. 1982. New higher primate fossils from the middle Miocene deposits at Majiwa and Kaloma, western Kenya. *Am. J. Phys. Anthropol.* 58:1–20.

Pilbeam, D. R. 1972. An idea we could live without—the naked ape. *Discovery* 7:63–70.

———. 1980. Major trends in human evolution. In *Current Argument on Early Man,* edited by Lars-Konig Konigsson, 261–285. London: Pergamon Press.

———. 1982. New hominoid skull material from the Miocene of Pakistan. *Nature* 295:232–234.

———. 1984. The descent of hominoids and hominids. *Sci. Amer.* 250(3):84–96.

Pilbeam, D. R., and S. J. Gould. 1974. Size and scaling in human evolution. *Science* 186:892–901.

Pilbeam, D. R., and A. Walker. 1968. Fossil monkeys from the Miocene of Napak, northeast Uganda. *Nature* 220:657–660.

Platt, J. R. 1964. Strong inference. *Science* 416:347–353.

Pollock, J. I. 1975. Field observations on *Indri indri*: a preliminary report. In *Lemur Biology,* edited by I. Tattersall and R. W. Sussman, 287–311. New York: Plenum Press.

———. 1977. The ecology and sociology of feeding in *Indri indri*. In *Primate Ecology,* edited by T. H. Clutton-Brock, 38–69. New York: Academic Press.

———. 1979. Female dominance in *Indri indri*. *Folia Primatol.* 31:143–164.

Pook, A. G., and G. Pook. 1982. Polyspecific association between *Saguinus fuscicollis, Saguinus labiatus, Callimico goeldii,* and other primates in north-western Bolivia. *Folia Primatol.* 38:196–216.

Portmann, A. 1939. Die Ontogenese der Säugetiere als Evolutionsproblem. *Biomorphol.* 1:109–126.

———. 1941. Die Trägzeit der Primaten und die Dauer der Schwängerschaft beim Menschen: ein Problem der vergleichende Biologie. *Rev. Suisse Zool.* 48:511–518.

———. 1965. Über die Evolution der Trägzeit bei Säugetieren. *Rev. Suisse Zool.* 71:658–666.

Post, D. 1978. Feeding and ranging behavior of the yellow baboon *(Papio cynocephalus).* Ph.D. diss., Yale University.

———. 1981. Activity patterns of yellow baboons *(Papio cynocephalus)* in the Amboseli National Park, Kenya. *Anim. Behav.* 29:357–374.

———. 1982. Feeding behavior of yellow baboons *(Papio cynocephalus)* in the Amboseli National Park, Kenya. *Int. J. Primatol.* 3:403–430.

Post, D., S. Goldstein, and D. Melnick. 1978. An analysis of cercopithecoid odontometrics. II. Relations between dental dimorphism, body size dimorphism and diet. *Am. J. Phys. Anthropol.* 49:533–543.

Post, D., G. Hausfater, and S. McCuskey. 1980. Feeding behavior of yellow baboons *(Papio cynocephalus):* relationship to age, gender, and dominance rank. *Folia Primatol.* 34:170–195.

Powell, G. V. N. 1974. Experimental analysis of the social value of flocking by starlings *(Sturnus vulgaris)* in relation to predation and foraging. *Anim. Behav.* 22:501–505.

Prasad, K. N. 1971. Ecology of the fossil Hominoidea from the Siwaliks of India. *Nature* 232:413–414.

———. 1975. Observations on the paleoecology of South Asian tertiary primates. In *Paleoanthropology: Morphology and Paleoecology,* edited by R. H. Tuttle, 21–30. The Hague: Mouton.

Preslock, J. P., S. H. Hampton, and J. K. Hampton, Jr. 1973. Cyclic variations of serum progestins and immunoreactive estrogens in marmosets. *Endocrinology* 92:1096–1101.

Preuss, T. M. 1982. The face of *Sivapithecus indicus:* description of a new, relatively complete specimen from the Siwaliks of Pakistan. *Folia Primatol.* 38:141–157.

Price, M. 1978. The role of microhabitat in structuring desert rodent communities. *Ecology* 59:910–921.

Price, P. W. 1975. *Insect Ecology.* New York: Wiley.

Prost, J. 1965. A definitional system for the classification of primate locomotion. *Am. Anthropol.* 67:1198–1214.

Puget, A. 1971. Observations sur le macaque rhesus, *Macaca mulatta* (Zimmerman 1780) en Afghanistan. *Mammalia* 35:199–203.

Pusey, A. 1979. Intercommunity transfer of chimpanzees in Gombe National Park. In *The Great Apes*, edited by D. A. Hamburg and E. R. McCown, 465–480. Menlo Park, Calif.: Benjamin/ Cummings.

Pyke, G. H., H. R. Pulliam, and E. L. Charnov. 1977. Optimal foraging: a selective review of theory and tests. *Annu. Rev. Biol.* 52:137–154.

Quick, H. F. 1963. Animal population analysis. In *Wildlife Investigational Techniques*, edited by H. S. Mosby and H. Hewitt, 190–228. Washington, D.C.: The Wildlife Society.

Quris, R. 1975. Ecologie et organisation sociale de *Cercoceebus galeritus agilis* dans le nord-est Gabon. *Terre Vie* 29:337–398.

———. 1976. Données comparatives sur la socioécologie de huit éspèces de cercopithécidae vivant dans une zone de forêt primitive périodiquement inondée (nord-est du gabon). *Terre Vie* 30:193–209.

Rabinowitz, P. D., M. F. Coffin, and D. Falvey. 1983. The separation of Madagascar and Africa. *Science* 220:67–69.

Raemaekers, J. J. 1978. Competition for food between lesser apes. In *Recent Advances in Primatology*, vol. 1, *Behaviour*, edited by D. J. Chivers and J. Herbert, 327–330. London: Academic Press.

Raemaekers, J. J., F. P. G. Aldrich-Blake, and J. B. Paynes. 1980. The forest. In *Malayan Forest Primates: Ten Years' Study in Tropical Rain Forest*, edited by D. J. Chivers, 29–61. New York: Plenum Press.

Raemaekers, J. J., and D. J. Chivers. 1980. Socio-ecology of Malayan forest primates. In *Malayan Forest Primates: Ten Years' Study in Tropical Rain Forest*, edited by D. J. Chivers, 279–315. New York: Plenum Press.

Rahaman, H., K. Srihari, and R. V. Krishnamoorthy. 1975. Polysaccharide digestion in cheek pouches of the bonnet macaque. *Primates* 16:175–180.

Rahm, U. 1960. The pangolins of West and Central Africa. *Afr. Wildl.* 14:271–275.

Rainier III, H.S.H., Prince of Monaco, and G. H. Bourne, eds. 1977. *Primate Conservation.* New York: Academic Press.

Ralls, K. 1976. Mammals in which females are larger than males. *Q. Rev. Biol.* 51:245–276.

Ramirez, M. F., C. H. Freese, and C. J. Revilla. 1978. Feeding ecology of the pygmy marmoset, *Cebuella pygmaea*, in northeastern Peru. In *The Biology and Conservation of the Callitrichidae*, edited by D. G. Kleiman, 91–104. Washington, D.C.: Smithsonian Institution.

Ransom, A. B. 1967. Reproductive biology of white-tailed deer in Manitoba. *J. Wildl. Manage.* 31:114–123.

Ransom, T. W., and T. E. Rowell. 1972. Early social development of feral baboons. In *Primate Socialization*, edited by F. E. Poirier, 105–144. New York: Random House.

Raven, P. H. 1972. Why are bird-visited flowers predominantly red? *Evolution* 26:674.

Reiter, R. J., and B. K. Follett, eds. 1980. *Seasonal Reproduction in Higher Vertebrates.* Vol. 5 of *Progress in Reproductive Biology.* Basel: Karger.

Reynolds, P. C. 1976. The emergence of early hominid social organization: 1. The attachment systems. *Yearb. Phys. Anthropol.* 20:73–95.

———. 1981. *On the Evolution of Human Behavior: The Argument from Animals to Man.* Berkeley: University of California Press.

Reynolds, V. 1967. *The Apes: The Gorilla, Chimpanzee, Orangutan, and Gibbon: Their History and their World.* New York: Dutton.

Reynolds, V., and F. Reynolds. 1965. Chimpanzees of the Budongo Forest. In *Primate Behavior: Field Studies of Monkeys and Apes,* edited by I. DeVore, 368–424. New York: Holt, Rinehart and Winston.

Rhine, R. J. 1975. The order of movement of yellow baboons *(Papio cynocephalus). Folia Primatol.* 23:72–104.

Rhoades, D. F., and R. G. Cates. 1976. A general theory of plant anti-herbivore chemistry. In *Biochemical Interaction between Plants and Insects,* edited by J. N. Wallace and R. L. Mansell, 168–213. Vol. 10 of *Recent Advances in Phytochemistry.* New York: Plenum Press.

Richard, A. F. 1974. Patterns of mating in *Propithecus verreauxi.* In *Prosimian Biology,* edited by R. D. Martin, G. A. Doyle, and A. C. Walker, 49–75. London: Duckworth.

———. 1976. Preliminary observations on the birth and development of *Propithecus verreauxi* to the age of six months. *Primates* 17:357–366.

———. 1977. The feeding behaviour of *Propithecus verreauxi.* In *Primate Ecology,* edited by T. H. Clutton-Brock, 72–96. New York: Academic Press.

———. 1978a. *Behavioral Variation: Case Study of a Malagasy Lemur.* Lewisburg, Pa.: Bucknell University Press.

———. 1978b. Variability in the feeding behavior of a Malagasy prosimian, *Propithecus verreauxi:* Lemuriformes. In *The Ecology of Arboreal Folivores,* edited by G. G. Montgomery, 519–533. Washington, D.C.: Smithsonian Institution.

———. 1981. Changing assumptions in primate ecology. *Am. Anthropol.* 83:517–533.

Richard, A. F., and S. R. Schulman. 1982. Sociobiology: primate field studies. *Annu. Rev. Anthropol.* 11:231–255.

Richard, A. F., and R. W. Sussman. 1975. Future of the Malagasy lemurs: conservation or extinction? In *Lemur Biology,* edited by I. Tattersall and R. W. Sussman, 335–350. New York: Plenum Press.

Richards, P. W. 1957. *Tropical Rain Forest.* Cambridge: Cambridge University Press.

Rick, C. M., and R. I. Bowman. 1961. Galápagos tomatoes and tortoises. *Evolution* 15:407–417.

Ricklefs, R. E. 1973. *Ecology.* Newton, Mass.: Chiron Press.

Ripley, S. 1967a. Intertroop encounters among Ceylon gray langurs *(Presbytis entellus).* In *Social Communication among Primates,* edited by S. A. Altmann, 237–253. Chicago: University of Chicago Press.

————. 1967b. The leaping of langurs: a problem in the study of locomotor adaptation. *Am. J. Phys. Anthropol.* 26:149–170.

————. 1970. Leaves and leaf-monkeys: the social organization of foraging in grey langurs, *Presbytis entellus thersites*. In *Old World Monkeys,* edited by J. R. Napier and P. H. Napier, 481–509. New York: Academic Press.

Rockwood, L. L., and K. E. Glander. 1979. Howling monkeys and leaf-cutting ants: comparative foraging in a tropical deciduous forest. *Biotropica* 11:1–10.

Rodgers, Q. R., and A. E. Harper. 1970. Selection of a solution containing histidine by rats fed a histidine-imbalanced diet. *J. Comp. Physiol. Psychol.* 72:66–71.

Rodman, P. S. 1973. Population composition and adaptive organisation among orang-utans of the Kutai Reserve. In *Comparative Ecology and Behaviour of Primates,* edited by R. P. Michael and J. H. Crook, 171–209. London: Academic Press.

————. 1977. Feeding behaviour of orang-utans of the Kutai Nature Reserve, East Kalimantan. In *Primate Ecology,* edited by T. H. Clutton-Brock, 384–413. London: Academic Press.

————. 1978. Diets, densities, and distributions of Bornean primates. In *The Ecology of Arboreal Folivores,* edited by G. G. Montgomery, 465–480. Symposia of the National Zoological Park. Washington, D.C.: Smithsonian Institution.

————. 1979. Individual activity patterns and the solitary nature of orangutans. In *The Great Apes,* edited by D. A. Hamburg and E. R. McCown, 235–255. Menlo Park, Calif.: Benjamin/Cummings.

Romer, A. S. 1966. *Vertebrate Paleontology,* 3d ed. Chicago: University of Chicago Press.

Roonwal, M. L. 1956. Macaque monkey eating mushrooms. *J. Bombay Nat. Hist. Soc.* 54:171.

Roonwal, M. L., and S. M. Mohnot. 1977. *Primates of South Asia: Ecology, Sociobiology, and Behavior.* Cambridge: Harvard University Press.

Root, R. B. 1967. The niche exploitation pattern of the blue-grey gnatcatcher. *Ecol. Monogr.* 37:317–350.

Rose, M. D. 1974. Postural adaptations in New and Old World monkeys. In *Primate Locomotion,* edited by F. A. Jenkins, 201–222. New York: Academic Press.

————. 1977a. Interspecific play between free ranging guerezas *(Colobus guereza)* and vervet monkeys *(Cercopithecus aethiops)*. *Primates* 18:956–964.

————. 1977b. Positional behaviour of olive baboons *(Papio anubis)* and its relationship to maintenance and social activities. *Primates* 18:59–116.

————. 1983. Miocene hominoid postcranial morphology: monkey-like, ape-like, neither, or both? In *New Interpretations of Ape and Human Ancestry,* edited by R. L. Ciochon and R. S. Corruccini, 301–324. New York: Plenum Press.

Rosenberger, A. L. 1979. Cranial anatomy and implication of *Dolichocebus,* a late Oligocene ceboid primate. *Nature* 179:416–417.

Ross, D. A., R. B. Whitmarsh, S. A. Ali, J. E. Boudreaux, R. Coleman, R. L. Fleischer, R. Girdler, F. Manheim, A. Malter, C. Nigrini, P. Stoffers, and P. R. Supko. 1973. Red Sea drillings. *Science* 179:377–380.

Ross, M. H., and G. Bras. 1975. Food preference and length of life. *Science* 190:165–167.

Rothe, H. 1978. Parturition and related behavior in *Callithrix jacchus* (Ceboidea, Callitrichidae). In *The Biology and Conservation of the Callitrichidae,* edited by D. G. Kleiman, 193–206. Washington, D.C.: Smithsonian Institution.

Roughgarden, J. 1983. Competition and theory in community ecology. *Am. Nat.* 122:583–601.

Rourke, J., and D. Wiens. 1977. Convergent floral evolution in South African and Australian Proteaceae and its possible bearing on pollination by nonflying mammals. *Ann. Mo. Bot. Gard.* 64:1–17.

Rowell, T. E. 1966. Forest living baboons in Uganda. *J. Zool. (Lond.)* 159:344–364.

———. 1967. A quantitative comparison of the behaviour of a wild and a caged baboon troop. *Anim. Behav.* 15:499–509.

———. 1969. Long-term changes in a population of Ugandan baboons. *Folia Primatol.* 11:241–254.

———. 1970. Baboon menstrual cycles affected by social environment. *J. Reprod. Fertil.* 21:133–141.

———. 1972a. Female reproduction cycles and social behavior in primates. *Adv. Study Behav.* 105:69–105.

———. 1972b. *The Social Behaviour of Monkeys.* Harmondsworth, England: Penguin Books.

———. 1973. Social organization of wild talapoin monkeys. *Am. J. Phys. Anthropol.* 38:593–597.

———. 1979. How would we know if social organization were *not* adaptive? In *Primate Ecology and Human Origins: Ecological Influences on Social Organization,* edited by I. S. Bernstein and E. O. Smith, 1–22. New York: Garland.

Rowell, T. E., and K. M. Hartwell. 1978. The interaction of behavior and reproductive cycles in patas monkeys. *Behav. Biol.* 24:141–167.

Rozin, P. 1969. Adaptive food sampling patterns in vitamin deficient rats. *J. Comp. Physiol. Psychol.* 69:126–132.

Rudran, R. 1973. Adult male replacement in one-male troops of purple-faced langurs *(Presbytis senex senex)* and its effect on population structure. *Folia Primatol.* 19:166–192.

———. 1978a. Intergroup dietary comparisons and folivorous tendencies of two groups of blue monkeys *(Cercopithecus mitis stuhlmanni).* In *The Ecology of Arboreal Folivores,* edited by G. G. Montgomery, 483–504. Washington, D.C.: Smithsonian Institution.

———. 1978b. Socioecology of the blue monkeys *(Cercopithecus mitis stuhlmanni)* of the Kibale Forest, Uganda. *Smithson. Contribu. Zool.* 249:1–88.

———. 1979. The demography and social mobility of a red howler *(Alouatta seniculus)* population in Venezuela. In *Vertebrate Ecology in the Northern Neotropics,* edited by J. F. Eisenberg, 107–126. Washington, D.C.: Smithsonian Institution.

Russell, R. J. 1975. Body temperatures and behavior of captive cheirogaleids. In *Lemur Biology*, edited by I. Tattersall and R. W. Sussman, 193–206. New York: Plenum Press.

———. 1977. The behavior, ecology, and environmental physiology of a noctornal primate, *Lepilemur mustelinus*. (Strepsirhini, Lemuriformes, Lepilemuridae). Ph.D. diss., Duke University.

Ryan, K. J., and B. R. Hopper. 1974. Placental biosynthesis and metabolism of steroid hormones in primates. In *Reproductive Biology of the Primates*, edited by W. P. Luckett, 258–283. Vol. 3 of *Contributions to Primatology*. Basel: Karger.

Sacher, G. A. 1970. Allometric and factorial analysis of brain structure in insectivores and primates. In *Advances in Primatology*, edited by C. Noback and W. Montagna, 245–287. Vol. 1 of *The Primate Brain*. New York: Appleton-Century-Crofts.

———. 1974. Maturation and longevity in relation to cranial capacity in hominid evolution. In *Primate Functional Morphology and Evolution*, edited by R. Tuttle, 417–442. The Hague: Mouton.

Sacher, G. A., and E. F. Staffeldt. 1974. Relation of gestation time to brain weight for placental mammals: implications for the theory of vertebrate growth. *Am. Nat.* 108:583–615.

Sade, D. S. 1965. Some aspects of parent-offspring and sibling relations in a group of rhesus monkeys with a discussion of grooming. *Am. J. Phys. Anthropol.* 23:1–18.

———. 1967. Determinants of dominance in a group of free-ranging rhesus monkeys. In *Social Communication Among Primates*, edited by S. Altmann, 99–114. Chicago: University of Chicago Press.

———. 1972a. Longitudinal study of social behavior of rhesus monkeys. In *The Functional and Evolutionary Biology of Primates*, edited by R. H. Tuttle, 378–398. Chicago: Aldine/Atherton.

———. 1972b. Sociometrics of *Macaca mulatta*. I. Linkages and cliques in grooming matrices. *Folia Primatol.* 18:196–223.

———. 1980. Population biology of free-ranging rhesus monkeys on Cayo Santiago, Puerto Rico. In *Biosocial Mechanisms of Population Regulation*, edited by M. N. Cohen, R. S. Malpass, and H. G. Klein, 171–187. New Haven: Yale University Press.

Sade, D. S., K. Cushing, P. Cushing, J. Dunaif, A. Figueroa, J. Kaplan, C. Lauer, D. Rhoades, and J. Schneider. 1976. Population dynamics in relation to social structure on Cayo Santiago. *Yearb. Phys. Anthropol.* 20:253–262.

Sadleir, R. M. F. S. 1969a. *The Ecology of Reproduction in Wild and Domestic Mammals*. London: Methuen.

———. 1969b. The role of nutrition in the reproduction of wild mammals. *J. Reprod. Fertil. (Suppl.)* 6:39–48.

Schaffer, W. M. 1974. Optimal reproductive effort in fluctuating environments. *Am. Nat.* 108:783–790.

Schaller, G. B. 1963. *The Mountain Gorilla: Ecology and Behavior*. Chicago: University of Chicago Press.

———. 1972. *The Serengeti Lion: A Study of Predator-Prey Relations*. Chicago: University of Chicago Press.

Schmidt-Nielsen, K. 1975. *Animal Physiology: Adaptation and Environment.* Cambridge: Cambridge University Press.

Schneider, D. G., L. D. Mech, and J. R. Tester. 1971. Movements of female raccoons and their young as determined by radio-tracking. In *Anim. Behav. Monogr.* 4(1):3–43.

Schoener, T. W. 1971. Theory of feeding strategies. *Annu. Rev. Ecol. Syst.* 2:369–404.

———. 1974. Resource partitioning in ecological communities. *Science* 185:27–39.

———. 1982. The controversy over interspecific competition. *Am. Sci.* 70:586–595.

———. 1983. Field experiments on interspecific competition. *Am. Nat.* 122:240–285.

Scholander, P. F. 1955. Evolution of climatic adaptation in homeotherms. *Evolution* 9:15–26.

Scholander, P. F., V. Walters, R. Hock, and L. Irving. 1950. Body insulation of some arctic and tropical mammals and birds. *Biol. Bull.* 99:225–236.

Schoonover, L. J., and W. H. Marshall. 1951. Food habits of the raccoon *(Procyon lotor hirtus)* in north-central Minnesota. *J. Mammal.* 32:422–428.

Schulman, S., and B. Chapais. 1980. Reproductive value and rank relations among macaque sisters. *Am. Nat.* 115:580–593.

Schultz, A. H. 1969. *The Life of Primates.* London: Weidenfeld and Nicolson.

Seyfarth, R. M. 1976. Social relationships among adult female baboons. *Anim. Behav.* 24: 917–938.

Seyfarth, R. M., D. L. Cheney, and R. A. Hinde. 1978. Some principles relating social interactions and social structure among primates. In *Recent Advances in Primatology,* vol. 1, edited by D. J. Chivers and J. Herbert, 39–51. London: Academic Press.

Seyfarth, R. M., D. L. Cheney, and P. Marler. 1980. Monkey responses to three different alarm calls: evidence of predator classification and semantic communication. *Science* 210:801–803.

Shackleton, N. J., and J. P. Kennett. 1975. Late Cenozoic oxygen and carbon isotopic changes at DSDP site 284: implications for glacial history of the Northern Hemisphere and Antarctica. *Initial Rep. Deep Sea Drill. Proj.* 29:801–807.

Sharon, N. 1980. Carbohydrates. *Sci. Amer.* 243(5):90–116.

Sharp, W. M., and L. H. Sharp. 1956. Nocturnal movements and behavior of wild raccoons at a winter feeding station. *J. Mammal.* 37:170–177.

Shaul, D. M. B. 1962. The composition of milk of wild animals. *Int. Zoo. Yearb.* 4:333–345.

Sheine, W. S. 1979. The effect of variations in molar morphology on masticatory effectiveness and digestion of cellulose in prosimian primates. Ph.D. diss., Duke University.

Sheine, W. S., and R. F. Kay. 1977. An analysis of chewed food particle size and its relationship to molar structure in the primates *Cheirogaleus medius* and *Galago senegalensis* and the insectivoran *Tupaia glis. Am. J. Phys. Anthropol.* 47:15–20.

Shelford, V. E., and A. C. Twomey. 1941. Tundra animal communities in the vicinity of Churchill, Manitoba. *Ecology* 22:47–69.

Short, R. V. 1976. The evolution of human reproduction. *Proc. R. Soc. Lond. B Biol. Sci.* 195:3–24.

Simberloff, D. 1983. Competition theory, hypothesis testing, and other community ecological buzzwords. *Am. Nat.* 122:626–635.

Simons, E. L. 1970. The deployment and history of Old World monkeys (Cercopithecidae, Primates). In *Old World Monkeys,* edited by J. R. Napier and P. H. Napier, 99–137. New York: Academic Press.

———. 1972. *Primate Evolution: An Introduction to Man's Place in Nature.* New York: Macmillan.

Simons, E. L., and D. R. Pilbeam. 1965. Preliminary revision of the Dryopithecinae (Pongidae, Anthropoidea). *Folia Primatol.* 3:81–152.

Simpson, G. G. 1945. The principles of classification and a classification of mammals. *Bull. Am. Mus. Nat. Hist.* 85:1–350.

———. 1951. *Horses.* New York: Oxford University Press.

———. 1953. *Evolution and Geography: An Essay on Historical Biogeography with Special Reference to Mammals.* Eugene: University of Oregon Press.

———. 1980. *Splendid Isolation: The Curious History of South American Mammals.* New Haven: Yale University Press.

Simpson, M. J. A. 1973. The social grooming of male chimpanzees. In *Comparative Ecology and Behaviour of Primates,* edited by R. P. Michael and J. H. Crook, 411–505. New York: Academic Press.

Slobodkin, L. B. 1961. *Growth and Regulation of Animal Populations.* New York: Holt.

Slobodkin, L. B., and A. Rapoport. 1974. An optimal strategy of evolution. *Q. Rev. Biol.* 49:181–200.

Slobodkin, L. B., F. E. Smith, and N. G. Hairston. 1967. Regulation in terrestrial ecosystems, and the implied balance of nature. *Am. Nat.* 101:109–124.

Smalley, R. L., and R. L. Dryer. 1967. Brown fat in hibernation. In *Mammalian Hibernation,* vol. 3, 325–345, edited by K. C. Fisher, A. R. Dawe, C. P. Lyman, E. Schonbaum, and F. E. South, Jr. Proceedings 3d International Symposium on Natural Mammalian Hibernation. New York: American Elsevier.

Smith, A. H., T. M. Butler, and N. Pace. 1975. Weight growth of colony-reared chimpanzees. *Folia Primatol.* 24:29–59.

Smith, C. C. 1977. Feeding behaviour and social organization in howling monkeys. In *Primate Ecology,* edited by T. H. Clutton-Brock, 97–126. London: Academic Press.

Smith, N. B., and F. S. Barkalow. 1967. Precocious breeding in the gray squirrel. *J. Mammal.* 48:328–330.

Smith, R. J. 1980. Rethinking allometry. *J. Theor. Biol.* 87:97–111.

Smythe, N. 1970. Relationships between fruiting seasons and seed dispersal methods in a neo-tropical forest. *Am. Nat.* 104:25–35.

Snodderly, D. M. 1978. Color discriminations during food foraging by a New World monkey. In *Recent Advances in Primatology,* vol. 1, 369–371, edited by D. J. Chivers and J. Herbert. London: Academic Press.

———. 1979. Visual discriminations encountered in food foraging by a neotropical primate: implications for the evolution of color vision. In *Behavioral Significance of Color,* edited by E. H. Burtt, Jr., 237–279. New York: Garland.

Snow, D. W. 1971. Evolutionary aspects of fruit-eating by birds. *Ibis* 113:194–202.

Soest, P. J. Van. 1981. *Nutritional Ecology of the Ruminant.* Corvallis, Oreg.: O & B Books.

Soper, J. D. 1941. History, range, and home life of the northern bison. *Ecol. Monogr.* 11:347–412.

———. 1944. The mammals of southern Baffin Island, Northwest Territories, Canada. *J. Mammal.* 25:221–254.

Soule, M., and B. R. Stewart. 1970. The "niche-variation" hypothesis: a test and alternatives. *Am. Nat.* 104:85–97.

Southwick, C. H. 1980. Rhesus monkey populations in India and Nepal: patterns of growth, decline, and natural regulation. In *Biosocial Mechanisms of Population Regulation,* edited by M. N. Cohen, R. S. Malpass, and H. G. Klein, 151–170. New Haven: Yale University Press.

Southwick, C. H., M. A. Beg, and M. R. Siddiqi. 1965. Rhesus monkeys in North India. In *Primate Behavior: Field Studies of Monkeys and Apes,* edited by I. DeVore, 111–159. New York: Holt, Rinehart and Winston.

Southwick, C. H., and M. F. Siddiqi. 1976. Demographic characteristics of semi-protected rhesus groups in India. *Yearb. Phys. Anthropol.* 20:242–252.

———. 1977. Population dynamics of rhesus monkeys in northern India. In *Primate Conservation,* edited by H.S.H. Prince Rainier III and G. Bourne, 339–362. New York: Academic Press.

Southwood, T. R. E. 1977. Habitat, the templet for ecological strategies. *J. Anim. Ecol.* 46:337–365.

Stacey, P. B. 1982. Female promiscuity and male reproductive success in social birds and mammals. *Am. Nat.* 120:51–64.

Stearns, S. C. 1976. Life-history tactics: a review of the ideas. *Q. Rev. Biol.* 51:3–47.

———. 1977. The evolution of life history traits. *Annu. Rev. Ecol. Syst.* 8:145–171.

Stern, J. T., Jr., and C. E. Oxnard. 1973. Primate locomotion: some links with evolution and morphology. *Primatologia, Handbook of Primatology* 4(11):1–93.

Stoltz, L. F., and M. E. Keith. 1973. A population survey of chacma baboons in the Northern Transvaal. *J. Hum. Evol.* 2:195–212.

Strong, D., L. Szyska, and D. Simberloff. 1979. Tests of community-wide character displacement against null hypotheses. *Evolution* 33:897–913.

Struhsaker, T. T. 1967a. Ecology of vervet monkeys *(Cercopithecus aethiops)* in the Masai–Amboseli Game Reserve, Kenya. *Ecology* 48:891–904.

———. 1967b. Social structure among vervet monkeys *(Cercopithecus aethiops). Behaviour* 29:83–121.

———. 1969. Correlates of ecology and social organization among African cercopithecines. *Folia Primatol.* 11:80–118.

———. 1973. A recensus of vervet monkeys in the Masai–Amboseli Game Reserve, Kenya. *Ecology* 54: 930–932.

———. 1975. *The Red Colobus Monkey.* Chicago: University of Chicago Press.

———. 1976. A further decline in numbers of Amboseli vervet monkeys. *Biotropica* 8(3): 211–214.

———. 1977. Infanticide and social organization in the redtail monkey *(Cercopithecus ascanius schmidti)* in the Kibale Forest, Uganda. *Z. Tierpsychol.* 45:75–84.

———. 1978a. Food habits of five monkey species in the Kibale Forest, Uganda. In *Recent Advances in Primatology,* edited by D. J. Chivers and J. Herbert. New York: Academic Press.

———. 1978b. Interrelations of red colobus monkeys and rain forest trees in the Kibale Forest, Uganda. In *The Ecology of Arboreal Folivores,* edited by G. G. Montgomery, 397–422. Symposia of the National Zoological Park. Washington, D.C.: Smithsonian Institution Press.

———. 1982. Polyspecific associations among tropical rain-forest primates. *Z. Tierpsychol.* 57:268–304.

Struhsaker, T. T., and J. S. Gartlan. 1970. Observations on the behaviour and ecology of the patas monkey *(Erythrocebus patas)* in the Waza Reserve, Cameroon. *J. Zool. (Lond.)* 161:49–63.

Struhsaker, T. T., and L. Leland. 1979. Socioecology of five sympatric monkey species in the Kibale Forest, Uganda. *Adv. Study Behav.* 9:159–228.

Struhsaker, T. T., and J. F. Oates. 1975. Comparison of the behavior and ecology of red colobus and black-and-white colobus monkeys in Uganda: a summary. In *Socioecology and Psychology of Primates,* edited by R. H. Tuttle, 103–123. The Hague: Mouton.

Strum, S. C. 1981. Processes and products of change: baboon predatory behavior at Gilgil, Kenya. In *Omnivorous Primates: Gathering and Hunting in Human Evolution,* edited by R. Harding and G. Teleki, 255–302. New York: Columbia University Press.

———. 1982. Agonistic dominance in male baboons: an alternative view. *Int. J. Primatol.* 3:175–202.

Strum, S. C., and J. D. Western. 1982. Variations in fecundity with age and environment in olive baboons *(Papio anubis). Am. J. Primatol.* 3:61–76.

Stubbs, M. 1977. Density dependence in the life-cycles of animals and its importance in *K*- and *r*-strategies. *J. Anim. Ecol.* 46:677–688.

Stuewer, F. W. 1943. Raccoons: their habits and management in Michigan. *Ecol. Monogr.* 13(2):203–257.

Sugawara, K. 1979. Sociological study of a wild group of hybrid baboons between *Papio anubis* and *P. hamadryas* in the Awash Valley, Ethiopia. *Primates* 20:21–56.

Sugiyama, Y. 1964. Group composition, population density and some sociological observations of hanuman langurs *(Presbytis entellus). Primates* 5:7–38.

———. 1965a. Behavioral development and social structure in two troops of hanuman langurs *(Presbytis entellus). Primates* 6:213–248.

———. 1965b. On the social change of hanuman langurs *(Presbytis entellus)* in their natural condition. *Primates* 6:381–418.

———. 1968. Social organization of chimpanzees in the Budongo forest, Uganda. *Primates* 9:225–258.

———. 1969. Social behavior of chimpanzees in the Budongo Forest, Uganda. *Primates* 10:197–225.

———. 1973. The social structure of wild chimpanzees: a review of field studies. In *Comparative Ecology and Behaviour of Primates,* edited by R. P. Michael and J. H. Crook, 375–410. London: Academic Press.

Sugiyama, Y., and H. Ohsawa. 1982. Population dynamics of Japanese monkeys with special references to the effect of artificial feeding. *Folia Primatol.* 39:238–263.

Sussman, R. W. 1974. Ecological distinctions in sympatric species of lemurs. In *Prosimian Biology,* edited by R. D. Martin, G. A. Doyle, and A. C. Walker, 75–108. London: Duckworth.

———. 1977. Feeding behaviour of *Lemur catta* and *Lemur fulvus.* In *Primate Ecology,* edited by T. H. Clutton-Brock, 1–37. London: Academic Press.

———. 1978. Nectar-feeding by prosimians and its evolutionary and ecological implications. In *Recent Advances in Primatology,* vol. 3, *Evolution,* edited by D. J. Chivers and K. A. Joysey, 119–124. New York: Academic Press.

———. In press. *Introduction to Primate Ecology.* New York: McMillian.

Sussman, R. W., and W. G. Kinzey. In press. The ecological role of the Callitrichidae: a review. *Am. J. Phys. Anthropol.*

Sussman, R. W., and P. H. Raven. 1978. Pollination by lemurs and marsupials: an archaic coevolutionary system. *Science* 200:731–736.

Suzuki, A. 1965. An ecological study of wild Japanese monkeys in snowy areas, focused on their food habits. *Primates* 6:31–72.

———. 1969. An ecological study of chimpanzees in savanna woodland. *Primates* 10:103–148.

———. 1971. Carnivory and cannibalism observed among forest-living chimpanzees. *J. Anthropol. Soc. Nippon* 79:30–48.

Swain, T. 1977. Secondary compounds as protective agents. *Annu. Rev. Plant Physiol.* 28:479–501.

Swift, R. W. 1948. Deer select most nutritious forages. *J. Wildl. Manage.* 12:109–110.

Szalay, F. S. 1968. The beginnings of primates. *Evolution* 22:19–36.

———. 1972. Paleobiology of the earliest primates. In *The Functional and Evolutionary Biology of Primates,* edited by R. Tuttle, 3–35. Chicago: Aldine/Atherton.

Szalay, F. S., and E. Delson. 1979. *Evolutionary History of the Primates.* New York: Academic Press.

Tansley, A. G. 1935. The use and abuse of vegetational concepts and terms. *Ecology* 16:284–307.

Tappen, N. C. 1960. Problems of distribution and adaptation of the African monkeys. *Curr. Anthropol.* 1:91–120.

Tarling, D. H. 1978. Plate tectonics: present and past. In *Evolution of the Earth's Crust,* edited by D. H. Tarling, 361–408. New York: Academic Press.

———. 1980. The geologic evolution of South America with special reference to the last 200 million years. In *Evolutionary Biology of the New World Monkeys and Continental Drift,* edited by R. L. Ciochon and A. B. Chiarelli, 1–41. New York: Plenum Press.

Tattersall, I. 1969a. Ecology of north Indian *Ramapithecus. Nature* 221:451–452.

———. 1969b. More on the ecology of north Indian *Ramapithecus. Nature* 224:821–822.

———. 1982. *The Primates of Madagascar.* New York: Columbia University Press.

Taub, D. M. 1977. Geographic distribution and habitat diversity of the Barbary macaque *Macaca sylvanus* L. *Folia Primatol.* 27:108–133.

Taylor, C. R., K. Schmidt-Nielsen, and J. L. Raab. 1970. Scaling of energetic cost of running to body size in mammals. *Am. J. Physiol.* 219:1104–1107.

Taylor, W. P. 1935. Ecology and life history of the porcupine *(Erethizon epixanthum)* as related to the forests of Arizona and the south-western United States. *Univ. Ariz. Bull. Biol. Sci. Bull.* 3.

Teas, J., T. L. Richie, H. G. Taylor, M. F. Siddiqi, and C. H. Southwick. 1981. Natural regulation of rhesus monkey populations in Kathmandu, Nepal. *Folia Primatol.* 35:117–123.

Teleki, G. 1973. *The Predatory Behavior of Wild Chimpanzees.* Lewisburg, Pa.: Bucknell University Press.

Teleki, G., E. E. Hunt, Jr., and J. H. Pfifferling. 1976. Demographic observations (1963–1973) on the chimpanzees of Gombe National Park, Tanzania. *J. Hum. Evol.* 5:559–598.

Temerin, L. A., and J. G. H. Cant. 1983. The evolutionary divergence of Old World monkeys and apes. *Am. Nat.* 122:335–351.

Temple, S. 1977. Plant-animal mutualism: coevolution with dodo leads to near extinction of plant. *Science* 197:885–886.

Tenaza, R. R. 1975. Territory and monogamy among Kloss's gibbons *(Hylobates klossi)* in Siberut Island, Indonesia. *Folia Primatol.* 24:60–80.

Terborgh, J. 1983. *The Behavioral Ecology of Five New World Primates.* Princeton: Princeton University Press.

Terborgh, J., and A. C. Wilson. In press. Tamarins: new evidence on the social system. *Nature.*

Thorington, R. W., Jr. 1967. Feeding and activity of *Cebus* and *Saimiri* in a Colombian forest. In *Progress in Primatology,* edited by D. Starck, R. Scheider, and H. J. Kuhn, 180–184. Stuttgart: Fischer.

Thorington, R. W., Jr., and C. P. Groves. 1970. An annotated classification of the Cercopithecoidea. In *Old World Monkeys: Evolution, Systematics, and Behavior,* edited by J. R. Napier and P. H. Napier, 629–647. New York: Academic Press.

Tlukherjee, A. K., and S. Gupta. 1965. Habits of the rhesus macaque, *Macaca mulatta* (Zimmerman), in the Sunderbans, 24-Parganas, West Bengal. *J. Bombay Nat. Hist. Soc.* 62:145–146.

Tokura, H., F. Hara, M. Okada, F. Mekata, and W. Ohsawa. 1975. Thermoregulatory responses at various ambient temperatures in some primates. In *Contemporary Primatology,* edited by S. Kondo, M. Kawai, and A. Ehara, 171–176. Basel: Karger.

Trivers, R. L. 1971. The evolution of reciprocal altruism. *Quart. Rev. Biol.* 46:35–37.

———. 1972. Parental investment and sexual selection. In *Sexual Selection and the Descent of Man,* edited by B. G. Campbell, 136–179. Chicago: Aldine.

Tucker, V. A. 1970. Energetic cost of locomotion in animals. *Comp. Biochem. Physiol.* 34:841–846.

Tutin, C. E. G. 1979. Mating patterns and reproductive strategies in a community of wild chimpanzees. *Behav. Ecol. Sociobiol.* 6:29–38.

Tutin, C. E. G., and P. R. McGinnis. 1981. Chimpanzee reproduction in the wild. In *Reproductive Biology of the Great Apes: Comparative and Biomedical Perspectives,* edited by C. E. Graham, 239–264. New York: Academic Press.

Vaitl, E. 1977a. Experimental analysis of the nature of social context in captive groups of squirrel monkeys *(Saimiri sciureus). Primates* 18:849–859.

———. 1977b. Social context as a structuring mechanism in captive groups of squirrel monkeys *(Saimiri sciureus). Primates* 18:861–874.

———. 1978. Nature and implications of the complexly organized social system in nonhuman primates. In *Recent Advances in Primatology,* vol. 1, edited D. J. Chivers and J. Herbert. London: Academic Press.

Van Couvering, J. A. 1980. Community evolution in East Africa during the late Cenozoic. In *Fossils in the Making: Vertebrate Taphonomy and Paleoecology,* edited by A. K. Behrensmeyer and A. P. Hill, 272–298. Chicago: University of Chicago Press.

Van Couvering, J. A., and J. Van Couvering. 1976. Early Miocene mammal fossils from East Africa. In *Human Origins,* edited by G. Isaac and E. McCown, 155–208. Menlo Park, Calif.: Benjamin/Cummings.

Vandermeer, J. H. 1975. Interspecific competition: a new approach to the classical theory. *Science* 188:253–255.

Van der Pijl, L. 1972. *Principles of Dispersal in Higher Plants.* New York: Springer-Verlag.

Van Horn, R. N. 1975. Primate breeding season: photoperiodic regulation in captive *Lemur catta*. *Folia Primatol.* 24:203–220.

———. 1980. Seasonal reproductive patterns in primates. In *Seasonal Reproduction in Higher Vertebrates,* edited by R. J. Reiter and B. K. Follett, 181–221. Vol. 5 of *Progress in Reproductive Biology.* Basel: Karger.

Vanzolini, P. E. 1973. Paleoclimates, relief and species multiplication. In *Tropical Forest Ecosystems in Africa and South America: A Comparative Review,* edited by B. J. Meggers, E. S. Ayensu, and W. D. Duckworth, 255–258. Washington, D.C.: Smithsonian Institution Press.

Vaughan, T. A. 1972. *Mammalogy.* Philadelphia: Saunders.

Vehrencamp, S. L. 1979. The roles of individual, kin, and group selection in the evolution of sociality. In *Handbook of Behaviorial Neurobiology: Social Behavior and Communication,* edited by P. Marler and J. C. Vandebergh, 351–394. New York: Plenum Press.

Vincent, F. 1968. La sociabilité du Galago de Demidoff. *Terre Vie* 22:51–56.

———. 1969. Contribution à l'étude des prosimiens africains. Le Galago de Demidoff. Thèse de Doctorat d'Etat, Paris. CNRS No. AO3575.

Vuilleumier, B. S. 1971. Pleistocene changes in the flora and fauna of South America. *Science* 173:771–780.

Wade, M. J. 1979. The evolution of social interactions by family selection. *Am. Nat.* 113:399–417.

Wagenen, G. van, and Simpson, M. E. 1965. *Embryology of the ovary and testis* Homo sapiens *and* Macaca mulatta. New Haven: Yale University Press.

Walker, A. C. 1967. Patterns of extinction among the subfossil Madagascan lemuroids. In *Pleistocene Extinctions: The Search for a Cause,* edited by P. S. Martin and H. E. Wright, 425–432. New Haven: Yale University Press.

Walker, A., and R. E. F. Leakey. 1978. The hominids of East Turkana. *Sci. Amer.* 239(2):54–66.

Walker, E. P., F. Warnick, S. E. Hamlet, K. I. Lange. M. A. Davis, H. E. Uible, and P. F. Wright. 1964. *Mammals of the World.* 2 vols. Baltimore: Johns Hopkins University Press.

Walker, P., and P. Murray. 1975. An assessment of masticatory efficiency in a series of anthropoid primates with special reference to the Colobinae and Cercopithecinae. In *Primate Functional Morphology and Evolution,* edited by R. H. Tuttle, 135–150. The Hague: Mouton.

Ward, S. C., and D. R. Pilbeam. 1983. Maxillofacial morphology of the Miocene hominoids from Africa and Indo-Pakistan. In *New Interpretations of Ape and Human Ancestry,* edited by R. L. Ciochon and R. S. Corruccini, 211–238. New York: Plenum Press.

Ward, S. C., and R. W. Sussman. 1979. Correlates between locomotor anatomy and behavior in two sympatric species of *Lemur. Am. J. Phys. Anthropol.* 50: 575–590.

Waser, P. M. 1975. Monthly variations in feeding and activity patterns of the mangabey, *Cercocebus albigena* (Lyddeker). *East Afr. Wildl. J.* 13:249–263.

———. 1976. *Cercocebus albigena:* site attachment, avoidance, and intergroup spacing. *Am. Nat.* 110:911–935.

————. 1977a. Feeding, ranging and group size in the mangabey *Cercocebus albigena*. In *Primate Ecology*, edited by T. H. Clutton-Brock, 183–222. New York: Academic Press.

————. 1977b. Individual recognition, intragroup cohesion, and intergroup spacing: evidence from sound playback to forest monkeys. *Behavior* 60:28–74.

————. 1982. Primate polyspecific associations: do they occur by chance? *Anim. Behav.* 30:1–8.

Waser, P. M., and O. Floody. 1974. Ranging patterns of the mangabey *Cercocebus albigena* in the Kibale Forest, Uganda. *Z. Tierpsychol.* 35:85–101.

Waser, P. M., and R. H. Wiley. 1980. Mechanism and evolution of spacing in animals. In *Handbook of Behavioral Neurobiology*, vol. 3, edited by P. Marler and J. G. Vandenbergh, 159–233. New York: Plenum Press.

Washburn, S. L., and D. A. Hamburg. 1965. The implications of primate research. In *Primate Behavior: Field Studies of Monkeys and Apes*, edited by I. DeVore, 607–622. New York: Holt.

Watt, K. E. F. 1966. The nature of systems analysis. In *Systems Analysis in Ecology*, edited by K. E. F. Watt, 1–14. New York: Academic Press.

Weir, B. J., and I. W. Rowlands. 1973. Reproductive strategies of mammals. *Annu. Rev. Ecol. Syst.* 4:139–163.

Western, D. 1979. Size, life history and ecology in mammals. *Afr. J. Ecol.* 17:185–204.

Western, D., and D. M. Sindiyo. 1972. The status of the Amboseli rhino population. *East Afr. Wildl. J.* 10:43–57.

Western, D., and C. Van Praet. 1973. Cyclical changes in the habitat and climate of an East African ecosystem. *Nature* 241:104–106.

Westoby, M. 1974. An analysis of diet selection by large generalist herbivores. *Am. Nat.* 108:290–304.

Wheatley, B. P. 1978. Foraging patterns in a group of longtailed macaques in Kalimantan Timur, Indonesia. In *Recent Advances in Primatology*, vol. 1, *Behaviour*, edited by D. J. Chivers and J. Herbert, 347–350. New York: Academic Press.

Wheelwright, N. T., and G. H. Orians. 1982. Seed dispersal by animals: contrasts with pollen dispersal, problems of terminology, and constraints on coevolution. *Am. Nat.* 199:402–413.

White, T. D., D. C. Johnson, and W. H. Kimbel. 1981. *Australopithecus africanus*: its phyletic position reconsidered. *S. Afr. J. Sci.* 77:445–470.

Whittaker, R. H. 1967. Gradient analysis of vegetation. *Biol. Rev.* 42:207–264.

————. 1975. *Communities and Ecosystems*, 2d ed. New York: Macmillan.

Whittaker, R. H., and S. A. Levin. 1976. *Niche: Theory and Application*. Stroudsburg, Pa.: Dowden, Hutchinson and Rose.

Wiegert, R. G. 1974. Competition: a theory based on realistic general equations of population growth. *Science* 184:539–552.

Wiens, J. A. 1977. On competition and variable environments. *Am. Sci.* 65:590–597.

Williams, G. C. 1975. *Sex and Evolution*. Princeton: Princeton University Press.

Wilson, E. O. 1975. *Sociobiology: The New Synthesis*. Cambridge, Mass.: Belknap.

Wilson, E. O., and W. H. Bossert. 1971. *A Primer of Population Biology*. Sunderland, Mass.: Sinauer.

Wilson, J. T., ed. 1976. *Continents Adrift, Continents Aground*. San Francisco: W. H. Freeman and Company.

Wimsatt, W. A. 1960. An analysis of parturition in Chiroptera, including new observations on *Myotis lucifugus*. *J. Mammal.* 41:183–200.

Wimsatt, W. C. 1980. Reductionistic research strategies and their biases in the units of selection controversy. In *Scientific Discovery: Case Studies,* edited by T. Nickko, 213–259. Dordrecht, Netherlands: Reidel.

Winterhalder, B. 1980. Hominid paleoecology: The competitive exclusion principle and determinants of niche relationships. *Yearb. Phys. Anthropol.* 23:43–63.

———. 1981. Hominid paleoecology and competitive exclusion: limits to similarity, niche differentiation and the effects of cultural behavior. *Yearb. Phys. Anthropol.* 24:101–121.

Wittenberger, J. F. 1980. Group size and polygamy in social mammals. *Am. Nat.* 115:197–222.

Wolf, K., and J. Fleagle. 1977. Adult male replacement in a group of silvered leaf monkeys *(Presbytis cristata)* at Kuala Selangor, Malaysia. *Primates* 18:949–955.

Wolfe, J. A. 1978. A paleobotanical interpretation of Tertiary climates in the Northern Hemisphere. *Am. Sci.* 66:694–703.

Wolfe, J. A., and D. M. Hopkins. 1967. Climatic changes recorded by Tertiary land floras in northwestern North America. In *Tertiary Correlations and Climatic Changes in the Pacific,* edited by K. Hatai, 67–76. Sendai, Japan: Sasaki.

Wolfheim, J. H. 1983. *Primates of the World*. Seattle: University of Washington Press.

Wondoleck, J. 1978. Forage-area separation and overlap in heteromyid rodents. *J. Mammal.* 59:510–518.

Woolfenden, G. E., and J. W. Fitzpatrick. 1978. The inheritance of territory in group-breeding birds. *Bioscience* 28:104–108.

Wrangham, R. W. 1974. Artificial feeding of chimpanzees and baboons in their natural habitat. *Anim. Behav.* 22:83–93.

———. 1976. Aspects of feeding and social behavior in gelada baboons. Science Research Council, London: Mimeo.

———. 1977. Feeding behaviour of chimpanzees in Gombe National Park, Tanzania. In *Primate Ecology,* edited by T. H. Clutton-Brock, 504–538. London: Academic Press.

———. 1979. Sex differences in chimpanzee dispersion. In *The Great Apes,* edited by D. A. Hamburg and E. R. McCown, 481–490. Menlo Park, Calif.: Benjamin/Cummings.

———. 1980. An ecological model of female-bonded primate groups. *Behaviour* 75:262–300.

———. 1981. Drinking competition in vervet monkeys. *Anim. Behav.* 29:904–910.

Wrangham, R. W., and B. B. Smuts. 1980. Sex differences in the behavioral ecology of chimpanzees in the Gombe National Park, Tanzania. *J. Reprod. Fert. Suppl.* 28:13–31.

Wrangham, R. W., and P. G. Waterman. 1981. Feeding behaviour of vervet monkeys on *Acacia tortilis* and *Acacia xanthophloea:* with special reference to reproductive strategies and tannin production. *J. Anim. Ecol.* 50:715–731.

Wright, P. C. 1978. Home range, activity pattern, and agonistic encounters of a group of night monkeys *(Aotus trivirgatus)* in Peru. *Folia Primatol.* 29:43–55.

Yerkes, R. M. 1916. Provision for the study of monkeys and apes. *Science,* n.s., 43:231–234.

Yerkes, R. M., and J. H. Elder. 1936. The sexual cycle of the chimpanzee. *Anat. Rec.* 67:119–143.

Yerkes, R. M., and A. W. Yerkes. 1929. *The Great Apes: A Study of Anthropoid Life.* New Haven: Yale University Press.

Yoshiba, K. 1968. Local and intertroop variability in ecology and social behavior of common Indian langurs. In *Primates: Studies in Adaptation and Variability,* edited by P. C. Jay, 217–242. New York: Holt, Rinehart and Winston.

Young, W. C., and R. M. Yerkes. 1943. Factors influencing the reproductive cycle in the chimpanzee; the period of adolescent sterility and related problems. *Endocrinology* 33:121–154.

Youngson, R. W. 1970. Rearing red deer calves in captivity. *J. Wildl. Manage.* 34:467–470.

Zuckerman, S. 1932. *The Social Life of Monkeys and Apes.* New York: Harcourt, Brace.

———. 1933. *Functional Affinities of Man, Monkeys, and Apes.* New York: Harcourt, Brace.

Zuckerman, S., and A. S. Parkes. 1932. The menstrual cycle of primates. V. The cycle of the baboon. *Proc. Zool. Soc. Lond.* 102:139–191.

Zuckerman, S., G. van Wagenen, and F. H. Gardiner. 1938. The sexual skin of the rhesus monkey. *Proc. Zool. Soc. Lond.* 108:385–401.

Glossary

Adaptive radiation: A set of species that exploit an environment in many different ways, having evolved from a single ancestral species.

Alkaloid: A class of secondary compounds, some of which are toxic to certain animals.

Altricial infants: Those born at an early stage of development, who require care after birth.

Amino acids: The constituents of protein molecules, composed of carbon, hydrogen, oxygen, nitrogen, and, in some cases, sulfur and phosphorus.

Angiosperm: A flowering plant.

Basal metabolic rate (BMR): The rate at which energy is exchanged in basic life processes.

Biomass: The weight of organic matter present at a given time per unit area of the earth's surface (also called standing crop).

Biome: The largest land community unit that is convenient to recognize.

Browse: Plant food that is high in fiber and low in nutrients (e.g., bark, woody tissue, mature leaves).

Cambium: The site of new growth in a woody stem, producing wood toward the inside and phloem toward the outside.

Carnivore: (1) a member of the order Carnivora; (2) an animal that eats primarily or exclusively meat.

Cellulose: The most abundant of all carbohydrates, cellulose is the major structural polysaccharide in plants.

Character displacement: A morphological change in one or two species that enables them to coexist sympathetically.

Chitin: A polysaccharide found in the shell of insects, with characteristics intermediate between those of structural and nonstructural carbohydrates.

Circulation: The movement of animals between social groups within a study population.

Climax: The stable vegetation type that eventually characterizes an area in the absence of disturbance.

Coevolution: The process by which species with close ecological relationships exert selective pressures on one another so that their evolution is interdependent.

Community: An assemblage of populations of plants, animals, bacteria, and fungi that together form a distinctive system with its own composition, structure, development, and function.

Competitive release: Niche expansion or population growth that occurs when one species is freed from competition with another.

Complementary fraction: The portion of a food item or items made up of minerals and insoluble sugars except cellulose (i.e., predominantly the least digestible part of a food item or items).

Congeners: Species belonging to the same genus.

Consortship: A close relationship between a male and an estrous female.

Digestion: The breaking down of large molecules into smaller components that can be absorbed across the lining of the gut.

Diurnal animals: Those active by day.

Ecocline: A gradual spacial change in the characteristics of an ecological community, which is a result of a gradual change in the abiotic environment.

Ecology: The scientific study of relations between organisms and between organisms and their nonliving environment.

Ecosystem: A community of organisms and their inorganic environment.

Emergent features: Properties of a level of organization that are not predictable from the properties of lower levels of organization.

Endozoochory: Transport of seeds in the gut.

Environment: The living and nonliving (e.g., climate, soil, topography) features of an animal's surroundings.

Epiphytes: Plants that grow on branches of trees and catch nutrients from the rain and the air.

Essential amino acids: Constituents of protein that cannot be synthesized by most animals and therefore must be supplied in the diet.

Estrus: The intensification of behaviors in female mammals (except higher primates) associated with mating and characterized by changes in sexual organs.

Eutheria: Infraclass of mammals possessing a placenta in which most of the development of young takes place in the uterus.

Exozoochory: Transport of seeds without ingestion.

Fecundity: The number of female offspring to which a female is biologically capable of giving birth over a specified interval.

Fertility: The number of female offspring to which a female gives birth over a specified interval.

Fundamental niche: The total range of conditions under which a species can exist.

Generation length: The elapsed time between the birth of a female and the birth of her median offspring.

Gross primary production: The total energy bound by green plants.

Haplorhines: The suborder of primates including tarsiers, monkeys, apes, and humans.

Home range: The area in which an animal or group of animals habitually moves, feeds, and rests.

Infanticide: The killing of infants.

Life history pattern: The timing of the major phases of development and reproduction in the life of an organism.

Limiting resource: An essential resource that is in short supply and, consequently, reduces the rate at which a population can grow.

Metabolism: The chemical changes that go on in the body's tissue to maintain normal life functions.

Metatheria: The infraclass of mammals in which most of the development of the young takes place in a pouch.

Monosaccharide: A simple, water-soluble sugar.

Natal group: The social group in which an animal is born.

Net primary production: The gross primary production minus the energy used by plants in respiration.

Nocturnal animals: Those active by night.

Nonessential amino acids: Constituents of protein molecules that are essential to an animal's wellbeing but that can be synthesized by the animal: they are not essential in the diet.

Nulliparous females: Those who have not yet bred.

Nutritional content: The proportions of nutrients in a food item or diet. A food item or diet's nutritional content can be specified without reference to the animals that eat it.

Nutritional value: The nutrients that an animal can extract from a food item or diet. A food item or diet's nutritional value can only be assessed in relation to a specified set of animals.

Orbital convergence: An evolutionary trend toward frontal orientation of the eye sockets.

Oligosaccharide: A molecule containing from 2 to 10 monosaccharides (simple sugars).

Parous females: Those who have bred.

Peptide links: Links composed of carbon, nitrogen, and oxygen molecules that bind amino acids together to form large protein molecules.

Phloem: The layer of wood stems in which nutrients and waste products are transported.

Polysaccharide: A large, complex molecule containing from 11 to as many as 26,000 monosaccharides (simple sugars).

Postpartum amenorrhea: The period after giving birth when a female does not cycle sexually.

Preadaptation: A feature that evolved to fulfill one function and is subsequently used for another.

Precocial infants: Those born at an advanced stage of development.

Primary productivity: The rate at which energy is bound or organic material is created by photosynthesis.

Primitive feature: One that has undergone little or no evolutionary change away from the corresponding feature in the presumed ancestral condition.

Prototheria: The subclass of mammals whose members produce eggs from which their young hatch.

Realized niche: The actual range of conditions under which a species exists.

Reductionism: The belief that complex data and phenomena can be explained in terms of something simpler.

Ruminant: An animal that digests food through microbial fermentation in a four-chambered stomach.

Seasonally polyestrous females: Those who cycle twice or more in one season if they do not conceive in the first cycle.

Secondary productivity: The rate at which animals convert plant material for their own use.

Sexual dimorphism: Differences in size and shape between the males and females of a species.

Social dynamics: Changes in social structure over time.

Social group: A set of animals that interact regularly, know one another individually, spend most of their time nearer to one another than to nonmembers, and usually behave aggressively toward nonmembers.

Social interaction: A single behavioral act involving communication between two or more individuals. Usually interactants are members of the same social group or network and are familiar with one another.

Social network: A set of animals with the properties of a social group except that members of a network do not necessarily spend most of their time nearer to one another than to nonmembers.

Social organization: A heterogeneous set of spacing and social behaviors that characterize a population of animals of known age and sex.

Social relationship: The sum of the social interactions between two individuals over time.

Social structure: The content and quality of social relationships among all the members of a social group or network.

Specialized: Describes a feature that has undergone major evolutionary change away from the corresponding feature in the presumed ancestral condition.

Strepsirhines: The suborder of primates comprising lemurs and lorises.

Structural carbohydrates: Primary components of cell walls that give plants their fibrous texture consisting of cellulose, hemicellulose, and lignin.

Succession: The gradual transformation from a disturbed and unstable vegetation type to an undisturbed and stable vegetation type.

Sympatric species: (1) Species with an overlapping geographical distribution or (2) that belong to the same community.

Tannins: A class of secondary compounds that act as digestibility-reducing agents in animals that ingest them in food.

Territory: An area that an individual or social group uses exclusively and defends.

Theria: The subclass of mammals whose members give birth to live young.

Toxin: A molecule that disrupts the metabolism of an animal that eats or inhales it. The toxicity of a molecule can be specified only in relation to a particular animal or set of animals.

Ungulates: Hoofed mammals in the orders Atiodactyla and Perissodactyla.

Wood: The innermost layer of a woody stem providing structural support.

Xerophyte: A drought-preferring plant.

Index